REFRACTORY MATERIALS

REFRACTORY MATERIALS

G.B. Rothenberg

NOYES DATA CORPORATION
Park Ridge, New Jersey, U.S.A.
1976

Published in the United States of America by
Noyes Data Corporation
Noyes Building, Park Ridge, New Jersey 07656

FOREWORD

The detailed, descriptive information in this book is based on U.S. patents issued since early 1973 that deal with refractory materials.

This book serves a double purpose in that it supplies detailed technical information and can be used as a guide to the U.S. patent literature in this field. By indicating all the information that is significant, and eliminating legal jargon and juristic phraseology, this book presents an advanced, technically oriented review of refractory materials as depicted in U.S. patents. To round out the complete technological picture, we have included one patent application published under the trial voluntary protest program initiated by the Commissioner of Patents and trademarks in January 1975.

The U.S. patent literature is the largest and most comprehensive collection of technical information in the world. There is more practical, commercial, timely process information assembled here than is available from any other source. The technical information obtained from a patent is extremely reliable and comprehensive; sufficient information must be included to avoid rejection for "insufficient disclosure." These patents include practically all of those issued on the subject in the United States during the period under review; there has been no bias in the selection of patents for inclusion.

The patent literature covers a substantial amount of information not available in the journal literature. The patent literature is a prime source of basic commercially useful information. This information is overlooked by those who rely primarily on the periodical journal literature. It is realized that there is a lag between a patent application on a new process development and the granting of a patent, but it is felt that this may roughly parallel or even anticipate the lag in putting that development into commercial practice.

Many of these patents are being utilized commercially. Whether used or not, they offer opportunities for technological transfer. Also, a major purpose of this book is to describe the number of technical possibilities available, which may open up profitable areas of research and development. The information contained in this book will allow you to establish a sound background before launching into research in this field.

Advanced composition and production methods developed by Noyes Data are employed to bring our new durably bound books to you in a minimum of time. Special techniques are used to close the gap between "manuscript" and "completed book." Industrial technology is progressing so rapidly that time-honored, conventional typesetting, binding and shipping methods are no longer suitable. We have bypassed the delays in the conventional book publishing cycle and provide the user with an effective and convenient means of reviewing up-to-date information in depth.

The Table of Contents is organized in such a way as to serve as a subject index. Other indexes by company, inventor and patent number help in providing easy access to the information contained in this book.

15 Reasons Why the U.S. Patent Office Literature Is Important to You —

1. The U.S. patent literature is the largest and most comprehensive collection of technical information in the world. There is more practical commercial process information assembled here than is available from any other source.

2. The technical information obtained from the patent literature is extremely comprehensive; sufficient information must be included to avoid rejection for "insufficient disclosure."

3. The patent literature is a prime source of basic commercially utilizable information. This information is overlooked by those who rely primarily on the periodical journal literature.

4. An important feature of the patent literature is that it can serve to avoid duplication of research and development.

5. Patents, unlike periodical literature, are bound by definition to contain new information, data and ideas.

6. It can serve as a source of new ideas in a different but related field, and may be outside the patent protection offered the original invention.

7. Since claims are narrowly defined, much valuable information is included that may be outside the legal protection afforded by the claims.

8. Patents discuss the difficulties associated with previous research, development or production techniques, and offer a specific method of overcoming problems. This gives clues to current process information that has not been published in periodicals or books.

9. Can aid in process design by providing a selection of alternate techniques. A powerful research and engineering tool.

10. Obtain licenses — many U.S. chemical patents have not been developed commercially.

11. Patents provide an excellent starting point for the next investigator.

12. Frequently, innovations derived from research are first disclosed in the patent literature, prior to coverage in the periodical literature.

13. Patents offer a most valuable method of keeping abreast of latest technologies, serving an individual's own "current awareness" program.

14. Copies of U.S. patents are easily obtained from the U.S. Patent Office at 50¢ a copy.

15. It is a creative source of ideas for those with imagination.

CONTENTS AND SUBJECT INDEX

INTRODUCTION

Refractories primarily are materials which can withstand high temperatures. Their usefulness depends on their ability to maintain their mechanical functions at high temperatures, quite often in contact with corrosive liquids and gases. There is no well-defined dividing line between refractories and nonrefractories; however, most generally recognized refractories have softening temperatures well over 1500°C (2732°F).

Refractories are comprised of ceramic materials and are chiefly used to line furnaces and high-temperature vessels. Also, they are needed to line passages or chambers remote from the combustion region, but which contain high-temperature gases. When refractories are used to line a vessel in which a fluid such as molten steel or molten glass is to be contained, the chemical reaction of the refractory is important.

Refractories are provided in a variety of physical forms such as shapes and monoliths which include plastics, ramming mixes, gunning mixes, casting mixes, etc. Shapes, usually brick, may be ceramically bonded by burning at elevated temperatures or may be chemically bonded with various chemical binders which set upon drying or curing at relatively low temperatures. Monolithic refractory products, those shaped within the furnace, almost always have a chemical binder. Binders used in refractory products include, for example, hydraulic cements, waste sulfite liquors, epsom salts, sodium silicate, phosphoric acid, and sodium phosphate salts.

Most industrially used refractories are composed of a mixture of metal oxides. In addition, carbon, graphite, and silicon carbide have found widespread use as refractories. Other carbides, nitrides, silicides, and borides are used in special applications. The same applies to most refractory metals. The principal advantage of oxidic refractories, apart from their refractoriness, is their stability under oxidizing conditions. The most commonly used refractory oxides are silica, zirconia, alumina, chromia, magnesia and calcia. Refractories containing mainly silica or zirconia are referred to as acid, those containing mainly alumina or chromia as neutral, and those containing mainly magnesia or calcia as basic. From the six refractory oxides mentioned above, the following classes of brick are presently made.

1

Silica Brick: These brick consist mainly of silica (in the form of christobalite, tridymite, and quartz), with minor amounts of lime, alumina, and iron [in the form of wollastonite ($CaSiO_3$) and glass]. They are made from naturally occurring quartzites and silica gravels.

Semisilica Brick: These brick are made from siliceous clays and contain mainly cristobalite grain bonded with a glassy phase rich in alumina and silica. They contain 80% silica and 20% alumina.

Fireclay Brick: These brick are made from naturally occurring fireclays. Their refractoriness increases with increasing alumina content and decreasing impurity content (alkalies, Fe_2O_3, TiO_2). Their alumina content ranges from 25 to 45%.

High-Alumina Brick: These brick contain more alumina than calcined kaolin (45%). They are generally made from varying combinations of calcined bauxites, clays, silica sand, and synthetic aluminum oxide. Their alumina content ranges from 45 to 100%.

Chrome Brick: These brick are made from naturally occurring chrome ore, $(Fe,Mg)O \cdot (Cr,Al)_2O_3$. Calcined magnesite is often added to the brick. The brick are then called chrome-magnesite brick.

Magnesite Brick: These brick consist mainly of the mineral periclase, MgO. When minor amounts of chrome ore are added, the brick are called magnesite-chrome brick. To increase their resistance to attack by basic slags, magnesite brick are sometimes impregnated with pitch.

Forsterite Brick: This class of brick consists mainly of the mineral forsterite, Mg_2SiO_4.

Zirconia-Containing Brick: The most common refractories in this class consist mainly of the mineral zircon, $ZrSiO_4$ or a mixture of zircon and alumina. The latter type of brick is usually formed by casting from a melt. Refractories for use at temperatures above 3500°F are made from synthetic, high-purity zirconia; these brick contain small quantities of lime, magnesia, or yttria to stabilize the cubic crystal form of zirconia.

Other oxides such as beryllia and thoria are used only in small quantities. Their price is generally too high for large-scale application. In use, refractories must successfully withstand shrinkage, fusion, spalling, slag erosion, abrasion, water absorption, modulus of rupture, heat conductivity and electrical conductivity.

This book describes the synthesis of hundreds of refractories and presents formulation and evaluation data for all type bricks, fibers, gunning mixes, ramming mixes, refractory coatings, cements, furnace linings, and metal casting refractories. In all, some 203 processes and compositions are presented from the U.S. patent literature during 1973 to 1976. As this technology has evolved, many of these refractories can be used in various applications depending on composition, binder, process conditions or impurities present. In this book, the refractories are discussed in the section relating to the major apparent component or use recognizing that many of these refractories may be used for other purposes as well as in a variety of special applications.

SILICA REFRACTORIES

HIGH SILICA REFRACTORIES

Lightweight Silica Foam

Foam glass products have not been useful as insulation at elevated temperatures. Most glasses have a substantial coefficient of thermal expansion and the existence of a temperature gradient of any magnitude across the glass causes cracking or crazing. Refractory glasses are generally those which withstand high temperatures and have a low coefficient of thermal expansion. Refractory glasses, however, require melting and processing at very elevated temperatures, thus consuming a considerable amount of energy to melt and causing rapid degradation of the refractory linings of the melting and processing tanks.

J.D. Johnson; U.S. Patent 3,945,816; March 23, 1976; assigned to Environ Control Products, Inc. describes a process whereby a refractory foam or cellular product substantially of silica may be produced in a few steps from a glass composition without requiring the glass to be subjected to temperatures above 850°C in any step subsequent to melting of the glass-forming ingredients to form a leachable, phase-separable glass.

The process is carried out by first melting a leachable, phase-separable glass, particularly a borosilicate glass, and then foaming the glass at elevated temperatures to form a lightweight phase-separable foam glass. The foam glass is then phase-separated at elevated temperatures to form a silica-rich phase and a silica-poor phase, which in the case of a borosilicate glass is a borate-rich phase. The silica-poor phase is preferably present as a substantially continuous phase.

The phase-separated foam glass is then leached with warm water or other aqueous leaching solution, e.g., dilute mineral acids, to remove the silica-poor phase from the glass to provide a cellular product consisting of silica. The refractory foam consists of cells resulting from the foaming action and interconnecting pores resulting from leaching of the silica-poor phase. The density of the resulting refractory cellular product may be 0.075 g/cc to 0.5 g/cc or higher. The refractory foam product has a very low coefficient of thermal expansion.

3

Example 1: A refractory foam was formed from the following materials: SiO_2, 40% by weight; B_2O_3, 50% by weight: and Na_2O, 10% by weight. A glass was formed by melting these materials in a conventional manner, fritted by rapid cooling, and further crushed to a fine particle size. The powdered glass was thoroughly admixed with 1% by weight of $CaCO_3$.

The $CaCO_3$ glass powder mixture was then introduced to a stainless steel mold having substantially greater volume than the volume of the $CaCO_3$ glass powder mixture. The mold was then heated (fired) in a kiln for $1\frac{1}{2}$ hours to 750°C. Thereafter, the mold was allowed to cool gradually in the kiln.

The foam article was then heated at 500°C for four hours to induce phase-separation of the sodium borate phase from the silica phase. The resulting foam had cells of one-fourth inch as an average dimension. The foam was fragile after leaching.

Example 2: A second composition substantially identical to that of Example 1 was prepared and foamed in a similar manner to that of Example 1. The foam was then divided into two pieces which were phase-separated by different techniques. Sample II(a) was heat treated at about 500°C for 8 hours while Sample II(b) was heat treated at 500°C for about 2 hours.

The samples were then leached at 90°C in distilled water. No significant difference was noted in the appearance and properties of the two samples. Each had cell dimensions of $\frac{1}{16}$ inch and each was fragile after leaching. The thermal coefficient of expansion of each sample was 5.0×10^{-7} m/m/°C.

Hydraulic Bonding Agent for Fused Silica

J.J. Capellman, J.C. Stultz and J.E. Cooper; U.S. Patent 3,824,105; July 16, 1974; assigned to PPG Industries, Inc. describe a hydraulic-setting bonding agent consisting of a hydraulic-setting calcium aluminate and nepheline syenite. The nepheline syenite is present in a concentration of 5 to 50% by weight based on the weight of the hydraulic-setting calcium aluminate and nepheline syenite. The nepheline syenite increases the fired cold flexural strength provided by the hydraulic-setting calcium aluminate. The bonding agent can be combined with an aggregate such as fused silica to provide a refractory such as the following, on a weight basis:

	Percent
Fused silica	60 – 80
Hydraulic-setting calcium aluminate	15 – 38
Nepheline syenite	2 – 15

This refractory composition when combined with 9 to 18% by weight water forms a fluid castable refractory composition which can be cast and hardened to form a refractory body such as a gas hearth module.

Nepheline syenite comprises a mixture of several different sodium aluminum silicates and potassium aluminum silicates. It is an igneous rock composed chiefly of the mineral nepheline mixed with albite ($NaAlSi_3O_8$) and orthoclase ($KAlSi_3O_8$). The native rock does not contain quartz.

Example: A gas hearth bed was prepared as follows: 1500 pounds of a fused silica grog having the following particle size distribution:

Tyler Standard Sieve Series, Held on	Percent by Weight
8 mesh	3.0
10 mesh	7.0
16 mesh	4.0
20 mesh	72.0
30 mesh	95.0
40 mesh	99.0

were split into two portions. One portion of about 500 pounds was further particulated in a ball mill for 180 to 240 minutes to give a silica grog having the following particle size distribution:

Tyler Standard Sieve Series, Held on	Percent by Weight
40 mesh	0.2
100 mesh	3.2
140 mesh	10.8
170 mesh	21.9
200 mesh	32.2
325 mesh	79.6
Through 325 mesh	21.6

The two portions of the silica grog were then recombined and mixed with 150 pounds of nepheline syenite having the following particle size distribution:

Tyler Standard Sieve Series, Held on	Percent by Weight
40 mesh	0.2
100 mesh	1.9
140 mesh	8.4
170 mesh	21.1
200 mesh	33.3
325 mesh	83.2
Through 325 mesh	16.8

in a mixer for 10 minutes. To this mixture were added 350 pounds of calcium aluminate cement. The calcium aluminate was mixed with the fused silica and nepheline syenite for about 10 minutes. The mixture of fused silica, nepheline syenite and calcium aluminate had the following particle size distribution:

Tyler Standard Sieve Series, Held on	Percent by Weight
10 mesh	3.0
16 mesh	17.5
20 mesh	30.0
30 mesh	41.0
40 mesh	48.0
100 mesh	56.2
140 mesh	64.5
170 mesh	71.5

(continued)

Tyler Standard Sieve Series, Held on	Percent by Weight
200 mesh	76.0
325 mesh	95.9
Through 325 mesh	4.5

The screen analyses given above were cumulative screen analyses. The above composition contained 75% by weight fused silica, 17.5% calcium aluminate cement and 7.5% nepheline syenite. To this mixture were added 250 pounds of water (12½% by weight water based on total weight of the dry mixture) at 70°F. Mixing was continued for an additional 5 minutes until the hydraulic mixture had the proper flow characteristics, that is, until it had a flow of 5 inches in 30 seconds as determined by measurement on the vibrating flow meter.

The mixture was then cast into a vibrating 4 foot by 6 foot by 8 inch wooden mold which had been coated on the interior surfaces with an axle grease release agent. The mold had brass inserts which corresponded to the hollow portions of the gas hearth modules. The hydraulic mixture was set for two hours at room temperature in the mold.

The mold was then covered with damp cloths to prevent rapid moisture loss at the exposed surface and the hydraulic mixture was set for an additional 24 hours. The casting was then removed from the mold and the inserts were removed from the casting to give a refractory shape suitable for use as a gas hearth module.

Before use, the casting was fired to increase its abrasion resistance and its cold flexural strength. The casting was fired by heating it in a kiln from room temperature to about 120°F at a rate of 50°F an hour. This slow rate of heating insures the complete removal of excess water without cracking the refractory. The temperature was then raised at 100°F per hour to 2100°F, where it was maintained for eight hours. The kiln was then slowly cooled at furnace rate to room temperature.

Bricks Resistant to Molten Glass

B.E. Yoldas; U.S. Patent 3,816,163; June 11, 1974; assigned to Owens-Illinois, Inc. describes impregnated refractory structures, such as, for example, bricks which have increased resistance to the destructive action of molten glass. The structures are produced by an impregnation and densification process in which conventional, prefabricated, solid, porous, refractory structures are first treated with a solution of a polysilicon compound which is thermally decomposable to silica, and the solution-treated structure is then heat treated, to cause a deposition of amorphous silica in the brick pores.

Because of the silica deposition and pore impregnation, a densification and decrease in porosity is effected which greatly decreases molten glass penetration and internal destruction of the silica impregnated refractory structure. The use of these impregnated structures also produces a glass with substantially fewer seeds. A particularly suitable polysilicon compound which may be employed is a hydrolyzed and condensed ethyl silicate.

Example: (A) Forming Hydrolyzed and Partially Condensed Ethyl Silicate — Four mols (832 g) of distilled ethyl orthosilicate (tetraethoxysilane) were added to a reactor and 540 g of anhydrous isopropanol then added to the ethyl silicate. The resultant mixture was agitated for several minutes. To this mixture there was then added 7.2 mols (130 g) of water. The water addition is generally done slowly and preferably the addition will be successively made in small aliquots until the total water charge has been added.

The water was added in 10-ml aliquots and, after each addition, the mixture stirred until clear; this procedure was successively repeated until the entire charge of water had been added. HCl (1 N solution) was then added to this ethyl silicate-isopropanol-water mixture such that there was a concentration of 10 parts HCl per million parts of total water and ethyl orthosilicate.

The above mixture was then subjected to a hydrolysis and partial condensation by heating the mixture to 50° to 60°C to generally effect hydrolysis and then refluxing the mixture for 2 hours at 80°C to effect partial condensation. The resultant solution containing a hydrolyzed and partially condensed polysilicon reaction product was then concentrated by evaporatory removal, at 80°C, of 750 g of the alcohol by-product and water.

Thus, there remained a gel-free hydrolyzed and condensed ethyl silicate polymer in an alcoholic solution, the solution containing the equivalent of 30 to 35 weight percent of silica (SiO_2). This solution was then stored in an enclosed glass receptacle and allowed to cool to room temperature.

(B) Densification and Impregnation — A commercially available silica refractory brick was then impregnated according to the following general procedure. This brick is composed of 99%+ fused silica and is generally recognized as having excellent resistance to the corrosive and erosive action of molten glass compositions. A large brick was cut to an acceptable size and the specific brick sample employed was a generally rectangular brick having an initial density of 1.82 g/cc, a porosity of 17.6%, and a size of 1 inch x 1 inch x 3 inches. By microscopic investigation, this sample was characterized by intercommunicating pores of a nominal size of about 50 to 100 microns.

The 1 inch x 1 inch x 3 inch sample had a porosity of 17.6% and density of 1.82 g/cc. The silica refractory brick sample while at room temperature was then immersed in the solution produced according to Step A above and allowed to remain and soak in this solution overnight to become thoroughly impregnated and saturated. This was done in the ambient environment. The saturated brick sample was then removed and the excess alcoholic solvent was allowed to evaporate therefrom by setting in open air until a brick sample with a substantially dry appearance resulted.

The brick sample, which is now impregnated with the decomposable polysilicon compound, was then heated to 325° to 350°C in an air environment for 1½ to 3 hours and held at that temperature for 3 to 4 hours. After this heating cycle, the sample was again heated, this time to 550° to 600°C and held there for several hours to effect a final decomposition of the hydrolyzed ethyl silicate polymer. As a result of this decomposition, amorphous silica was deposited in the brick pores, thus producing a silica impregnated brick.

Silica was deposited, apparently without the formation of any residual carbon as determined by chemical analysis. Since carbon is known to discolor certain glass compositions as well as set up undesirable electrochemical reactions, this carbon-free silica impregnation makes the bricks well suited for use in the glass-melting operation.

The density of the silica impregnated brick after this first treatment was 1.87, that is, a densification of 3.01% resulted. Likewise, the porosity was 15.1%, which is the equivalent of a pore reduction of 14.2%. When microscopically examined, it was found that the silica impregnated brick had silica deposited in a uniform and homogeneous manner throughout the entire thickness of the brick which is attributable to the solution treatment and soaking. The pretreatment pore intercommunication was also substantially reduced.

The silica impregnated brick which had been treated as set forth above was then successively treated five additional times according to the same procedure. The table below summarizes the resultant densification and pore reduction of this brick after each of these successive treatments.

Densification and Pore Reduction of Example 1

	Before Treatment	First Treatment	Second Treatment	Third Treatment	Fourth Treatment	Fifth Treatment	Sixth Treatment
Density g/cc	1.82	1.87	1.90	1.92	1.93	1.93	1.94
Porosity (%)	17.6	15.1	13.6	12.4	12.3	12.2	11.9
Densification (%)	0	3.01	4.8	6.1	6.4	6.6	6.9
Pore reduction (%)	0	14.2	22.6	28.8	30.1	31.0	32.4

Rare Earth Oxide-Silica Ceramics

G.J. Kamin and W.T. Kiger; U.S. Patent 3,798,040; March 19, 1974; assigned to GTE Sylvania Incorporated describe refractory compositions having low thermal expansions and high melting point characteristics comprising rare earth oxide-silica polycrystalline ceramics having a rare earth oxide to silica mol ratio of 3:7 to 9:1.

The compositions are prepared by blending a rare earth oxide and silica for 1 to 3 hours to form a substantially uniform admixture having a rare earth oxide to silica mol ratio of 3:7 to 9:1. The admixture is then heated above 1500°C for at least 1 hour to form a rare earth oxide-silica polycrystallic ceramic. The compound may then be ground to a particle size so that 100% passes through a USS-60 mesh screen, formed into the desired shape, and thereafter heated above 1500°C for at least 1 hour to densify the compound to form a shaped body.

Compositions that are especially preferred include compositions having mol ratios of yttria to silica of 1:1, 1:2, and 2:3 respectively, a composition having a mol ratio of lanthanum oxide to silica of 1:1, and a composition having a mol ratio of misch metal to silica of about 1:1.

The final ceramic materials show low coefficients of thermal expansion ranging from 28×10^{-7} to 77×10^{-7} in/in/°C from 25° to 800°C. The materials have high melting points ranging from 1700°C for the high silica compositions to 2500°C for the high yttria compositions. The ceramics show high resistance to fracture due to thermal cycling. They also show excellent compatability with molten ferrous alloys and many nonferrous alloys. The ceramic bodies can be used in oxidizing or reducing atmospheres.

The ceramics find application in those areas where high-temperature, thermal-shock resistant materials are needed, such as crucibles (for melting metals or performing chemical reactions), kiln or furnace parts, heat shields for aerospace applications, or even automobile parts such as turbines, manifold liners, and the like. The prereacted powders can be used for impact formed crucibles, spray coating, or any other usage generally associated with refractory powders.

Example: A blend of 50 mol percent yttria of 99.9% purity, and 50 mol percent silica of greater than 98% purity is prepared by mixing the blend for two hours in a V-blender. The material is then passed through a number 80 sieve. The blended powders are then isostatically pressed into one-half inch diameter bars, 2 x 2 x 4 inch bricks and small crucibles 2 inches in diameter and 4 inches high, at 40,000 psi. The pressed pieces are then fired at 1500°C for 48 hours in an air ambient atmosphere.

The final pieces are of 80% theoretical density. The pieces show coefficients of thermal expansion of 30×10^{-7} in/in/°C over 25° to 800°C. The material is rapidly cycled from room temperature to 1100°C in excess of a hundred times without any serious flaw development. Test pieces heated to 1200°C and plunged into cold water also show no thermal shock damage. The compound melts congruently at 1980°C.

Pieces of the ceramic material are put in contact with various molten metals by dropping cold pieces of the ceramic into the molten metal phase. In none of the tests do the ceramic pieces crack due to the thermal shock. Examination of the interfaces between the molten metal and ceramic after cooling shows no reaction between the materials in the case of low-carbon steel, iron-nickel-cobalt alloys, nickel, aluminum or copper.

Crucibles of the ceramic are used to fire phosphor materials of the halophosphate type generally used in fluorescent lamps. The phosphor shows no darkening or loss of brightness. The crucibles were repeatedly cycled during heating and showed no thermal shock damage or deterioration due to the contact with the chloride or fluoride gases evolved from the phosphor. The crucibles are also used to fire rare earth type television phosphors with no sign of reaction or loss of brightness.

Silica Brick with Tridymite Structure

According to *H. Flood and A. Seltveit; U.S. Patent 3,788,866; January 29, 1974; assigned to Elkem A/S, Norway* refractory bricks in which all of the silica is in the tridymite form can be prepared from finely divided silicas, especially silica dust recovered from waste gases of furnaces producing silicon metal and silicon alloys. This dust which is of colloidal nature, is very finely divided and has a surface area of 15 to 20 square meters for each gram. The dust is moistened with water and then lime is added thereto. The lime may be present in an amount of 1 to 5% by weight of the mixture and may be in the form of unslaked lime, slaked lime, lime milk, etc., of which the preferred is slaked lime.

In addition to the catalytic effect for tridymite formation, the addition of lime

also establishes a slightly basic condition (pH values 8 to 9) which is highly beneficial for obtaining a proper plasticity of the mixture. When the blend has been
thoroughly admixed, a pulverulent ammonium salt is added to the mixture in
an amount approximately equal to or greater than the amount of lime that was
added. The ammonium salt must be one which is capable of precipitating calcium as a substantially insoluble salt. Suitable ammonium salts are ammonium
carbonate, ammonium bicarbonate, ammonium oxalate, ammonium phosphate,
etc.

After the mixture is prepared, it is formed into bricks and then fired. The insoluble calcium precipitate seems to act as a very good catalyst for the cristobalite to tridymite conversion. Preferably, the brick is fired up to 800° to 900°C,
fired at this temperature for a period of time, after which the temperature is
raised to 1300° to 1500°C and the firing is continued. Where a porous refractory
material is desired, pore-forming materials such as sawdust, expanded polystyrene
and the like can be added to the refractory mixture.

The amount of silicon dioxide material should be by weight 7 to 20 parts per
10 parts of water. The amount of pore-forming material, if used, will be from
3 to 8 parts per 10 parts of water where the pore-forming material is sawdust
or a similar material. It is important to note that the total firing cycle of 50
hours for these bricks is much shorter than the firing cycle for silica bricks made
from conventionally raw materials which is 200 to 250 hours. This means lower
production costs and increased capacity on the kilns.

For such short firing periods it must be stressed that if sawdust or other organic
materials are added to the raw material mixture, the temperature rise around
500°C must be slow. Preferably the bricks should be kept at 500°C for about
10 hours to ensure complete removal of carbonaceous matter without detrimental
temperature increase inside the bricks due to the heat of oxidation and cracking
due to fast gas evolution.

Likewise it has been found that the firing curve should not be too steep in the
temperature range 800° to 900°C and best results can be obtained if the temperature of the bricks can be kept within this range for a period of 4 to 10 hours.
Good results have sometimes been obtained with even faster heating in this range
but very often faster heating has resulted in extensive shrinkage of the bricks.
The main purpose of relatively slow heating rate at 800° to 900°C is to convert
amorphous silica to highly disordered cristobalite which can be easily converted
to tridymite at higher temperatures. For firing in large kilns where temperature
gradients can hardly be avoided, the following firing curve is recommended.

°C	Hours
0 - 500	4
500	10
500 - 800	4
800 - 900	10
900 - 1400	5
1400	10
1400 - 2000	10

Example: A mixture comprising 1500 g silica dust, 570 g sawdust and 1.35 liters
of water was admixed for ten minutes in a laboratory scale mixer. The silicon
dioxide dust was colloidal silica which has been precipitated from the waste gases

of a metallic silicon furnace. To this admixture was added 40 g of finely divided calcium hydroxide and admixing was continued for an additional ten minutes. Thereafter, 60 g of ammonium bicarbonate was added and mixing was again continued for ten minutes.

The resulting mixture was relatively plastic and was formed by hand in a simple iron mold to bricks of 230 x 115 x 65 mm. The bricks were dried at ambient temperature for fifteen hours and then were dried in an oven at 50°C for 48 hours. Thereafter, the bricks were transferred to a laboratory furnace with Kanthal elements.

The bricks were baked at 450°C for two hours, raised to 500°C for an additional four hours and then the temperature was raised to 830°C. The bricks were baked between 830° and 900°C for fifteen hours after which the temperature was raised to 1300° to 1400°C at which temperature the bricks were cooked for ten hours. The furnace was then shut off and the bricks were allowed to cool overnight. Two of the bricks were analyzed with the following results.

	Brick 1	Brick 2
Porosity, percent	67	67
Weight per volume (kg/dm^2)	0.74	0.74
t_a, °C	1,595	1,605
t_e, °C	1,630	1,625

t_a and t_e are a measure of the comparison strength of the refractory material. t_a is defined as that temperature limit which corresponds to 0.6% longitudinal shrinkage when the sample is heated under a pressure of 1 kg/cm^2 and the temperature is raised according to a standard temperature/time curve. t_e is defined as that temperature which corresponds to 40% longitudinal shrinkage under the same conditions. At t_e, the brick is considered to be totally broken down.

SILICA-ALUMINA REFRACTORIES

Cordierite Coating on Ceramics

F.W. Martin; U.S. Patent Application (Published) B 554,655; February 24, 1976; assigned to Corning Glass Works describes sinterable powdered glasses consisting, on the oxide basis, of 22.2 to 26 mol percent MgO, 23.6 to 27.8 mol percent Al_2O_3, 44.3 to 52.7 mol percent SiO_2, 0.7 to 5.6 mol percent total of one or more modifying oxides selected from the group consisting of BaO, PbO, SrO and CaO, and at least 23.6 mol percent total of modifying oxides plus MgO.

Glassy phases in the crystallized glass-ceramic product are minimized by increasing the alumina content of the composition in approximate correspondence with modifying oxide additions, to approach stoichiometry. Thus the Al_2O_3 content of the composition is maintained at a level which exceeds the total modifying oxide content by at least 21 but not more than about 23 mol percent.

The sintering and crystallization of glass powders is accomplished by heating to 900° to 1050°C for ¼ to 12 hours. The sole crystal phase consists of silicate crystals corresponding to hexagonal cordierite, this crystal phase comprising the oxides of magnesium, aluminum, silicon, and the modifying ions in the approximate molar proportions $4MgO \cdot (4+n)Al_2O_3 \cdot (10-2n)SiO_2 \cdot (n)MO$, where MO is one or more modifying oxides selected from the group consisting of PbO, BaO, SrO and CaO, and where n is 0.25 to 1. The use of the glasses to provide crystallized electrical barrier layers on glass-ceramic articles is also described.

Example: A batch for a glass having a composition, on the oxide basis, of about 22.3 mol percent MgO, 25.0 mol percent Al_2O_3, 49.9 mol percent SiO_2, and 2.8 mol percent PbO is compounded, ball-milled, and melted at 1650°C in a platinum crucible for seven hours. The glass melt is then cast into a glass slab and annealed at 700°C. The glass slab is crushed and ground to provide a glass powder having a maximum particle size of 150 microns (passing 100 mesh). This powder is then pressed into a cylindrical sample 0.5 inch in diameter and 0.5 inch in height.

The pressed cylinder is then heated to 950°C and maintained for ½ hour to sinter and crystallize the powdered glass into a highly crystalline glass-ceramic cylindrical sample. A linear shrinkage on firing of 5.5% which is indicative of good sintering is observed. The sample is then subjected to high-temperature electrical resistivity measurements to determine the DC volume resistivity at 400° to 825°C. The following log resistivity (ρ) values (ohm-centimeters) are recorded.

Temperature	Log ρ (ohm-centimeters)
425°C	12.93
500°C	11.64
625°C	10.42
712°C	9.39
825°C	8.37

X-ray diffraction examination of the sample shows a high degree of crystallization to crystals corresponding in structure to hexagonal cordierite, but with the presence of lead being reflected by altered line intensities.

Masking with Bentonite Clay and Inorganic Oxide

R.B. Forker, Jr. and J.N. Panzarino; U.S. Patent 3,932,681; January 13, 1976; assigned to Corning Glass Works describe masking compositions which will provide a mask which is both impermeable to most sprayed coating compositions and readily removable from nonporous ceramic surfaces even after firing to 700°C or higher. The masking compositions comprise 40 to 90% of bentonite clay and 10 to 60% of a finely divided refractory metal oxide release agent component, which is essential to the removability of the composition after exposure to elevated temperatures above 500°C. A bentonite clay rather than a kaolinite, beidellite, or other clay species should be employed if impermeable, adherent, yet removable masks are to be obtained.

The finely divided (less than 100 mesh U.S. Standard Sieve) refractory metal oxide release agents may be silica, titania, iron oxide, tin oxide, or calcium oxide, metal oxide products of fusion which provide the required degree of refractoriness such as 96% silica glasses, or metal oxide mixtures, compounds, or combinations such as alundum cement, a mixture typically composed of 90 parts alumina, 5 parts silica and 5 parts lime by weight.

Among the ceramic materials which may be usefully masked with these compositions are vitreous ceramics (glasses) and semicrystalline ceramics commonly referred to as glass-ceramics. Both vitreous and semicrystalline ceramics typically have nonporous surfaces which permit ready removal of the masking composition after firing. Hence, very mild removal methods which do not affect the quality of the functional ceramic coating on unmasked surface portions of the ceramic may be utilized to remove these masks. Crystalline ceramic articles having increasing amounts of surface porosity present increasingly difficult removal problems, but the ready removability of these masking compositions permits their use even on some porous ceramic surfaces where masking coatings have not previously been employed.

A convenient method of applying the masking compositions comprises mixing a selected composition with a volatile liquid vehicle to form a fluid suspension thereof. This suspension may then be applied by spraying, brushing, dipping, rolling or other conventional means to form a coating and the volatile vehicle may then be driven off by heating the coated ceramic surface to a temperature above the boiling point of the vehicle. A particularly preferred method comprises applying the masking composition in the form of a slip or slurry by spraying onto the ceramic surface.

Removal of the masking composition after the application of the functional decorative, conductive, reflective or protective coating to the ceramic surface is extremely simple, being readily accomplished by water rinsing with occasional hand rubbing even when the masking composition has been subjected to firing temperatures approaching 700°C.

Example: Two glass-ceramic dishes to be provided with electrically conductive tin oxide coatings on their base portions only are selected for treatment. The dishes are composed of a lithium aluminosilicate glass-ceramic material, being semicrystalline in nature but having surfaces essentially free of porosity. A masking composition in the form of a slurry is prepared by adding 40 g of a bentonite clay, 30 g of a finely divided alundum cement metal oxide release agent, and 10 g of a CaO metal oxide release agent to 200 ml of a methanol-water vehicle consisting of 50% methanol and 50% water by volume.

The alundum cement had a maximum particle size of 100 mesh and the CaO also had a maximum particle size of 100 mesh. The resulting slurry is flowable and has a viscosity suitable for spraying according to conventional methods.

The slurried masking composition is applied at room temperature by spraying to the edge portions of one of the glass-ceramic dishes selected for treatment to provide a mask thereon. The edge portions of the other glass-ceramic dish are masked by spraying with a sprayable slurry consisting only of bentonite and water, the thicknesses of the masking layers on each of the two dishes being approximately equivalent.

Following the application of masking layers, the two dishes are heated to 600°C in a kiln, removed, and sprayed while hot with a solution of tin chloride. Upon contact with the heated dishes, the solution is volatilized and the tin chloride is converted by pyrolysis to tin oxide (SnO_2).

After coating, the tin-oxide-coated dishes are cooled to room temperature and attempts are made to remove the masking compositions. The mask which is composed of bentonite alone is very difficult to remove, requiring the use of an abrasive detergent composition and vigorous scrubbing to obtain complete removal. The complete avoidance of damage to areas of the tin oxide coating adjacent to the mask is impossible, and some scratching of the ceramic surface is incurred.

In contrast, the masking composition, being composed of bentonite plus the alundum cement-lime release agent, is found to be rather friable upon cooling, and most of the mask can be removed by vigorous spraying with water. Only light brushing or hand rubbing with additional water is required to remove the remainder. The edges of the tin oxide coating adjacent the masked areas are sharp and well-defined, being equivalent in thickness to the remainder of the coating and undamaged by the mask removal process.

Anisotropic Cordierite Monolith

I.M. Lachman and R.M. Lewis; U.S. Patent 3,885,977; May 27, 1975; assigned to Corning Glass Works describe a polycrystalline monolithic cordierite honeycomb product having a microstructure of oriented cordierite crystals with sufficient orientation of c-axis in the web planes of the honeycomb to contribute to a coefficient of thermal expansion of as low as 5.5×10^{-7} in/in/°C or lower over 25° to 1000°C in the compositional range, on the oxide basis, of 41 to 56.5% SiO_2, 30 to 50% Al_2O_3, and 9 to 20% MgO. The product excels in its resistance to thermal shock when exposed to wide deviations in temperature because of low thermal expansion.

The cordierite crystallites' a-axis contribution produces a high expansion direction perpendicular to the low expansion direction in the monolith but the product geometry obviates any problem by providing the high expansion direction transverse to the smallest dimension of very thin webs, which additionally have internal space within the monolith in which to expand.

Specifically, 25% of the MgO may be replaced by an equal number of formula weights of NiO (in the form of oxide, sulfate, carbonate, etc.). Similar replacements may be made for 15% of the MgO by CoO, 40% of the MgO by FeO, 98% of the MgO by MnO or 15% of the MgO by TiO_2. The product is particularly adapted to use as a catalytic support matrix for emissions control.

Ceramics from Red Mud

In the commercial treatment of bauxite ore to extract the oxides of aluminum therefrom, such as in the well known Bayer process, the bauxite is digested with a sodium hydroxide solution and filtered. The dissolved aluminum compounds are separated from the undissolved iron-containing residue, the latter commonly being termed red mud. The red mud has in the past been piped to settling ponds once the available alumina has been leached out and most of the alkali has been washed out for recycling. These ponds are most unsightly and take up a great amount of land. Also, the residual caustic content of the red mud in the pond poses a serious threat to animal and plant life.

Accordingly *J.M. Klotz; U.S. Patent 3,879,211; April 22, 1975* describes a process for preparing from 81 to 97% of red mud containing at least 40% iron expressed as Fe_2O_3, 2 to 9% of sodium bentonite or 2 to 25% of calcium bentonite, and 0 to 10% of calcium sulfate. The product is then vitrified by firing at 1950°

to 2500°F. The product of the above composition may be prepared wherein the ingredients mentioned above are blended together with sufficient water to provide a cohesive but formable mass. After forming into a desired shape, this mass is exposed to an atmosphere which is substantially saturated with water and then fired at the aforesaid temperature range for a period of time to cause the mass to be vitrified.

The red mud can vary somewhat in its chemical content. A typical range of composition is as follows:

Element	Expressed As	Weight Percent Dry Basis
Fe	Fe_2O_3	42 – 50
Al	Al_2O_3	13 – 22
Si	SiO_2	7 – 8
Ti	TiO_2	5 – 7
Na	Na_2O	5 – 9

Example: A mixture was made to contain 95% of dried red mud, 4% sodium bentonite, and 1% calcium sulfate (plaster of Paris). After sufficient water was added to give the mixture a workable consistency, the mixture was allowed to stand for 30 minutes. It was then placed in a chamber and steam was added at such a rate that the chamber temperature was 150°F. After 20 minutes of steaming, the mixture was removed and then fired for 10 hours at 2100°F. The resulting vitrified product had the following properties.

Water absorption, %	0.735
Specific gravity	
Bulk saturated surface dry	2.686
Bulk oven dried	2.667
Apparent	2.720
Compressive strength, psi	26.825

Ceramics and Glass-Ceramics from β-Spodumene

Ceramic structures to be used for high-temperature applications should be quite low in thermal expansion in order to minimize thermal stress, so that useful service life may be realized. They should also be nonreactive with respect to their environment at high temperatures. In automotive exhaust applications, for example, high temperatures are accompanied by reducing and oxidizing exhaust gases, sulfur dioxide, water vapor, and oxides of nitrogen. If the effect of these conditions is to partly modify the composition of the ceramic material, rapid degradation of the ceramic structure frequently results.

According to *D.G. Grossman and H.L. Rittler; U.S. Patent 3,834,981; Sept. 10, 1974; assigned to Corning Glass Works* low-expansion ceramic and glass-ceramic articles can be manufactured from beta-spodumene $Li_2O–Al_2O_3–SiO_2$ compositions using a $H^+ \rightleftharpoons Li^+$ ion exchange process which produces a hydroxy-aluminosilicate phase convertible by appropriate heat treatment to desirable aluminosilicate phases. The low-expansion ceramic and glass-ceramic articles are useful for certain high-temperature applications where the reactivity of lithium aluminosilicates precludes their use.

The beta-spodumene crystals involved in the ion-exchange reaction may arise from essentially any source, including alpha-spodumene, a lithium aluminosilicate glass, a sintered batch of appropriate composition, or a sintered beta-spodumene ceramic article. Also, any of the known beta-spodumene solid solutions may be treated.

The ion-exchange treatment of the process comprises the initial step of contacting the beta-spodumene-containing ceramic articles with strong mineral acids at 25° to 320°C, for a period of time to permit the exchange of hydrogen ions for lithium ions in the crystals of the ceramic article. The process is time and temperature dependent so that, at higher temperatures, shorter treatments are required.

The product of this treatment is a ceramic article containing hydroxy-aluminosilicate crystals having the unit formula $H_2O \cdot Al_2O_3 \cdot nSiO_2$. Sufficiently long ion-exchange treatments permit essentially complete replacement of lithium in the crystals by hydrogen, resulting in an article wherein the described hydroxy-aluminosilicate crystals constitute the principal crystalline phase.

Following completion of the ion-exchange step, residual acid and lithium-containing residues are removed from the surface of the article, and it is then heated. Heating removes the water from the crystalline lattice and converts the hydroxy-aluminosilicate crystals to nonreactive, low-expansion aluminosilicate crystals. At 350° to 1000°C, water is driven off, but the extent of removal is proportional to the temperature of the heat treatment, and temperatures of at least 1000°C are required to remove all of the water from the structure.

Examination of a structure from which all of the water has been removed by heat treatment at 1000°C discloses the presence of aluminosilicate crystals of unestablished composition and structure which have an X-ray diffraction pattern closely resembling that of keatite. Keatite is a synthetic hydrothermal silica form, also referred to as silica-K. The aluminosilicate crystals produced by this treatment have a negative coefficient of thermal expansion as evidenced by the fact that treatment as described to produce essentially total conversion to keatite-type aluminosilicate crystals produces a negative-expansion body. The crystals are quite probably composed of alumina-silica solid solutions which are iso-structural with keatite, and are referred to as aluminous keatite.

Articles composed essentially entirely of these aluminous keatite crystals are useful not only for their low thermal expansion characteristic, but also because they are refractory and alkali-free. Hence, the articles contain crystals which consist of silica and alumina in a molar ratio $Al_2O_3:nSiO_2$, wherein n is 3.5 to 10, contain only minor or trace amounts of lithium, have solidus temperatures well in excess of the spodumene-containing structures from which they are produced, and are essentially nonreactive at high temperatures even in moist oxidizing or reducing atmospheres.

The utility of such aluminosilicate articles may be improved for certain applications through the use of further heat treatment above 1000°C. Such heating causes phase transformations from the aluminous keatite structure to mullite $(3Al_2O_3 \cdot 2SiO_2)$ and, above 1200°C, cristobalite (SiO_2). The extent to which these transformations occur may be controlled by controlling the temperature and time of the heat treatment. Mullite and cristobalite crystals are higher in thermal expansion than is aluminous keatite; thus, a controlled thermal treatment can be used to produce partial transformations which result in a ceramic article of specified crystalline composition and thermal expansion characteristics. For example, articles exhibiting essentially zero average coefficients of thermal expansion at 0° to 800°C may be obtained.

Example: A glass-ceramic honeycomb structure formed by thermal crystallization of a glass consisting essentially, in weight percent on the oxide basis as calculated from the batch, of about 64.5% SiO_2, 18.9% Al_2O_3, 3.7% Li_2O, 2.5% ZnO, 4.4% TiO_2, 5.0% B_2O_3, 0.8% As_2O_3 and 0.2% F is selected for treatment. The article has a principal crystal phase consisting essentially of beta-spodumene solid solution, and has an average coefficient of thermal expansion of about 45 x 10^{-7}/°C over the range from 25° to 800°C.

The structure is ion-exchange treated by immersion in concentrated sulfuric acid at 90°C for 32 hours. It is then removed from the acid and rinsed with distilled water to remove all lithium-containing residues of the ion exchange treatment. Following the ion-exchange treatment the article is heated in air to a temperature of about 1050°C, maintained at that temperature for about 16 hours, and finally cooled to room temperature.

Examination of the product of the above treatment discloses a strong integral article essentially identical in size and configuration with the initial structure. Thermal expansion measurements show a decrease in the average coefficient of thermal expansion over the range from 25° to 800°C to about -1.0 x 10^{-7}/°C. The article has a deformation temperature in excess of about 1400°C, as compared to about 1250°C for the original beta-spodumene structure from which it is produced.

Forming Mullite with Fluorine Compound

T.C. Shutt; U.S. Patent 3,761,294; September 25, 1973; assigned to Fiberglas Canada Limited, Canada describes a method of promoting mullite formation in clay comprising the step of admixing a minor proportion of a fluorine compound and heating the mixture. The fluorine compound may be fluorine, hydrogen fluoride and compounds which form one or both of them on heating to 600° to 900°C in the mullite formation process. The fluorine compound will normally be added in an amount of 0.5 to 5%, or preferably 1 to 3%, based on the weight of the clay.

Although mullite normally forms on heating clay to 1200°C, the process provides forming mullite at lower temperatures, for example 700° to 1000°C or more, usually 700° to 950°C. Mullite ($3Al_2O_3 \cdot 2SiO_2$) contains 72% Al_2O_3. It is a colorless, acicular mineral of the orthorhombic system which is only rarely found in nature.

Example 1: Experiments were carried out using barium fluoride, lithium fluoride and sodium silicofluoride as mineralizer. Owen clay (a kaolinite) was the only mineral in the mix apart from the mineralizers. The mix compositions, in parts by weight, were as follows: Owen clay (–300 U.S. sieve), 60; water, 40; and fluoride, 1.

The mixes were dry blended, water added, mixed and cast and dried at 90°C. Two specimens of each fluoride mix were prepared and fired at 800° and 1000°C. Study of thin sections of the resulting material showed that in every case except the 800°C-fired barium fluoride mix, mullite had formed without causing harmful effects such as cracking or spalling.

Example 2: Sodium silicofluoride was used as the mineralizer. Seven mixes were made; the only component was clay which, although it would give low strength values, would make the study of mullite easier because of its abundance. Water and monoaluminum phosphate were used as bonding agents.

Composition	Mix Number						
	1	2	3	4	5	6	7
Owen clay (green)	700	700	700	700	700	700	700
Sodium silicofluoride	7	7	$3\frac{1}{2}$	7	–	–	$3\frac{1}{2}$
50% solution M.A.P.	500	–	–	250	500	–	500
Water	–	400	400	250	–	400	–

Note.—NB:
1. All readings in grams.
2. Treatment—Air dried; dried at 60°C for 4 hours; 2 hour soak on temperature.

Modulus of Rupture — In both the phosphate and water-bonded samples, the highest strengths are found in the 1% fluoride additions, followed by the $\frac{1}{2}$% fluoride additions and the weakest is without fluoride. For the purpose of simplicity only the water-bonded specimens will be discussed, since these specimens illustrate best the effect of fluoride addition. With any given starting composition, the initial variation in green strength is due to fabrication technique. At the 500°C firing the strength of No. 2 is 75% greater than No. 6 and at 1000°C it is 110% greater. The drop in No. 2 at 540°C could be prevented by a slightly longer firing time.

Thin Sections — Thin sections were made of every specimen. Photomicrographs of the 800°C test specimens were made, from which it could be seen that the 1% fluoride mixes produced approximately 5% of mullite. The $\frac{1}{2}$% fluoride seemed to produce about 1% mullite and the mixes without fluoride have produced no mullite.

X-Ray Results — 5% is the lower limit of detection by X-ray determination. Discounting the specimens that are fired at 1200°C (which must form mullite) the only other specimens that contained measurable amounts of mullite, all had fluoride addition.

High Purity Mullite

R.E. Farris and J.C. Hicks; U.S. Patent 3,758,318; September 11, 1973; assigned to Kaiser Aluminum & Chemical Corporation describe a method of making high purity synthetic mullite from finely divided hydrated aluminous material. The material contains, on the ignited basis, at least 95% Al_2O_3, substantially all of the aluminous material passing a 325 mesh screen and having an average particle size less than 10 microns, and a finely divided hydrated aluminosilicate material containing, on the ignited basis, at least 95% Al_2O_3 plus SiO_2, substantially all of the aluminosilicate material passing a 325 mesh screen and having an average particle size less than 5 microns.

The amounts of the two materials are chosen so that the two materials together contain, on the ignited basis, 65 to 80 parts by weight Al_2O_3 for each 20 to 35 parts by weight SiO_2. The materials are lightly calcined at 750° to 1150°C, milled to an average particle size of less than 5 microns, admixed, and fired to form a refractory material containing a predominant proportion of mullite and less than 5% material other than Al_2O_3 and SiO_2.

Example: The aluminum trihydrate had an average particle size of 40 microns, substantially all of it being coarser than 2 microns. Its chemical analysis was as follows: 64.8% Al_2O_3, 0.4% Na_2O, 0.1% CaO, 34.7% ignition loss, together with trace amounts (i.e., less than 0.05% total) of other impurities. One thousand grams of the aluminum trihydrate, slurried in 1,000 grams of water, were milled for 8 hours in a 1-gallon porcelain ball mill filled half full with alumina balls. After this milling, substantially all of the aluminum trihydrate was finer than 20 microns and it had an average particle size of less than 5 microns.

The kaolin was substantially all finer than 20 microns and had an average particle size of about 1.3 microns. Its chemical analysis was as follows: 38.3% Al_2O_3, 45.5% SiO_2, 0.5% Fe_2O_3, 1.5% TiO_2, 0.3% CaO, 0.3% MgO, and 13.6% ignition loss, together with trace amounts (i.e., about 0.05% total) of other impurities.

Fifty-five parts by weight of the aluminum trihydrate and 45 parts by weight of the kaolin were mixed as a slurry by a propeller-type mixer. This slurry was dried and the dried cake pulverized in a hammer mill. From the foregoing analyses, it can be calculated that the overall admixture, on the ignited basis, would have the following chemical analysis: 70.6% Al_2O_3, 27.4% SiO_2, 0.3% Fe_2O_3, 0.9% TiO_2, 0.3% Na_2O, 0.2% MgO, and 0.3% CaO. Thus, it can be seen that the alumina and silica account for 98% of the admixture, and that they are of a weight ratio of 72 parts Al_2O_3 to 28 parts SiO_2, corresponding to the 3 mols of alumina to 2 mols of silica found in mullite.

The pulverized admixture was placed in 1" deep silica trays which were then placed in an electrically heated furnace at a temperature of 800°C. After the furnace returned to a temperature of 800°C, the material was soaked (i.e., held at that temperature) for 1 hour. The lightly calcined material was then wet milled for 8 hours in a 1-gallon porcelain ball mill half filled with alumina balls. Water was used in the proportion of 1,000 grams to 1,000 grams of the lightly calcined admixture.

After milling, the material was dried at 105°C, and subsequently blended with 10% water and 2% of a lignosulfonate temporary binder. The wetted admixture was granulated by forcing it through a 20 mesh screen. Pellets ½" in diameter by ½" high were formed from the granulated powder under uniaxial pressure of 20,000 psi. After drying, the formed pellets were sintered at 1750°C for 20 minutes, being raised to that temperature at 875°C/hr. After firing, the compacts had a bulk density of 3.00 g/cc.

Petrographic examination of the fired compacts showed them to comprise 92 volume percent mullite, 3 volume percent glass, and 5 volume percent total pores. The pores were fewer and smaller than those in compacts made the same way except for omission of the light calcination step. The mullite crystals were one micron in size. No residual alpha-alumina was seen.

ALUMINUM, ZIRCONIUM
AND CHROMIUM REFRACTORIES

HIGH ALUMINA REFRACTORIES

High Alumina Brick

P.G. Kihlstedt and K.S.E. Forssberg; U.S. Patent 3,953,563; April 27, 1976; assigned to Advanced Mineral Research, Sweden describe the preparation of high alumina refractory bricks containing more than 50% by weight alumina. The process is carried out by (1) forming a damp mixture of an alumina rich material containing at least 50% by weight of nonactivated alumina and a binding agent amounting to 5 to 50% by weight of the total mixture in the dry state and comprising finely divided particles of activated alumina; (2) pressing the moist mixture into brick shapes; (3) drying the shapes to prevent crack formation during steam hardening; and (4) steam hardening the shapes at 160° to 230°C and 5 to 70 atm for 1 to 24 hours to convert all of the activated alumina into boehmite.

The binder may be slaked lime in quantities of 1 to 5% by weight of the total mixture in the dry state together with activated alumina; $Mg(OH)_2$ in quantities of 1 to 25% by weight of the total mixture in the dry state together with activated alumina; or finely divided SiO_2 in quantities of 1 to 10% by weight of the total mixture in the dry state and having a specific surface exceeding 20,000 square centimeters per cubic centimeter together with activated alumina.

Example 1: Sixty-five percent by weight of comminuted nonactivated alumina having a particle size of $K_{80} = 0.8$ mm (that is, 80% by weight of the alumina particles pass through a mesh having an opening size of 0.8 mm) was admixed with 20% by weight of a nonactivated alumina having particle size of $K_{80} = 0.05$ millimeters and 15% by weight activated alumina having a specific surface of 54 m^2/g and a particle size of $K_{80} = 0.01$ mm.

The materials were carefully mixed and a quantity of water amounting to 5% by weight of the mix was added thereto. Mixing was continued until it could be seen visually that the mix was homogeneous. The mix was then charged to a hydraulic press and slugs having a diameter of 40 mm and a height or length of 30 mm were produced therein at 1,500 kp/cm². The slugs were then removed from the press and dried to steady weight for 16 hours at 105°C, so that the water content of the slugs was less than 1% by weight. The dried slugs were then autoclaved for 16 hours at 205°C and corresponding pressure in a saturated steam atmosphere. It was found that the autoclaved slugs had a porosity of 18% and a cold compression strength of 2,100 kp/cm², when a load was applied to the opposing end surfaces of the slugs.

Example 2: In the manner of Example 1, 65% by weight of nonactivated alumina having a particle size of K_{80} = 0.8 mm, 17% by weight of nonactivated alumina having a particle size of K_{80} = 0.05 mm and 15% by weight activated alumina having a specific surface of 54 m²/g and a particle size of K_{80} = 0.01 mm were admixed with 3% by weight slaked lime [$Ca(OH)_2$] having an approximate particle size of K_{80} = 0.03 mm and a specific surface of about 10 m²/g.

Water was added to a content of 5% by weight of the mix. The mix was then treated in the manner described in Example 1, to form autoclaved slugs. The slugs were found to have a porosity of 17% and a cold compression strength of 2,300 kp/cm², when measured in accordance with Example 1.

Corundum Bonded with Aluminoborosilicate

R.G. LaBar; U.S. Patent 3,879,210; April 22, 1975; assigned to The Carborundum Company describes a substantially crack-free corrosion and thermal shock resistant fused-cast refractory article with a crystal structure comprising crystals of corundum bonded by a refractory aluminum borosilicate glass. The refractory comprises 98 to 99 weight percent alumina, 0.25 to 1 weight percent boric oxide and 0.25 to 1 weight percent silica. Small amounts of other metallic oxides may be present if they constitute less than 0.3 weight percent of the refractory.

The preferred composition comprises 99 weight percent alumina with 0.5 weight percent boric oxide and 0.5 weight percent silica. After fusion, the resulting castings may contain 0.4 to 0.9 weight percent silica and 0.3 to 0.7 weight percent boric oxide, depending on the conditions of melting and sampling of the casting. The refractory microstructure comprises crystalline corundum and is preferably composed of large interlocking crystals of α-alumina or corundum and nearly continuous intercrystalline bonding film of an aluminum borosilicate glass.

The refractories may be made by fusing the raw batch of material in a suitable furnace such as an arc-furnace. In carrying out the fusion, the furnace may consist of a water-cooled iron or steel shell, having no other lining than that built up by the material being fused as it is fed into the furnace. Fusion is effected initially by the heat from graphite starting rods between the tips of two or more graphite electrodes.

After sufficient molten material is formed around the starting rods, they are removed and the electricity flows through the molten oxide. The batch is continually fed in and the electrodes raised, as the quantity of molten material increases. When a sufficient amount of melt is produced, the molten oxide is poured into graphite or water-cooled iron molds by tilting the furnace. It is extremely important to avoid contamination of the melt by carbon from the starting rods, electrodes or any other source. Thus, the starting rods must be removed as soon as possible to avoid submergence of the electrodes. The molded articles are left in the mold for a brief period of time depending on the volume and surface area of casting. Once a solidified outer layer of refractory forms, the molds may be removed. The molded articles are then placed in a container and buried in a lightweight refractory powder, which insulates the articles from rapid cooling; or they may be placed in a hot kiln for controlled cooling. This annealing procedure is critical to the successful recovery of crack-free refractory articles.

High Purity β-Alumina

T.L. Francis and G. MacZura; U.S. Patent 3,859,427; January 7, 1975; assigned to Aluminum Company of America describe a process for preparation of β-alumina useful as a solid electrolyte. The process involves forming β-alumina by reacting a mixture of alumina hydrate or calcined alumina or both, a source of cationic oxide other than aluminum oxide and a different source of fluoride from that for the cationic oxide other than aluminum oxide.

The median crystal size of the β-alumina is 5 to 15 microns as determined by microscopic appraisal. The alumina hydrate starting material may be Bayer hydrate prepared according to the well-known Bayer process by caustic digestion of bauxite and precipitation of the alumina hydrate from the resulting sodium aluminate solution by seeding, carbonation or other suitable treatment. The starting alumina hydrate is preferably predominantly α-alumina trihydrate.

The calcined alumina which may be used as the starting material in place of or in addition to the alumina hydrate may be prepared by calcining a Bayer or other alumina hydrate. When calcined alumina is included in the starting material, it may be in the α- or γ-alumina forms or both, used either alone or with the hydrate. The median crystal size of the starting calcined alumina or alumina hydrate is 0.5 to 6 microns. The purity of the product may be as high as 99% by weight β-alumina. The density of the β-alumina product when fired is 1.3 to greater than 3.2 g/cc.

The cationic oxide may be supplied from any appropriate source other than the source of the fluoride which under the reaction conditions will provide available soda, for example, carbonate, sulfate, nitrate or the like, the carbonate being preferred. Preferred cations of the cationic oxide are the alkali metal ions; preferably sodium and potassium.

Preferred cations have a valence of not greater than two. The temperature for reacting the alumina hydrate, calcined alumina or both, cationic oxide source, and fluoride source may be as high as 1600°, preferably, 900° to 1400°C. It is preferred that the reaction take place in the solid state.

The reaction may take place in a static or in a moving bed such as a rotary kiln, and the reaction may be completed in 10 minutes to 30 hours. There is a tendency for the rotary kiln to require less fluoride than a static bed. Materials, such as aluminum fluoride, sodium fluoride, hydrogen fluoride, cryolite, chiolite and the like may be used as the separate source of the fluoride, aluminum fluoride being preferred.

Example 1: A mixture of 90% calcined Bayer alumina and 10% Na_2O added as sodium carbonate when heated statically at 1675°C (25°C above the melting point of sodium aluminate) for 15 minutes yielded a substantially pure sodium β-alumina. The fired density was 3.2 g/cc.

Example 2: The same mixture used in Example 1 when heated statically at 1400°C for 15 minutes yielded a mixture of primarily sodium ζ-alumina ($Na_2O\cdot5Al_2O_3$) and the remainder principally sodium β-alumina ($Na_2O\cdot11Al_2O_3$).

Example 3: Five weight percent AlF_3 was added to the starting mixture in Example 1. The resulting mixture was then heated in a static bed at 1400°C for 15 minutes. A substantially pure sodium β-alumina was produced.

Example 4: The starting material of Example 3 was heated in a static bed at 1300°C for 15 minutes. A substantially pure sodium β-alumina ($Na_2O\cdot11Al_2O_3$) resulted.

Example 5: The starting material of Example 3 was heated at 1350°C in a rotary kiln. A substantially pure sodium β-alumina ($Na_2O\cdot11Al_2O_3$) resulted.

Alumina with Improved Smoothness and Electrical Properties

According to *K. Niwa, Y. Anzai, K. Hashimoto and H. Yokoyama; U.S. Patent 3,854,965; December 17, 1974; assigned to Fujitsu Limited, Japan* an alumina substrate with improved surface smoothness and electrical properties is prepared by the addition of 0.1 to 0.4 weight percent magnesium oxide and 0.001 to 0.05 weight percent chromium oxide to alumina powder, slip casting and firing at 1550° to 1670°C. The most desirable alumina substrate is produced by the addition of 0.25 weight percent of MgO and 0.02 to 0.03 weight percent of Cr_2O_3 and firing at 1600°C.

In addition, polyvinyl butyral is used as a binder, solubitane trioleate is used as a deflocculent, dibutyl phthalate is used as a plasticizer, and methyl ethyl ketone (MEK), methyl alcohol and n-butyl alcohol are used as solvents. These materials are added to 100 parts of alumina powder. The materials are 5.5 to 9.5 parts of polyvinyl butyral (4.0 to 6.0 parts for an average degree of polymerization of 250 to 500 and 1.5 to 3.5 parts for a degree of polymerization of 1,000 to 2,000), 0.8 to 1.5 parts of solubitane trioleate, 0.5 to 1.2 parts of dibutyl phthalate, 20 to 30 parts of MEK, 10 to 20 parts of methyl alcohol, 8 to 12 parts of n-butyl alcohol. The alumina powder consists of γ-alumina powder of 30 weight percent or less and residual α-alumina powder.

Chromic Orthophosphate Bonding Agent

According to *N.A. Hill; U.S. Patent 3,847,629; November 12, 1974; assigned to Morgan Refractories Limited, England* a plasticized moldable alumina-base composition hardenable on firing to give a refractory, consists of an alumina aggregate, a plasticizer, and a water-insoluble chromium phosphate as a bonding agent for the alumina, the phosphate forming 0.5 to 10% by weight of the composition.

Particularly chromic orthophosphate, or other water-insoluble phosphate is used as the bonding agent. The material used in the examples below approximates the formula $CrPO_4\cdot3H_2O$, with a Cr_2O_3 content of 35 to 37%. Orthophosphoric acid is preferably also added to the compositions in an amount that, while not itself sufficient to effect the bonding when the compositions are fired, nor to cause premature hardening in storage, gives a degree of air-hardening or stiffening in drying of the compositions at ambient temperatures before they are fired, and also some low temperature strength at 300° to 400°C. The plasticizer may be clay, at 3 to 15% by weight, or an organic plasticizer at 1 to 5%. Applications include soaking-pit linings, the bottoms of steel ingot molds, and critical areas in arc melting furnaces.

Example 1: The following composition was made up:

	Parts by Weight
Alumina grog mixture 4/100 mesh	64
Fine alumina (particle size about 10 microns	19
Ball clay/china clay (equal amounts)	10
Raw kyanite	7
Chromic orthophosphate powder	1.25
Phosphoric acid (H_3PO_4, 67% solution)	0.75
Water	9

Fused alumina grog was used but tabular alumina is also suitable. In a four month storage test with weekly determinations of workability the value, initially 26%, remained above 22% even after sixteen weeks of storage. In contrast in a similar composition containing two parts of 67% phosphoric acid and eight parts water (that is, no chromic orthophosphate), the initial workability, although higher at 30%, came down after sixteen weeks of storage to a value below 12%, which is regarded as unacceptable. The composition gave an excellent refractory when fired.

Example 2: A composition as in Example 1 but containing two parts of the chromic orthophosphate with eight parts of water (that is, no phosphoric acid) was similarly stored and tested. The workability, initially 34%, remained above 24% for ten weeks and after twelve weeks was still above 20%. The composition gave an excellent refractory. The crushing strength after firing at 1000°C was 9,000 psi.

Diammonium Phosphate Binder

J.R. Parsons and H.L. Rechter; U.S. Patent 3,832,193; August 27, 1974; assigned to Chicago Fire Brick Company describe a storable water-wet plastic refractory composition comprising: (a) refractory particles selected from clay, alumina, bauxite, and/or kyanite; (b) a bonding amount of diammonium phosphate; (c) uncombined water in amount sufficient to wet the composition; and (d) an acid or acid salt mixed therewith in an amount to give a pH of 7.0 to 3.0.

The acid includes phosphoric acid, sulfuric, acetic and citric acids, and the acid salt includes monoaluminum phosphate and aluminum sulfate. The compositions, by varying proportions and water content, may be used as ramming, gunning and patching mixes, mortar, and for plastic, brick, block and the like. They form phosphate bonded refractory products capable of high temperature service. It is not necessary to maintain the refractory-diammonium phosphate composition completely dry in order to prevent evolution of ammonia and loss of binding strength. Any acid will do, but aluminum phosphate and/or phosphoric acid will add some bond strength. A minimum pH of about 3.0 will insure that long shelf life with reactive clays can be maintained. A pH of 6.0 to 6.9 is recommended for a particularly wet product, such as a mortar, containing fine alumina and clay. A blend of diammonium and monoaluminum phosphate improves air setting properties in some cases and suppresses evolution of ammonia from the wet mixes. Any acid addition will suppress the odor of ammonia, such as acetic acid, sulfuric acid, hydrochloric acid or aluminum sulfate.

The particle sizing of the refractory products can be ⅜ inch or larger such as with use of calcined bauxite or flint clays, and as fine as submicron sized clays, but generally range from ¼ inch and finer for ramming mixes and plastics, and 28 mesh and finer for mortars.

The diammonium phosphate is used with the refractory particles in amount sufficient to bond the particles together on drying or on pressing and heating. The amount of diammonium phosphate may range from 0.5 to 20% or higher, but when the diammonium phosphate is used without aluminum phosphate bonding agent or other bonding agent, it is preferred to use at least 6% of the diammonium phosphate to obtain higher strength of bond.

Example 1: A high alumina (89% on a fired basis) plastic refractory was formulated as follows:

	Amount, wt %	Suitable Range
−6 Mesh alumina	46.9	20 - 60
−325 Mesh alumina	23.5	50 - 10
35 Mesh kyanite	7.8	0 - 15
Kaolin clay	3.9	15 - 0
Bentonite	2.9	0 - 10
Diammonium phosphate	5.9	4 - 12
Citric acid	2.9	1 - 5
Water	6.1	4 - 8

This material had an initial workability of over 40% and this was retained over a period of 5 months during which monitoring took place.

Example 2: Another high alumina plastic refractory formulation was made using a binder combination of monoaluminum phosphate with diammonium phosphate:

	Amount, wt %	Suitable Range
−6 Mesh alumina	48.7	20 - 60
−325 Mesh alumina	24.4	50 - 10
35 Mesh kyanite	8.1	0 - 15
Kaolin clay	4.1	15 - 0
Bentonite	3.0	0 - 10
$Al(H_2PO_4)_3$	4.1	½ - 8
$(NH_4)_2HPO_4$	2.0	4 - 12
Water	5.6	3 - 8

The aluminum phosphate used in this example contained 10% Al_2O_3, 52% P_2O_5 and 37% water by weight. This formulation exhibited an average flexural strength of over 2,500 psi and cold crushing strength in excess of 6,000 psi, with a density of 166 lb/ft³ after firing at 2000°F for five hours. The water content was adjusted for pressing brick, and more water will increase the plasticity for use as a plastic refractory.

Lignosulfonate Binder for High Alumina Brick

S.R. Pavlica and E.P. Weaver; U.S. Patent 3,832,194; August 27, 1974; assigned to Dresser Industries, Inc. describe a binder for high alumina refractory brick to provide an improvement in pressing or working characteristics, density and strength comprising 0.5 to 5% of an aluminum lignosulfonate or chromium lignosulfonate, preferably, 1 to 2%.

The sulfonated lignin materials comprise the aluminum or chromium salt of sulfonated lignin-containing materials such as those obtained from the pulping of wood and other lignocellulosic material, for example, spent sulfite liquor obtained from the pulping of wood by the neutral and acid bisulfite process, sulfonated kraft process lignin and sulfonated soda lignin; the metal ion of the salts preferably being present in an amount chemically equivalent to 2 to 15% on an oxide basis. Preferably, the chromium is present in amounts of 3 to 15% and the aluminum between 2 and 3%. Partial chemical analysis showing the inorganic metal oxides contained in the binders utilized in the examples is given in the table below.

	Binder Designation					
	A	B	C	D	E	F
Partial chemical analysis (percent):						
Alumina (Al_2O_3)	2.6	1.9	–	–	–	–
Iron oxide (Fe_2O_3)	0.08	0.02	–	–	–	–
Chromic oxide (Cr_2O_3)	–	–	3.4	5.3	6.6	14.4
Manganese oxide (MnO)	–	–	2.5	0.0	3.6	0.1
Soda (Na_2O)	1.79	0.01	6.61	3.92	7.25	0.4
Potash (K_2O)	0.20	0.19	0.21	0.17	0.20	0.02
Lithia (Li_2O)	0.01	0.01	0.02	0.02	0.02	0.02
Ash content (solids basis)	1.25	6.5	27.0	16.8	32.3	18.1
pH value	3.9	3.6	5.2	3.6	3.0	3.5

Examples 1 through 9: In the following examples, the standard power press method of making refractory brick was employed. The components were crushed and thoroughly blended together to give a typical brick making grind wherein 15 to 25% of the refractory material is minus 10 plus 28 mesh, 20 to 35% is minus 28 plus 65 mesh and balance is minus 65 mesh.

About 1 to 6% by weight of water was added to alumina mixtures consisting of 75% tabular alumina, 15% calcined alumina, 5% potter's flint and 5% volatilized silica. The particular binder utilized in each of the Examples 1 through 9 is set forth in the table below. After batch mixing, the material was pressed into brick, 9" x 4½" x 2½", at 8 and 10,000 psi. The shapes were removed from the press and subsequently fired for 10 hours at 2730°F.

	Mixes								
	1	2	3	4	5	6	7	8	9
Binder and percent	C-1%	C-2%	D-1%	E-2%	F-1%	A-2%	B-1%	D-1%	A-1%
Density, pcf	192	195	195	193	190	194	194	193	183
Modulus of rupture:									
At room temp	4,910	5,170	5,030	5,050	3,670	5,230	4,670	5,220	3,080
At 2700°F	2,450	2,720	2,810	2,710	2,380	2,530	2,970	2,970	1,600

As can be seen from the above, all of the brick can be characterized as having a high density and also a high strength as measured by the modulus of rupture test at both room temperature and elevated temperatures.

High Density Alumina

When the alumina refractory bodies are prepared, that is, compositions containing 90+% alumina, a reaction between the fine alumina and any free silica, which is present in the aluminum composition, takes place and tends to form mullite. This reaction causes a reduction in the density of the refractory to take place. The body also tends to be more porous and linear expansion takes place. When some compounds are added in sufficient quantity to reduce the linear expansion effectively, the fired body loses some of its refractoriness and therefore becomes unacceptable for high temperature service.

Accordingly *E.L. Manigault; U.S. Patent 3,808,013; April 30, 1974; assigned to The Chas. Taylor's Sons Company* has prepared a composition comprising 76 to 96.5% tabular alumina, 0 to 10% calcined alumina, 3 to 10% silica and 0.5 to 4% of rutile titanium dioxide or barium titanate.

This refractory composition produces fired ceramic bodies which have low porosity and absorption and high bulk density. In addition the linear expansion is also reduced. The tabular alumina should be of a size range of -¼ inch through 325 mesh while the calcined alumina should be -325 mesh. Both the barium titanate or rutile and the silica portions should be -325 mesh size.

To this mixture were added sufficient water and a binder to properly temper the mixture. The mix was then formed into bricks and fired at 1500° to 1650°C. The resulting bricks had the following ranges of properties.

Modulus of rupture (psi)	3,171 - 6,126
Porosity (percent)	7.6 - 12.0
Absorption (percent)	2.3 - 3.8
Bulk density (g/cc)	3.18 - 3.27
Linear change (percent)	-0.16 - +0.14

Example 1: In this example the following ingredients were thoroughly blended together to form a ceramic mixture.

Ingredient	Size Range (mesh)	Weight (kg)
Tabular alumina	-4+14	33.5
Tabular alumina	-14+60	19.5
Tabular alumina	-60+325	21.0
Tabular alumina	-325	14.0
Calcined alumina	-325	7.0
Fumed silica	-325	5.0
Barium titanate	-325	1.5
Lignin liquor	--	1.0
Water	--	2.4

The above ingredients were blended in a muller-type mixer, then formed into bricks (9" x 4½" x 3") and, after drying, were fired at 1560°C for 5 hours. The resulting bricks had the following properties.

Modulus of rupture (psi)	5,415
Porosity (percent)	11.0
Absorption (percent)	3.4
Bulk density (g/cc)	3.22
Linear change (percent)	-0.25

Examples 2 through 5: The procedure of Example 1 was repeated except that varying amounts of barium titanate were used in these examples. The results obtained are recorded along with those of Example 1 in the following table. For control purposes the same procedure described in Example 1 was repeated except that no barium titanate was added. Again the results are recorded in the following table.

	Example Number					
Ingredients	1	2	3	4	5	Control
Tabular alumina, kg	88	88	88	88	88	88
Calcined alumina, kg	7	7	7	7	7	7
Silica, kg	5	5	5	5	5	5
Barium titanate, kg	1.5	0.5	1.0	2.0	3.0	–
Modulus of rupture, psi	5,415	5,185	6,101	6,002	6,126	3,870
Porosity, %	11.0	12.0	10.6	10.2	9.4	13.7
Absorption, %	3.4	3.8	3.3	3.1	2.9	4.4
Bulk density, g/cc	3.22	3.18	3.22	3.24	3.26	3.13
Linear change, %	-0.25	-0.16	-0.36	-0.30	-0.42	+0.22

Polycrystalline Abrasion Resistant Alumina

W.T. Kiger and W.E. Blodgett; U.S. Patent 3,802,893; April 9, 1974; assigned to GTE Sylvania Incorporated describe polycrystalline alumina compositions having a uniform fine grain structure which contributes to the strength, toughness and durability of the composition when formed into articles capable of being polished to obtain an extremely smooth surface for uses where abrasion resistance is of primary importance such as cutting tool inserts.

The composition consists of 99.5 to 99.9% alumina oxide, 0.01 to 0.25% magnesium oxide and 0.01 to about 0.25% samarium oxide. The composition has an average grain size of 3 microns and at least 99% of the grains are within the 2 to 5 micron range.

The polycrystalline alumina ceramic is prepared by blending in a noncontaminating medium the aluminum oxide, magnesium oxide, samarium oxide and at least 2 parts per part of the oxide of an aqueous solution of a binder-lubricant system for 16 hours; drying and screening the mixture to form a free-flowing powder having a particle size smaller than the openings in a U.S. Standard 45 mesh screen; cold pressing the powder to form a shaped article; presintering the shaped article to remove the organics from the article; and firing in a hydrogen atmosphere above 1500°C for 5 hours.

The organic binders which can be used are those which are water-soluble at least to the extent of 5% by weight, upon drying at 110°C form cakes, and upon being heated to above 500°C decompose without leaving any appreciable amount of residue or ash. Suitable materials for the binders include dextrin, glycerin, the relatively long chain polyhydric alcohols such as PVA and the like.

While saturated alcohols such as polyethylene glycol are preferred as lubricants, any organic which is water-soluble and forms a film which aids in the blending operation and retards further reduction of particle size can be used. Additionally, the film which is formed after the water is evaporated must give a relatively high green density that is 30% of theoretical. The lubricant must also decompose at

500°C without forming an appreciable amount of ash or residue. The density of the polycrystalline alumina is at least 3.90 g/cm³, preferably 3.93 to 3.97 grams per cubic centimeter. Typical hardness of the fired ceramic is 91.5 to 94.5 on the Rockwell A scale, generally 92 to 94.

Example: Twelve hundred parts of 99.98% pure agglomerate-free alumina, 0.6 part of 99.99% pure samarium oxide and 0.6 part of pure magnesium oxide are weighed and placed in a high purity alumina jar mill containing a high purity alumina grinding medium. To this mixture 2,400 parts of water solution containing 0.5% by weight of polyvinyl alcohol and 2.0% by weight of polyethylene glycol are added. The mixture is then milled for 16 hours.

After milling, the solution is dried at 110°C for at least 24 hours. The dried cake is broken up and passed through a U.S. Standard 45 mesh screen. The resulting powder is pressed into a desired shape at 25,000 psi using a hydraulic press. The pressed parts are then air sintered at 500°C to remove the organic binders and lubricant.

The final sintering is carried out in hydrogen at 1550°C for 5 hours. The resulting parts have a density of 3.96 g/cc with an average grain size of 3 microns. The grains range in size from 2 to 5 microns. The parts exhibit a hardness of 94 on the Rockwell A scale. The parts are finished by diamond lapping to a surface finish of 2 to 3 rms and a flatness of 1 light band.

Substantially similar results are achieved when the magnesium oxide and samarium oxide content is varied throughout the range of from 0.01 to 0.25% by weight. The materials when used as cutting tool inserts on finishing brake drums exhibit wear characteristics exceeding those of other alumina ceramic inserts that do not contain samarium oxide as an additive. About 50% more drums were finished using the ceramic of this example.

In other alumina ceramics in which other rare earth materials such as yttrium oxide, ytterbium oxide, gadolinium oxide, and europium oxide are substituted for samarium oxide, however, at the firing temperatures used, excessive grain growth occurred.

Dense High Purity Alumina from Gibbsite

D.E. Paul and L.M. Housh; U.S. Patent 3,795,724; March 5, 1974; assigned to Kaiser Aluminum & Chemical Corporation describe a method of making dense alumina refractory grain containing at least 95% by weight Al_2O_3 by (a) selecting an aluminum trihydrate of the gibbsite crystal structure containing at least 95% Al_2O_3 on the ignited basis and less than 3% by weight water removable by drying at 110°C; (b) placing particles of the aluminum trihydrate in a confined zone; (c) applying a pressure of at least 60,000 psi to the aluminum trihydrate in the confined zone to produce compacts of high green strength having a green density at least equal to 88% of the true specific gravity of the aluminum trihydrate; and (d) firing the compacts to at least 1650°C.

An apparatus particularly suited for this compaction is the well known roll-type briquetting press, widely used in the refractory and other industries. In addition to subjecting material to high compaction pressures, roll presses impart shear stresses during compaction.

As is common practice in operating roll presses, it will generally be found advantageous to recycle a certain fraction of the compacted material back to the roll feed so that it is subjected to a second compaction. As is also customary in roll press compaction, the hydrate raw material can be admixed before pressing with a lubricant such as aluminum stearate or magnesium stearate.

In addition, when a material such as magnesium stearate is used as lubricant, it decomposes upon firing to yield magnesium oxide (MgO) in the compacts and the presence of this foreign material in the alumina is believed to inhibit crystal growth of the alumina and assist in producing a grain having smaller crystallites and lower porosity.

It is sometimes advantageous to operate the roll press so that the two rolls are turning at slightly different speeds, thus increasing the shear stresses on the compacted material. It is also sometimes useful to operate the roll press with the recesses in the two rolls out of alignment so that the compacts formed are not the usual regular ellipsoid briquettes, but are more irregular in shape, and by this means also increase shearing stresses on the material.

It is possible to produce green, unfired compacts whose green or unfired density is 93% of the theoretical density of gibbsite itself using the gibbsite tertiary thickener seed, an undersize product from a Bayer process, which passes a 65 mesh screen. Raw compacts can have, in themselves, less than 10 vol % porosity.

Aluminum trihydrate compacts have the following characteristics: (1) a green density of at least 2.12 g/cc; (2) the ability in the 1" x ⁵⁄₈" x ³⁄₈" size to withstand, when heated to 1000°C, the impact of a sharpened 56 gram weight dropped from 5 cm or higher; and (3) the ability to be dropped into a rotary kiln at 1000°C and emerge from that kiln after firing above 1650°C sufficiently intact so that at least 80% of the material is retained on a 20 mesh screen.

Tertiary thickener seed as described above was dried to a water content of less than 0.5 weight percent and 50 pounds of this material admixed with one-half weight percent magnesium stearate in a dry mixer. Because the amount of material used was not sufficient to run the roll press continuously, the effect of recirculation of the compacted material was achieved by compacting half the material between smooth rolls 12 inches in diameter by 2 inches wide at a roll separating force of 37 tons, granulating this compacted material so that it all passed a 4 mesh screen, and admixing the precompacted material with the remaining raw material.

This mixture, at 90°C, was fed to briquetting rolls at a roll separating force of 27 tons. These rolls were 12 inches in diameter and 2 inches wide, their surfaces containing recesses 1 inch wide by ⁵⁄₈ inch long and ³⁄₁₆ inch deep, whereby the rolls formed ellipsoidal briquettes about 1" x ⁵⁄₈" x ³⁄₈".

The roll separating force of 37 tons corresponds to an equivalent force of 18.5 tons per linear inch of roll length. It will be understood that, for smooth cylindrical rolls, it is essentially impossible to determine the area of compaction between the rolls, and hence not possible to convert the rolls separating force to an equivalent pressure. The 27 ton roll separating force corresponds to 36.5 tons per square inch (tsi) briquetting pressure. With briquetting rolls, the cross sectional area of the pockets formed by the two rolls at the line of closest contact can be used to calculate an equivalent pressure. The cross-sectional area of a single briquette pocket is calculated by considering it to be elliptical and using the usual formula for the area of an ellipse, $A = \pi ab$, where a and b are the semimajor and semiminor axes of the ellipse. Using the dimensions given above, the cross-sectional area of a single pocket is calculated to be 0.492 square inch.

The cross sectional area of a single briquette pocket is multiplied by a geometrical factor which depends on the number of pockets across the width of the rolls. For the 2" rolls used in the example, which have two staggered rows of pockets, the geometrical factor is 1.5. Thus, the total area on which the 27 ton roll separating force is applied is determined to be 0.74 square inch, giving an equivalent pressure of 36.5 tons per square inch, or 73,000 psi.

The resulting green (that is, unfired) briquettes produced had a density of 2.25 grams per cubic centimeter, 93% of the theoretical density of gibbsite. These briquettes were sintered at 1800°C with a one hour soak. They remained coherent during passage through a rotary kiln, and had a fired porosity of 2.7 volume percent. The Al_2O_3 crystallites in the grain averaged 30 microns in size. The grain analyzed over 99% Al_2O_3 and about 0.1% each of Na_2O, MgO, Fe_2O_3 and K_2O.

Fine Grain Alumina Ceramics

R.L. Clendenen; U.S. Patent 3,776,744; December 4, 1973; assigned to Shell Oil Company describes a ceramic product consisting of 50 to 85% by weight of aluminum oxide and 15 to 50% by weight ferric oxide, chromium oxide and/or titanium oxide. The composition has an average grain size of not more than 3 microns in diameter, a density of at least 95% of the theoretical maximum and is further characterized as being deformable without fracture at strain rates of at least 10% per minute at 55 to 75% of its melting temperature.

Example: A series of alumina-transition metal oxide ceramic compositions was produced by coprecipitating aluminum and transition metal hydroxides with aqueous ammonia from aqueous solutions of the metal nitrates of calculated concentration. Each precipitate was filtered and dried at 50°C to remove free water and placed in a piston-cylinder graphite die. The die was pressurized and heated at a controlled heating rate to calcine and sinter the ceramic.

The ceramic products thus formed had a very fine grain structure, of the order of one micron. The compositions of ceramic products thus produced are listed below, together with the pressure and final temperature of their production and the heating rate employed. The term T/T_m measures the ratio of the temperature employed to the melting temperature and the term d_c/d_{theor} measures the ratio of the actual density of the ceramic to the theoretical density.

Product	Composition, % wt	Pressure, psig	Temp °C	Heating Rate, °C/hr	T/T_m	d_c/d_{theor}
A	50% Al_2O_3 50% Fe_2O_3	4,000	1100	500	0.70	>0.99

(continued)

Product	Composition, % wt	Pressure, psig	Temp °C	Heating Rate, °C/hr	T/T_m	d_c/d_{theor}
B	50% Al_2O_3 50% Fe_2O_3	8,000	1100	300	0.70	>0.99
C	50% Al_2O_3 50% Fe_2O_3	4,000	1050	400	0.68	0.99
D	90% Al_2O_3 10% TiO_2	4,000	900	400	0.55	0.85
E	85% Al_2O_3 15% Fe_2O_3	4,000	1150	400	0.63	0.95
F	50% Al_2O_3 50% Cr_2O_3	4,000	1300	400	0.65	0.99
G	75% Al_2O_3 25% Fe_2O_3	4,000	1200	400	0.74	0.99
H	65% Al_2O_3 35% Fe_2O_3	4,000	1200	400	0.74	>0.99
I	95% Al_2O_3 5% Fe_2O_3	4,000	1300	400	0.70	0.95
J	84% Al_2O_3 14% Fe_2O_3 2% TiO_2	4,000	1100	400	0.60	0.96
K	85% Al_2O_3 15% Fe_2O_3	3,000	950	400	0.55	0.85
L	74% Al_2O_3 24% Fe_2O_3 2% TiO_2	4,000	1150	400	0.64	0.99
M	50% Al_2O_3 30% Fe_2O_3 20% Cr_2O_3	4,000	1400	400	0.75	0.99
N	64% Al_2O_3 34% Fe_2O_3 2% TiO_2	4,000	1100	400	0.62	0.99

Transparent Sintered Alumina

A. Muta, T. Noro, G. Toda and C. Yamazaki; U.S. Patent 3,769,049; October 30, 1973; assigned to Hitachi, Ltd., Japan describe a process for preparing a transparent sintered alumina body having an in-line transmission of not less than 10% per 0.5 mm thickness for radiant energy of all wavelengths from 0.21 to 1.1 micron.

The process is carried out by adding substances capable of inhibiting the grain growth of alumina to high purity alumina or aluminum compounds capable of being converted to high purity alumina by calcination, and then subjecting the mixture to molding and sintering. The additives may be zinc oxide, cadmium oxide, zinc compounds convertible to zinc oxide by calcination and/or cadmium compounds capable of being converted to cadmium oxide by calcination and a small amount of magnesium oxide and/or magnesium compounds capable of being converted to magnesium oxide by calcination.

Magnesium oxide inhibits the discontinuous grain growth of crystal grains of alumina and facilitates the removal of pores present in the crystal grains and crystal grain boundary of alumina, thereby densifying the resulting sintered body. When magnesium oxide and zinc oxide or cadmium oxide are used in combination, the two compounds synergistically act, even when added in very small amounts, to give excellent polycrystalline sintered alumina.

The magnesium, cadmium and zinc oxides added to alumina individually vaporize, in major proportions, at the sintering stage, and the amounts left in the resulting sintered body are no more than trace amounts.

The properties of zinc oxide and cadmium oxide at the sintering stage are such that the zinc oxide starts to vaporize at 1300°C and, when the temperature reaches 1700°C, almost all of the added zinc oxide is vaporized, while the cadmium oxide begins to vaporize at 1100°C and, when the temperature becomes 1300°C or above, almost all of the added cadmium oxide is vaporized. On the other hand, the alumina is rapidly densified at 1200° to 1600°C to show a high shrinkage. That is, the shrinking step of alumina and the vaporization step of the above-mentioned additives proceed at temperatures within the same range. The role of zinc or cadmium oxide is considered ascribable to the fact that the surfaces of alumina crystal grains are activated at the vaporization step of each added oxide.

The sintered alumina is higher in in-line transmission than sintered alumina prepared without the addition of cadmium or zinc oxide. The thus-obtained sintered bodies have wide uses as materials for metal gas sealing tubes of high pressure metal vapor discharge lamps, furnace materials, electronics materials and other heat-resisting and corrosion-resisting materials.

Strengthening Alumina and Spinel

H.P. Kirchner and R.E. Walker; U.S. Patent 3,758,328; September 11, 1973 describe a method of strengthening alumina or spinel of the $Al_2O_3 \cdot MgO$ type by packing the alumina or spinel body in chromium oxide, either with or without a fluorinating pretreatment, which chromium oxide includes a fluoride or chloride taken from the group consisting of CrF_3, $CrCl_3$, AlF_3, CaF_2 and NH_4F, and then refiring the body. It has been found that a low expansion coating of a solid solution is formed which greatly increases the strength of the body as compared with the method in which there is no fluoride or chloride included in the packing material. The preferred ratio of chromium oxide to fluoride is 20 to 70% fluoride. Thereafter, the packed body is refired at 1000°C to a temperature just below the melting point, but preferably 1500 to 1725°C for 1 to 4 hours.

Examples 1 through 6: 94% alumina bodies in the form of tiles 3 x 3 x ¼" were cut into test bars 3 x ¼ x %4" by diamond sawing. A first group of 5 samples was used as controls, and a second group of 5 samples was packed in Cr_2O_3 only and refired at 1500°C in a gas kiln for 4 hours. Six further groups of 5 samples each were then packed in various mixtures of Cr_2O_3 and $CrF_3 \cdot 3\frac{1}{2}H_2O$, except for the last group in which the Cr_2O_3 was omitted. The samples were refired at the same conditions as the second group of samples.

It will be seen, in the table on the following page, that except for Example 1 where the proportion of fluoride is low, and Example 6 where there is no Cr_2O_3 present, the flexural strengths are higher than when Cr_2O_3 is used alone.

Flexural Strength of Alumina Packed in Various Mixtures of Cr_2O_3 and $CrF_3 \cdot 3\frac{1}{2}H_2O$

Number	Treatment	Number Samples	Average Flexural Strength, psi	Strength Difference, psi
	As cut, controls	5	34,200	–
	Packed in Cr_2O_3, gas kiln	5	41,800	+7,600
1	Packed in 90% Cr_2O_3 plus 10% $CrF_3 \cdot 3\frac{1}{2}H_2O$, gas kiln	5	36,600	+2,200
2	Packed in 70% Cr_2O_3 plus 30% $CrF_3 \cdot 3\frac{1}{2}H_2O$, electric kiln	5	49,500	+15,300
3	Packed in 50% Cr_2O_3 plus 50% $CrF_3 \cdot 3\frac{1}{2}H_2O$, electric kiln	5	50,300	+16,100
4	Packed in 50% Cr_2O_3 plus 50% $CrF_3 \cdot 3\frac{1}{2}H_2O$, gas kiln	5	41,900	+7,700
5	Packed in 30% Cr_2O_3 plus 70% $CrF_3 \cdot 3\frac{1}{2}H_2O$, electric kiln	5	50,800	+16,600
6	Packed in 100% $CrF_3 \cdot 3\frac{1}{2}H_2O$, gas kiln	5	41,700	+7,500

High Shear Rolled Alumina Substrate

K.A. Kappes and S. Bateson; U.S. Patent 3,740,243; June 19, 1973 describe a high density high shear alumina ceramic substrate which is especially suitable for use as a substrate for the deposition of thin metallic films used in electronic applications. The substrate is formed from a viscoelastic mass and subjected to high shear rolling. It has an average bulk density of at least 90% of the theoretical density of crystalline alumina and a surface layer of randomly oriented close packed alumina crystals of tabular hexagonal prismatic form providing the surface layer with a density substantially that of the crystalline alumina forming the surface layer, substantially zero porosity and a surface smoothness of less than 8 microinches. The body for thin film microcircuitry desirably has a thickness of 0.004 to 0.40 inch and usually has a bulk density of at least 3.93 and preferably at least 3.96.

This crystal structure is obtained by preparing a viscoelastic mass comprising reactive alumina, a viscoelastic thermoplastic binder and a solvent in particular proportions, forming the mass into a sheet, rolling the sheet preferably in both directions, under high shear between very hard steel rolls to form a sheet having a yield value, subsequently heating the rolled sheet in stages in which the temperature is raised at a controlled rate to remove any residual solvent and then decomposing the binder under oxidizing conditions and firing the sheet at sintering temperature to cause grain growth of 1½ to 5 times the original grain size.

The thermoplastic polymer includes homopolymers or copolymers such as polyethylenes, polycellulosic polymers, nylons or polystyrenes and is preferably a vinyl resin such as a vinyl chloride/vinylidene chloride, vinyl chloride/acetate or vinyl chloride/butyrate copolymer or an acrylic resin such as butyl methacrylate polymers or a vinyl butyral polymer. The particular plasticizer depends on the thermoplastic polymer employed and the capability of the plasticizer to dissolve in the organic solvent. A particularly suitable plasticizer for vinyl chloride copolymer is triethylene glycol di-2-ethylhexoate.

In order to prevent the mixture from sticking to the rollers and to aid in plasticizing the polymer it is desirable to add a wax to the polymer as a lubricant, preferably a methoxypolyethylene glycol. The polymer plasticizer/wax composition is dissolved in a volatile organic solvent capable of forming a solution with the binder composition such toluene, xylene, methylethyl ketone, or acetone, preferably toluene.

Sealing High Density Alumina with Al_2O_3–Ga_2O_3

R.H. Arendt; U.S. Patent 3,736,222; May 29, 1973; assigned to General Electric Company describes a method of forming a seal between high-density alumina bodies by placing between the portions of the bodies to be sealed an aqueous slurry comprising an inorganic mixture of 20 to 60 mol percent Ga_2O_3 and Al_2O_3 and an organic binder, drying the aqueous slurry, firing the composite bodies above the melting point of Ga_2O_3 and below the deformation temperature of the alumina bodies, preferably 1750° to 1850°C, whereby the organic binder is removed and at least a portion of the inorganic mixture initially forms a liquid phase that is transformed to a solid solution of Ga_2O_3–Al_2O_3 at the interface between the bodies, and cooling the sealed bodies. This process produces a gas tight seal. Organic binders include polyvinyl alcohol and polyimides.

Example 1: An inorganic mixture of 1.7613 grams of alumina and 3.2418 grams of gallia was prepared to give a 50–50 molar composition. The powdered oxides were dry mixed overnight in a polyethylene container to form a homogeneous mixture. To this mixture was added an aqueous solution containing 2% by weight of polyvinyl alcohol such that a viscous slurry was formed. The slurry was then applied between a polycrystalline alumina cylinder and a polycrystalline alumina disk to form a coating between the two bodies. Thereafter the slurry was air-dried.

The composite material was placed in a molybdenum element furnace and fired in a hydrogen atmosphere at 1750° to 1850°C for 5 minutes to 2 hours. The heating rate was in excess of 2000°C per hour. The composite body was then rapidly cooled to room temperature.

The sealed alumina composite article was found to be leaktight to gaseous helium. Microstructural analysis of the sealing area indicated that the bond was a crystalline solid solution of gallia and alumina containing aggregate particles of alumina embedded in the gallia-alumina solid solution bond. The portion adjacent to the alumina bodies indicated that the sealing composition had wetted and dissolved a portion of the alumina bodies. Attempts to separate the two bodies indicated that a very strong bond had been formed between them.

Examples 2 through 4: Following the procedure of Example 1, the ratio of a alumina to gallia in the inorganic mixture was varied as shown in the table below:

	Example 2	Example 3	Example 4
Ingredient (grams)			
Ga_2O_3	2.752	2.205	1.577
Al_2O_3	2.245	2.800	3.425
Molar ratio—Al_2O_3:Ga_2O_3	60:40	70:30	80:20

Slurries of these mixtures were also placed between two alumina bodies and fired under the same conditions as described in Example 1. The bonds obtained using each of these slurries was substantially identical in strength and in structure to those previously obtained and a vacuum tight seal was obtained in each instance.

Yttria-Magnesia Sintering Aid for Transparent Alumina

A. Muta, G. Toda, T. Noro and C. Yamazaki; U.S. Patent 3,711,585; January 16,

1973; assigned to Hitachi, Ltd., Japan describe a process for preparing a transparent alumina sintered body by adding a small amount of substances that can
inhibit the alumina grain growth, to the main component comprising highly pure
alumina and/or aluminum compounds that can be converted to highly pure alumina by calcination in an oxidizing atmosphere, and then subjecting the mixture
to molding and to sintering. The grain-growth inhibitor may be a small amount
of at least one member selected from yttrium oxide and yttrium compounds
that can be converted to yttrium oxide by calcination in an oxidizing atmosphere,
and at least one member selected from magnesium oxide and magnesium compounds that can be converted to magnesium oxide by calcination in an oxidizing
atmosphere.

Magnesium oxide and yttrium oxide inhibit the discontinuous grain growth of
alumina crystalline powder and facilitate the removal of pores in the alumina
crystalline grains and on the crystalline grain boundary, whereby they increase
the density of the resulting sintered body.

Almost all of the magnesium oxide added to alumina vaporizes at the sintering
stage and only a trace of magnesium oxide remains in the resulting sintered body.
Yttrium oxide, in contrast thereto, remains in the alumina crystalline grains even
after the final sintering, and thus it plays a role of inhibiting the discontinuous
grain growth of the crystalline grains, even after the vaporization of the magnesium oxide.

Raw alumina is preferably as pure and as fine in particle size as possible. Ordinarily, alumina having an average particle size up to 0.5 μ is used. Particularly
from the viewpoints of heat resistance and corrosion resistance to metal vapor
or the like, the use of alumina which contains as little silica as possible is desired.
The employment in place of alumina of such an aluminum compound as aluminum sulfate which can be converted to α-alumina by calcination may provide
substantially the same results.

As the above-mentioned additives, such soluble compounds as chlorates, fluorates
or carbonates of magnesium and yttrium which can finally be converted to the
oxides by calcination may preferably be used with a wet method. Alternatively,
the powders of magnesium oxide and yttrium oxide as such may be mixed with
alumina in accordance with a wet or dry method under such conditions that the
alumina can be mixed thoroughly and homogeneously.

The mixture of alumina and the additives in accordance with a wet method is
dried to vaporize the solvent mainly by stirring at room temperature and removing the residual volatiles completely in an air bath kept at 100° to 200°C. The
pressurization at the molding step is preferably 1 to 3 t/cm^2.

The presintering step is intended to convert the additives to the oxides, to remove the unnecessary volatile components from the mixture and, if desired, to
give preferable properties for working which may be required for the reforming
to the desired shape. Therefore, the presintering is carried out mainly in an
oxidizing atmosphere at not less than 1000°C, and heating for 1 hour is adequate.

These transparent polycrystalline alumina sintered bodies have an in-line transmission of at least 40% per 0.5 mm thickness of a radiant energy of wavelengths
of 320 to 1,100 mμ.

ZIRCONIA REFRACTORIES

Fired Zirconia-Zirconium Silicate Shaped Elements

H.G. Schwarz; U.S. Patent 3,899,341; August 12, 1975; assigned to Didier-Werke AG, Germany describes zirconium-based material for shaped elements to be inserted into closures in casting units for steel, iron and metal having high resistance to wear and high resistance to thermal shock.

The composition comprises zirconium oxide at 10 to 50% by weight, preferably 35% by weight and zirconium silicate at 50 to 90% by weight, preferably 65% by weight. Moreover, 30 to 60% by weight of the entire amount of zirconium oxide present must be in stabilized form and 70 to 40% by weight must be in a nonstabilized form. The zirconium oxide is employed in a grain size of at most 0.1 mm, and preferably smaller than 0.06 mm, and the zirconium silicate is introduced in a grain size exceeding 0.2 mm, preferably 0.2 to 0.5 mm.

The composition of zirconium silicate in coarse grains and zirconium oxide in fine powder produces a fired element with a framework of zirconium silicate comprising an intermediate matrix of sintered zirconium oxide.

The shaped elements are manufactured both by the ceramic slurry casting process and the pressing process. In order to avoid contamination of the starting material, a zirconium-oxide-containing slurry is used in the casting process. In the pressing process, this slurry or organic binding materials are employed as binders. In order to avoid contamination of the zirconium material through abrasion of foreign substances, the slurry is produced in a wet mill whose lining consists of shaped elements of zirconium oxide, which elements in suitable shape also perform the function of grinding elements.

Example: Slurry-Casting Process — A mixture of 83 kg stabilized zirconium oxide, grain size less than 0.1 mm, and 83 kg nonstabilized zirconium oxide, grain size less than 0.1 mm, with 34 liters water and 5 ml aqueous solution of an alkali salt of a polycarboxylic acid, is ground for 200 hours in a wet mill charged with grinding elements of zirconium oxide, and provided with a lining of zirconium oxide bricks. After 200 hours, 60 parts by weight of zirconium silicate, grain size 0.2 to 0.5 mm, is added to 40 parts by weight of such a slurry and a mixture is made. The mixture is cast into gypsum molds, removed after 12 hours, dried and fired at 1620°C.

Pressing Process — In a muller mixer, there are mixed 650 kg of zirconium silicate, grain size 0.2 to 0.5 mm, while adding 15 liters sulfite waste liquor (30°B), with 175 kg stabilized zirconium oxide, grain size smaller than 0.06 mm, and 175 kg nonstabilized zirconium oxide, for 10 minutes. The completed mass is pressed on a hydraulic press under 500 kg/cm^2 pressure. The pressed blank is dried and then fired at 1620°C.

The steel-pouring nozzles produced possess the following properties:

Chemical Analysis

ZrO_2, % by weight	75
CaO, % by weight	1.5
SiO_2, % by weight	20

(continued)

Gross density	4.0 g/cc
Specific gravity (weight)	5.0 g/cc
Total porosity	20% by volume

When employed in steelworks, the nozzles withstood the tests of use with excellent results.

Zircon Adhesive for Investment Casting

C.P. Mabie, Jr.; U.S. Patent 3,892,579; July 1, 1975; assigned to American Dental Association describes an adhesive composition useful in investment casting, particularly in the preparation of metal dental devices. For example, in casting alloys in the preparation of partial dentures, it is customary to make a pattern or model of that part of the dentition and mouth cavity to which the dental device would be attached, called a refractory cast or model. This cast or model is usually made from gypsum, phosphate or silica-bonded investment. The size and shape of the prospective dental device is determined by adding wax or plastic preforms to the surface of the refractory cast.

The adhesive composition is useful both as a tacky adhesive for securing wax or plastic preforms to a pattern or model and as a liner or protective coat on the inner surface of an investment mold.

The composition consists of an organic solvent solution of an organic acid soap or salt, the cation of which is a metal taken from the group consisting of zirconium and/or cobalt and having dispersed therein very fine granules of a refractory ceramic material. A preferred composition is a salt solution of zirconium octoate in methylene chloride having dispersed refractory granules of zirconium orthosilicate (zircon). The anion portion of the salts are the usual carboxylic groups of organic acids having 3 to 10 carbon atoms.

The salt or soap is prepared in an organic solvent solution preferably methylene chloride, the soap comprising 40 to 70%, preferably 60% by weight of the solution. To the solution are added refractory granules such as refractory metal oxides or silicates to form a slurry comprising 40 to 60% by weight of the refractory granules. The granules should be of very fine particle size, preferably less than 5 μ but not greater than 100 μ. Preferably, zircon granules are used having a majority of particles of a size of 2 to 5 μ mean diameter.

The refractory granules may be metal silicates, aluminates, oxides and aluminosilicates. The metals may be tungsten, molybdenum, tantalum, hafnium, thorium and similar metals. The addition of the refractory oxide, silicate or the like, prohibits or restricts interfacial liquefaction and corrosion resulting from contact of the coating with the molten casting metal. In addition to methylene chloride, higher alcohols may be used as the solvent for the refractory soap, for example, butanol, pentanol or hexanol.

It is important that the coating composition be applied to the refractory cast or the wax or plastic preform in a very thin coat not more than 20 μ, preferably 10 μ. If the coating is too thick, it will induce blistering and pits on the casting surface. This thin coat should be applied by dipping. Brushing the slurry or pouring it over the surface to be coated is not a satisfactory method of application.

Example 1: A coating slip was initially prepared by precipitation of zirconium octoate from a 350 ml sample of a 22% mineral spirits solution (22% ZrO_2) by repeated absolute ethanol rinses. The salt obtained by precipitation, which resembles a white taffy, was dissolved in 350 ml of methylene chloride. This solution contained 60% of zirconium octoate (30% ZrO_2). To the solution was added 600 grams of precision casting grade zircon having a particle size of 2 to 5 μ. The slurry was mixed by repeated pouring back and forth between two polyethylene containers.

A refractory cast, consisting of a gypsum-quartz investment material of a partial denture was dipped into the slurry to coat one-half the surface and then permitted to dry for 30 minutes before application of plastic or wax preforms. Plastic preforms may be lifted off the refractory coated models with care. Engineered plastic and wax patterns may be dipped again in the coating mixture. Excess coat must be knifed around the edges of the pattern in order to eliminate fin formation in the final cast part. Outer investment was then applied and normal burn out and alloy casting was executed.

A low fusing chromium base alloy which fuses below 1400°C was melted and injected centrifugally into the casting investment. Subsequently, the resulting casting and its surrounding investment was let to cool on standing to room temperature. The resulting metal casting was knocked free of surrounding investment with a hammer and then sandblasted to eliminate adhering investment and scale. Comparative profilometer measurements were made over the sandblasted partial denture casting surface with and without this coating. The average center of line values (CTA) were measured and their significance computed.

On the outer investment side, the CTA value was 1.351 ± 0.123 μ for the surface which was cast against this coating. The CTA value for a surface cast against the investment was 2.002 ± 0.127. Profilometer values on the refractory-cast casting side were the same for both the coated and uncoated areas, 1.880 μ. Prior to sandblasting the surface cast against the coating was shiny and showed no sign of scaling. This was in contrast to the surface cast directly against the investment, which was a dull gray and exhibited abundant scaling. The absence of scaling, when this coat is used, is due to the fact that the zircon-zirconium octoate composition after burnout does not liquefy during the casting and thus react with oxides and suboxides of vaporized casted alloy constituents. Thermal and corrosive effects were not present when this composition was employed as a protective coating.

Example 2: The zirconium octoate solution of Example 1 was replaced in the amount of 50% by a methylene chloride solution of cobalt octoate. The cobalt octoate improved the flux capability of the coating although the refractoriness of the coating was reduced. The improved flux capability of the cobalt octoate addition aids in the adherence of the protective coat of this process to the investment both during and after burnout, thus reducing the chance of spalling into the mold.

Example 3: Partial denture plates were made from plastic triforms using a single dip of the composition of Example 1 as an adhesive coat on the refractory cast. Two castings were made utilizing low fusing chromium-nickel-base alloy which fuses below 1400°C and investing technique and three were made from the high fusing chromium-cobalt base alloy which fuses above 1440°C and phosphate bonded investment. In all castings high detail resolution and nearly complete scale removal were effected.

High Strength Shaped Zirconium Oxide

It is known that cubic zirconium oxide can be stabilized by heating the oxide to a high temperature in the presence of oxides of metals having an ionic radius similar to that of zirconium, such as calcium, magnesium, yttrium, lanthanum, cerium, ytterbium and other rare earth metals, as well as cadmium, manganese, cobalt and titanium, which diffuse into the zirconium oxide and are bound thereto. Sintered bodies prepared from stabilized zirconium oxide are coherent and strong enough to permit their use in laboratory equipment in grinding balls, extruder mandrels, yarn guides and the like.

H. Sturhahn; U.S. Patent 3,887,387; June 3, 1975; assigned to Feldmuhle Anlagen und Produktions GmbH, Germany has shown that shaped bodies combining the desirable thermal and chemical properties of zirconium oxide with superior mechanical strength can be obtained by sintering compacts prepared from intimately mixed pulverulent monoclinic zirconium oxide, pulverulent zirconium oxide stabilized as described above, and pulverulent magnesium oxide or calcium oxide.

The monoclinic zirconium oxide should amount to 30 to 90% of the compacted mixture, the stabilized zirconium oxide to 7.8 to 69.5%, and the magnesium or calcium oxide to 0.5 to 2.2%, the amount of stabilizing oxide in the stabilized zirconium oxide being chosen to make the amount of all oxides other than zirconium oxide in the sintered body 2.5 to 3.5%. When the other oxides are magnesium oxide and/or calcium oxide, the combined weight of magnesium oxide and calcium oxide should be 2.7 to 3.3%.

The sintering time and sintering temperature are selected in such a manner that 75 to 95% of the crystals in the sintered body are cubic. It is then found that the density of the sintered body is at least 5.5 g/cm^3, but may closely approach that of zirconia monocrystals, and the flexural strength of test rods prepared from the sintered material is at least 30 kp/mm^2.

When magnesium oxide and zirconium oxide are the only oxides present in amounts greater than incidental trace amounts or as impurities, a flexural strength of 45 kp/mm^2 is readily exceeded when the amount of magnesium oxide is between 2.5 and 3.0%, a strength greater than 55 kp/mm^2 is achieved within the range of 2.6 and 2.85% magnesium oxide, and optimum values of flexural strength are reached at magnesium oxide contents near 2.7%. Densities greater than 5.6 g/cm^3 and even beyond 5.7 g/cm^3 are found under preferred conditions.

When the oxides in the sintered body are magnesium oxide and calcium oxide in a weight ratio between 1:2 and 2:1, the shear strength of the material exceeds 55 kp/mm^2, the average grain size is greater than 35 $m\mu$, and the density is greater than 5.55 g/cm^3.

These zirconium oxide products have some mechanical properties closely similar to corresponding properties of pure, sintered aluminum oxide, and other mechanical properties which are superior to those of sintered, polycrystalline aluminum oxide.

Example: A pulverulent mixture was prepared from 4,200 grams (68.7%) mono-clinic zirconium oxide, 1,800 grams (29.4%) stabilized zirconium oxide and 117 grams (1.9%) magnesium oxide. The ingredients were ground in the presence of a small amount of water until the specific surface area of the powder reached 4.5 m^2/g.

The zirconium oxide powder, which was the major component of the mixture, was prepared by crushing molten zirconium oxide having a purity of more than 99% ZrO_2. The prestabilized fraction contained 3.8% magnesium oxide, based on zirconium oxide.

Approximately 1% polyvinyl alcohol in the form of a 10% aqueous solution was added to the finely-ground mixture to facilitate the shaping of green compacts in a conventional laboratory press. The compacts were gradually heated in a tunnel oven in air to 1770°C over 10 hours, held at temperature for 2 hours and cooled to ambient temperature in 10 hours.

The 30 sintered, rectangular test pieces so prepared had dimensions of 4 x 4 x 32 millimeters. They had a specific gravity of 5.75, a hardness of 1,200 kp/mm^2 (Knoop indenter, 300 g), and an average flexural breaking strength of 63 kp/mm^2 in the "as sintered" condition. The strength of test pieces prepared in the same manner, but polished after sintering, was 10% higher.

Tough ZrO_2 and CaO/ZrO_2 Eutectic

C.O. Hulse and J.A. Batt; U.S. Patent 3,887,384; June 3, 1975; assigned to United Aircraft Corporation describe an article exhibiting high toughness to very high temperatures comprising a matrix phase consisting of a compound of cal-cium oxide and zirconium oxide and a reinforcing phase consisting of zirconium oxide embedded therein in the form of lamellae substantially oriented in the direction of principal anticipated stress.

The eutectic is that existing between CaO/ZrO_2 and ZrO_2. This eutectic compo-sition melts at 2300°C and both phases are cubic. The coefficients of thermal expansion of the two phases are identical and, therefore, any possible effects due to this variable are eliminated. The eutectic also exhibits a unique ability to undergo plastic deformation at elevated temperatures without loss of strength.

The microstructure of the directionally-solidified eutectic is lamellar with the maximum spacing between the lamellae being 3 μ when solidified at a rate of 10 cm/hr.

Sintering Yttria-Stabilized Zirconia with Thoria

W.H. Rhodes; U.S. Patent 3,862,283; January 21, 1975; assigned to the U.S. Secretary of the Air Force describes the use of thorium oxide as a single phase sintering aid for oxide or zirconia refractories. The thoria, in finely-divided form, is introduced into a batch of zirconia as an additive material in 0.2 to 10.0% by weight of the final powder mix. The zirconia is chemically pure 6.5 mol percent yttria-stabilized zirconium dioxide having a particulate size of 10 μ or less. The powder mixture is then introduced into conventional mixing apparatus, such as a porcelain ball, and mixed intimately for an hour in order to assure proper mix-ing. The resultant dry mix is then formed into a desired shape by dry pressing.

The shaped body is then placed into a conventional furnace and fired or sintered using either air or oxygen as the sintering environment. Firing is accomplished at 1400° to 1900°C for 4 to 19 hours with 1500° to 1550°C being preferable. Densities of 95 to 98% with an impurity pickup of less than 0.1% were accomplished.

The thorium ions segregate to the grain boundaries limiting grain growth, thus allowing more time for pore removal prior to pore entrapping grain growth. Thus, higher relative densities are achieved during sintering without the formation of a discrete second phase. The zirconia products are single phase which maintain their intrinsic properties such as refractoriness, creep strength and optical properties.

Examples 1 through 13:

- - - - - - - - - - - - - - - Effect of Thoria on Sintering - - - - - - - - - - - - - - - -

Ex. No.	Weight Percent ThO_2	Cold Pressing Pressure	Sintered Relative Density
1	0.2	40	0.878
2	0.2	70	0.934
3	0.4	40	0.912
4	0.4	70	0.929
5	1.0	40	0.901
6	1.0	70	0.935
7	5.0	40	0.922
8	5.0	70	0.955
9	10.0	40	0.882
10	10.0	70	0.941
11	0.2	70	0.942
12	1.0	70	0.936
13	5.0	70	0.999

The samples of zirconia in Examples 1 through 12 were sintered at 1500°C for 4 hours in air, while the sample in Example 13 was sintered at 1550°C for 19 hours in oxygen.

In the examples, the thorium oxide was added as the nitrate in a water solution. The slip was wet ball milled using zirconia balls and a polyethylene jar. The dried powder was pressed into samples and fired in air, thus allowing for the conversion to ThO_2 during the initial stages of sintering.

The effectiveness of the ThO_2 with the undoped powder and the compositions tested are compared. An increase in sintered bulk density with up to 5 weight percent ThO_2 was realized. The fact that the 10 weight percent composition gave lower density was taken as proof that the additive was affecting the sintering characteristics of the powder and not just affecting the theoretical density of the mix. The sample in Example 13 containing 5% ThO_2 was given an extended sintering cycle. The sample was highly translucent and had a bulk density of 6.071 g/cc, the highest measured. However, it was cracked and the microstructure illustrated that mixing was inhomogeneous.

The microstructure of a 1 weight percent addition shows fully dense islands within a more porous matrix. The starting undoped powder gave uniform micro-structures, thus the observed dense regions are not a result of dominant intra-agglomerate sintering. The grain size was 2.6 μ as compared with 7.6 μ for un-doped zirconia powder receiving an identical sintering cycle. The results indicate that ThO_2 is an effective grain growth inhibitor and sintering aid. Further, the 5 weight percent addition appears to be near the proper level to gain optimum benefit.

Purifying Alumina or Zirconia for Electrophoretic Deposition

R.L. Lambert; U.S. Patent 3,859,426; January 7, 1975; assigned to GTE Sylvania Incorporated describes a method of purifying alumina or zirconia for use in electrophoretic deposition processes. The method comprises firing a quantity of the oxide in a muffle furnace for 14 to 16 hours at 1050° to 1100°C. After cooling the oxide in a dessicator to 20° to 25°C, it is repeatedly leached with hot deionized water until a rinse shows a pH of 7.0 to 7.5. Then the oxide is dried at 105°C and stored in a dessicator.

This process removes water and alcohol-soluble salts to the point where they are present at only a few parts per million and allows the use of a low-voltage, low-current cataphoretic process. The use of the cataphoretic process greatly reduces the surface oxidation of the heater core and permits high temperature sintering in a low dew point hydrogen atmosphere.

Examples 1 through 5: To purify a refractory oxide for use with a cataphoretic deposition process a quantity of electronic grade alumina or zirconia is first fired at 1050° to 1100°C in a muffle furnace for 14 to 16 hours. After the firing the oxide is placed in a dessicator to avoid contamination (it is very sensitive to adsorption after this treatment) and is cooled to room temperature, i.e., 20° to 25°C. When the oxide has reached the required temperature, it is repeatedly leached with separate rinses of hot, deionized water until a rinse shows a pH of 7.0 to 7.5. Thereafter, the oxide is dried at 105°C and then stored in a dessicator with fresh dessicant until it is to be used. This process removes residual salts such as sulfate and sodium compounds which not only adversely affect the cataphoresis bath but also adversely affect finished heaters by contributing to heater-cathode leakage.

Typical formulations utilizing the purified oxides are as follows:

Example 1:

Purified alumina	100 grams
1% B-1 binder in isopropanol*	1 ml
Anhydrous isopropanol	349 ml

*B-1 binder is essentially a basic aluminum nitrate

Example 2:

Purified zirconia	100 grams
1% B-1 binder in isopropanol	0.7 ml
Anhydrous isopropanol	159.3 ml

Example 3:

Purified alumina	100 grams
2% aluminum salt of 2,4-pentanedione in anhydrous methanol	100 ml
Anhydrous methanol	100 ml
Methyl ethyl ketone	400 ml

Example 4:

Purified zirconia	100 grams
Zirconium salt of 2,4-pentanedione	1.2 grams
Anhydrous methanol	200 ml
Methyl ethyl ketone	200 ml

Example 5:

Purified alumina	100 grams
Zirconium salt of 2,4-pentanedione	1.2 grams
Anhydrous methanol	100 ml
Anhydrous isopropanol	300 ml

In Example 1 the given concentration of binder would provide 10 ppm of Al^{+3} on a solids basis. In Example 2 the amount would be 7 ppm of Al^{+3}. In Example 3 the aluminum salt of 2,4-pentanedione would provide 160 ppm on a solids basis. Example 3 also illustrates the use of a mixed solvent system which permits multiple dips without washing away previous coatings as the salt is only very slightly soluble in the methyl ethyl ketone.

In Examples 1, 2 and 5 methyl alcohol can be substituted for the isopropanol if the heater coil being coated does not have too many turns per inch. If the number of turns per inch is great, then bridging will more likely occur with the methyl alcohol since the electrophoretic current will be higher and thus the amount of solvent codeposited will be increased.

All of the above-referenced coating suspensions can be mixed in a high speed stirrer or in a ball mill jar and are immediately ready for use. To be employed as a coating suspension the mixture selected is poured to the required depth in a stirred container suitable for electrophoretic deposition. A stainless steel or tantalum anode of proper configuration serves as the positive terminal in the system and the coil or other object to be coated is connected to the negative of the power supply. Obviously, the necessary time-controlled switches, automated dip and transfer devices, etc. can be used as with other coatings so long as proper polarities are observed.

With only slight modifications in respect to solution concentrations, dilutions, etc. coatings of practical weights can be deposited in 2 to 10 seconds at voltages of 10 to 40 volts and current densities of 10 to 600 $\mu A/cm^2$.

It will be seen from the above that insulator coatings of refractory oxides can effectively be applied cataphoretically at low current and low voltage. The treated refractory oxides are low in the residual salts which can cause heater-cathode leakage. By using a known, controlled amount of electrolyte to provide the desired ion and particle mobility a more favorable environment is presented to the core during the thermal decomposition of those salts during sintering. Further, coating density can be better controlled to minimize cracking and peeling, and bridging between turns of a coil is reduced.

Zircon Brick Modified with Chromic Oxide and Rutile

E.L. Manigault; U.S. Patents 3,804,649; April 16, 1974 and 3,791,834; February 12, 1974; both assigned to The Chas. Taylor's Sons Company describes a ceramic composition comprising 85.0 to 98.8% zircon, 1.0 to 8.0% chromic oxide and 0.2 to 5.0% titanium dioxide, all of the percentages expressed on a weight basis. Such a composition when fired at 1450° to 1600°C produced bricks and other shapes which have superior properties to those obtained from the well-known zircon bricks. This refractory is particularly useful as a backup, or subfloor, under the glass contact bottom blocks in a glass-melting tank.

In preparing the refractory composition, the zircon (zirconium silicate) is sized to –70 mesh. The chromic oxide and the titanium dioxide are ground to –325 mesh size. The three ingredients are thoroughly mixed with water and lignin liquor to temper the mixture and the mixture is then pressed into shapes and fired at 1450° to 1600°C.

The bulk density and the modulus of rupture are both increased while the porosity is decreased. The refractory product is also more resistant to iron oxide penetration than the prior art product.

Example 1: 93.5 kg of zircon (–70 mesh) were mixed with 6 kg of chromic oxide and 0.5 kg of rutile titanium dioxide. Both the chromic oxide and the rutile were –325 mesh size.

3 kg of an organic binder and 1.1 kg of water were mixed with the above ingredients to temper the mixture. The mixture was then formed into bricks 9 x 4½ x 3 inches which were fired at 1560°C for 5 hours. The fired bricks had the following properties:

Modulus of rupture, psi	3270
Porosity, %	15.5
Absorption, %	3.9
Bulk density, g/cc	3.95

These properties are superior to a typical standard zircon brick which possesses the following properties:

Modulus of rupture, psi	2300
Porosity, %	17.5
Bulk density, g/cc	3.72

In order to show the bricks' resistance toward penetration of tramp metal into the brick, the following test was run. A ¾-inch black steel nut was placed on the surface of the brick produced and the brick was heated to 1565°C for 5 hours. After cooling the depth of penetration of the iron oxide into the brick was measured as 0.39 inch. For comparison a prior art zircon brick was subjected to the same test and it was found that the depth of penetration was 2⅛ inches.

Examples 2 through 4: In these examples the procedure of Example 1 was repeated except that varying amounts of the ingredients were employed. The amounts of ingredients used and the results obtained are recorded in the following table along with those of Example 1 and the prior art zircon bricks.

Examples 2 and 3 contained chromic oxide in amounts of 2.0 and 4.2% respectively while in Example 4 an amount of 10.0% was used which lies outside the range of chromic oxide addition.

From the table it has clearly been shown that the bricks possessed superior modulus of rupture, increased bulk density, decreased porosity and greater resistance toward iron penetration than the zircon bricks of the prior art. Example 4, which had an excessive amount of chromic oxide, produced inferior modulus of rupture and porosity.

	-----Example Number-----				Prior-Art
	1	2	3	4	Zircon Brick
Zircon, kg	93.5	98	95.4	90	99
Chromic oxide, kg	6	2.0	4.2	10	–
Rutile, kg	0.4	0.4	0.4	0.4	1.0
Modulus of rupture, psi	3,270	3,336	3,514	2,026	2,300
Porosity, %	15.5	16.8	15.8	19.2	17.5
Absorption, %	3.9	4.3	4.1	5.0	3.72
Bulk density, g/cc	3.95	3.87	3.87	3.80	2.12
Nut test, depth of penetration, inches	0.39	0.78	–	–	2.13

Zirconium Oxide Spraying Material

M. Shiroyama and E. Noguchi; U.S. Patent 3,753,745; August 21, 1973; assigned to Nippon Tungsten Company, Ltd., Japan describe zirconium oxide series spraying material capable of providing a dense coating on a metallic substrate and excellent adhesion.

The zirconium oxide series spraying materials may be prepared by grinding and mixing zirconium oxide powder with nickel powder or nickel oxide powder in an amount of 20 to 60 Ni percent by weight of zirconium oxide, pressing the mixture into a block, sintering the block under an oxidizing or reducing atmosphere to form a briquette, grinding the same and size-grading to 140-325 mesh (ASTM) particles.

A very dense coating having a porosity of less than 5% and an adherence strength to a substrate of more than 300 kg/cm^2 is obtained by a flame spraying use of oxyacetylene, and a nonporous coating having an adherence strength of more than 500 kg/cm^2 may be obtained by a plasma spraying, thus providing for a high protective coating to the substrate and a high thermal shock resistance.

These materials are of high oxidation resistance at an elevated temperature giving a high protecting property, and hence they may also be used solely and/or in combination with other ceramic materials as an undercoating in furnace parts, tuyere, mold or nozzle for burning material effectively and in a variety of applications.

Example 1: Nickel oxide particles less than 200 mesh were mixed with zirconium oxide particles again less than 200 mesh in a weight ratio of 38:62, the mixture was ground by ball-milling, the resulting particles were compressed under a pressure of 300 kg/cm² into a block and then the block was sintered into a briquette at 1400°C in the atmosphere. The sintered block was then ground, and the resulting particles were size-graded to obtain zirconium oxide particles containing nickel oxide particles of 140-325 mesh.

These particles were sprayed on a substrate of mild steel, using the oxyacetylene flame, up to a thickness of 0.3 mm, providing a sprayed coating having a porosity of 3% and an adhesion strength of 370 kg/cm². A substrate was covered by a sprayed coating where nickel oxide was uniformly dispersed in zirconium oxide.

The weight increment versus oxidation time while maintaining the coating at 1000°C in the atmosphere proves that there is no weight increment after 20 hours. A series of thermal shock tests in which a test piece heated up to 1000°C in the atmosphere is immersed in cold water, and the sprayed coating observed after 30 repetitions of such heating and immersion, showed that there was no cracking in the coating and peeling off from the substrate. Another test in which a test piece is immersed in a molten pig iron at 1500°C showed no cracking and peeling off of the sprayed coating.

Example 2: In a similar manner to Example 1, nickel oxide powder and zirconium oxide powder both less than 200 mesh were mixed together in a weight ratio of 38:62. The mixture was ball-milled and compressed under a pressure of 300 kg/cm² into a block, the block was sintered into a briquette at 1400°C and hydrogen atmosphere, and then the briquette was ground. The particles were size-graded to obtain zirconium oxide containing nickel oxide of 140-325 mesh.

These particles were spray coated on a substrate of a copper plate using a plasma torch to form an undercoating having a thickness of 0.1 mm and an adherence strength of 500 kg/cm², on which aluminum oxide was coated up to a thickness of 0.3 mm. In the test of immersion into a molten pig iron at 1500°C for 5 seconds similar to Example 1, there was observed no cracking and peeling off in the coating as well as adhesion of the pig iron.

Example 3: Nickel powder and zirconium oxide powder both of less than 200 mesh were mixed together in a weight ratio of 32:68, the mixture was ball-milled and then compressed under a pressure of 300 kg/cm² into a block. The block was sintered into a briquette at 1400°C and in a hydrogen atmosphere. The briquette was then ground into size-graded particles to obtain zirconium oxide of 140-325 mesh containing nickel.

These particles were spray coated on a substrate of mild steel using the oxyacetylene flame up to a thickness of 0.3 mm providing a sprayed coating having a porosity of 5% and an adhesion strength over 300 kg/cm².

Zircon-Pyrophyllite Unfired Refractory Brick

N. Nameishi, H. Yoshino and S. Uto; U.S. Patent 3,752,682; August 14, 1973; assigned to Harima Refractories Co., Ltd., Japan describe an unfired zircon-pyrophyllite refractory having excellent corrosion resistance at elevated temperatures so as to enhance its resistance to both molten steel and slag, prevent adhesion of molten metal and prevent penetration of both metal and slag.

This refractory is manufactured by providing 15 to 30% by weight of naturally-occurring zircon sand having a grain size of less than 0.3 mm, 20 to 40% by weight of comminuted zircon flour having a grain size of less than 0.15 mm, the proportion of zircon sand and zircon flour being adjusted to 60:40 to 40:60, 30 to 35% by weight of raw pyrophyllite having a grain size of 1 to 4 mm, up to 20% by weight of raw pyrophyllite having a grain size of less than 1 mm, and up to 10% by weight of calcined pyrophyllite having a grain size of less than 1.5 mm; adding as a binder up to 3% by weight of powdered sodium silicate, or up to 4% by weight of aqueous sodium silicate or water glass, or both powdered and aqueous sodium silicate together, and 1 to 6% by weight of water so as to maintain the content of Na_2O up to 0.8% by weight; mixing, kneading and molding at 500 kg/cm² to shape; drying in the open atmosphere; and then drying at 80° to 100°C for 24 to 36 hours.

The addition of pyrophyllite to the mixture of zircon sand and zircon flour prevents metal adherence, an inherent defect of zircon refractories, and further, restricts the occurrence of lamination due to the original particle size of the zircon sand being as large as 0.3 mm during the molding step, so that the manufacture of a relatively large size of refractory brick is feasible.

Example 1:

	Percent by Weight
Zircon sand, less than 0.3 mm	29 to 30
Zircon flour, less than 0.15 mm	36 to 40
Raw pyrophyllite, 1 to 4 mm	35 to 30
Powdered sodium silicate No. 1	0.5 to 2.5
Water glass, No. 2	0.5 to 3.5
Water	1 to 5.5

The sum of the first three ingredients should be 100% by weight, and percent by weight of each of the other ingredients listed should be based on the 100% by weight composition.

All ingredients are mixed and kneaded, placed in a mold, molded with a molding pressure of 500 kg/cm² to shape, then dried in the open atmosphere and finally dried at 80° to 100°C for 24 to 36 hours. The manufacturing steps are the same in Examples 1 through 5.

Example 2:

	Percent by Weight
Zircon sand, less than 0.3 mm	29 to 30
Zircon flour, less than 0.15 mm	35 to 30
Raw pyrophyllite, 1 to 4 mm	35 to 30
Raw pyrophyllite, up to 1 mm	1 to 10
Powdered sodium silicate No. 1	2 to 3
Water glass No. 2	1 to 2
Water	2 to 5.5

Example 3:

	Percent by Weight
Zircon sand, less than 0.3 mm	20 to 15
Zircon flour, less than 0.15 mm	20 to 24
Raw pyrophyllite, 1 to 4 mm	30 to 35
Raw pyrophyllite, less than 1 mm	20 to 18
Calcined pyrophyllite, less than 1.5 mm	10 to 8
Water glass No. 2	3 to 4
Powdered sodium silica No. 1	0.5 to 1
Water	1 to 4.5

Example 4:

Zircon sand, less than 0.3 mm	18 to 22
Zircon flour, less than 0.15 mm	24 to 20
Raw pyrophyllite, 1 to 4 mm	33 to 35
Raw pyrophyllite, up to 1 mm	15 to 18
Calcined pyrophyllite, less than 1.5 mm	10 to 5
Powdered sodium silicate, No. 1	0.5 to 1
Water glass, No. 2	3 to 4
Water	1 to 4.5

Example 5:

Zircon sand, less than 0.3 mm	20 to 29
Zircon flour, less than 0.15 mm	30 to 20
Raw pyrophyllite, 1 to 4 mm	33 to 31
Raw pyrophyllite, less than 1 mm	15 to 10
Calcined pyrophyllite, less than 1.5 mm	2 to 10
Powdered sodium silicate No. 1	0.5 to 1
Water glass No. 2	3 to 4
Water	1 to 4.5

The table below shows the chemical analysis of principal ingredients employed in Examples 1 through 5.

	Ignition Loss	SiO_2	TiO_2	Al_2O_3	Fe_2O_3	ZrO_2	Na_2O	K_2O	Total, %
Zircon sand	0.03	32.80	0.08	0.02	0.07	66.90	–	–	99.90
Zircon flour	0.12	32.42	0.10	0.71	0.20	66.42	–	–	99.97
Raw pyrophyllite	3.30	77.00	0.16	17.64	0.60	–	0.25	0.50	99.45
Calcined pyrophyllite	0.28	80.02	0.31	17.40	1.17	–	0.12	0.49	99.79

The physical properties of the products of the Examples 1 through 5 are shown in the following table.

	Example Number				
	1	2	3	4	5
Apparent specific gravity	3.74	3.59	3.09	3.20	3.30
Bulk density	3.12	3.11	2.71	2.79	2.91
Apparent porosity, %	16.4	13.4	12.3	12.6	11.9
Compressive strength, kg/cm^2	678	851	376	430	404
Refractoriness, SK	29	28	26	27	27
Softening point under load at high temperatures, °C, 2 kg/cm^2:					
T_1	1,335	1,325	1,170	1,195	1,210
T_2	1,440	1,370	1,275	1,300	1,305
T_3	1,490	1,440	1,435	1,410	1,405
Corrosion rate, %	49	55	69	69	57

Note: The less the corrosion rate the better for the refractory.

ALUMINA-ZIRCONIA REFRACTORIES

Corrosion Resistant Al_2O_3-ZrO_2-SiO_2

G. Cevales; U.S. Patent 3,754,950; August 28, 1973; assigned to Refradige SpA, Italy describes a method for reducing the vitreous phase of electromelted material so as to render it more corrosion resistant.

The process is carried out by treating a batch comprising, on the basis of the oxides, 45 to 65% by weight of Al_2O_3, 10 to 40% by weight of ZrO_2, 12 to 20% by weight of SiO_2 and 0.8 to 1.4% by weight of Na_2O, and constituted initially of 30 to 60% by weight of corundum, 10 to 40% by weight of baddeleyite, 0 to 40% by weight of mullite as crystalline phases and 15 to 25% by weight of a vitreous phase at 1300° to 1600°C for 8 to 12 days.

After thermal treatment, a product is obtained the individual phases of which fall, according to the treatment temperatures, within the following ranges:

Phases	Treatment Temperatures			
	1300°C %	1400°C %	1500°C %	1600°C %
Corundum	30–50	30–40	20–30	10–22
Baddeleyite	10–40	9–38	9–35	9–35
Mullite	10–40	15–48	18–54	35–60
Vitreous phase	14–18	12–14	10–12	7–9

The vitreous phase may be reduced down to 7%; also its composition turns out to be modified as will be seen from the following tables. The determination of the single phases was carried out by means of chemical analysis and by diffractometric x-ray analysis. The method used for the analysis of the vitreous phase was the commonly used one of extraction with hydrofluoric acid.

For the thermal treatment there may be used any conventional kind of furnace operating with gaseous or liquid fuels, provided that it be equipped with means insuring the desired control of the processing temperature. At the end of the residence time in the furnace at the established temperature, the refractory material is left to cool down at an average rate of temperature drop of 120° to 150°C per day.

Example 1: A sample of electromelted refractory, having the following composition was used:

	Percent
Al_2O_3	47.7
ZrO_2	36.1
SiO_2	14.7
Na_2O	1.0
Fe_2O_3	0.06
TiO_2	0.04
CaO	0.22
MgO	0.05

It was heated in a furnace fired by oil burners, at 1300° to 1600°C. The duration of the treatment was 240 hours. The ranges of the crystalline and vitreous phases of the starting product as well as the samples after treatment for 240 hours at 1300°, 1400°, 1500° and 1600°C are shown below.

Phases	Starting product %	Product after treatment at the temperature of			
		1300°C %	1400°C %	1500°C %	1600°C %
Corundum	43	38	29	20	12
Baddeleyite	36	35	34	34	34
Mullite	3	15	25	36	46
Vitreous phase	18	14	12	10	8

The compositions of the vitreous phase of the starting material and of the samples subjected to thermal treatment are shown below.

Component	Vitreous phase extracted on the starting products %	Vitreous Phase Extracted after a treatment of 240 hours at the temperatures indicated			
		1300°C %	1400°C %	1500°C %	1600°C %
Al_2O_3	21.9	29.10	34.20	42.9	47.6
ZrO_2	4.9	6.91	9.07	12.0	15.7
SiO_2	67.05	57.10	49.52	38.44	30.99
Na_2O	5.45	5.93	6.10	5.20	3.87
Fe_2O_3	0.21	0.27	0.27	0.35	0.60
TiO_2	0.13	0.11	0.13	0.14	0.24
CaO	0.36	0.58	0.71	0.97	1.00

Alumina-Zirconia-Silica Fused Cast Refractory

Fused cast refractories are refractories which are formed by electrical fusion followed by casting and annealing. They have excellent resistance to the strongly corrosive action of the highly alkaline materials which are used to make glass and these refractories are therefore used in the construction of glass melting tanks.

F.T. Felice, R.E. Fisher and L.J. Jacobs; U.S. Patent 3,868,241; February 25, 1975; assigned to Combustion Engineering Incorporated describe a method of producing a molten refractory material for use in forming fused cast refractories comprising: (1) melting alumina-zirconia and silica in an electric arc furnace having carbon electrodes whereby some carbon is transferred to the molten refractory; (2) flowing the molten refractory from the electric arc furnace; and (3) directing a stream of oxygen onto the molten material as it is being poured.

Specifically, the molten refractory material is poured in a free falling stream. In addition, step (3) comprises (a) impinging a first stream of oxygen directly on the free falling stream to break up the stream and produce contact between the oxygen and the molten refractory material; and (b) directing a second stream of oxygen tangentially around the free falling stream to consolidate the broken-up free-falling stream. This procedure minimizes the exudation of the glossy matrix of the refractory during service.

Alumina and Yttria-Stabilized Zirconia Eutectic

C.O. Hulse and J.A. Batt; U.S. Patent 3,761,295; September 25, 1973; assigned to United Aircraft Corporation describe a directionally solidified refractory oxide eutectic existing between alumina and zirconia where the zirconia has been stabilized in a cubic crystallographic structure, the alumina matrix is prestressed in compression at room temperature and the cast microstructure is ordered in a regular morphology.

The replacement, of 15 weight percent zirconia with yttria stabilizes the zirconia in the cubic structure and thereby effects prestressing of the alumina matrix in compression upon cooling for optimum properties. If the stabilizing agent is not added, the zirconia will not have a sufficiently high thermal expansion to prestress the alumina in compression upon cooling and it will undergo a monoclinic to tetragonal inversion resulting in very poor mechanical properties.

The composition consists of 54.5 weight percent alumina, balance zirconia plus stabilizing agent, which stabilizing agent may be yttria, calcia, or any other refractory material capable of providing the stabilizing function. Magnesia is also a commercially utilized stabilizing agent for zirconia.

In the particularly preferred alumina, zirconia, yttria combination the material melting point is 3450°F and, because of the very low vapor pressure of the component oxides, it is extremely stable even in high vacuum at its melting point. At 2865°F in air the directionally solidified material has demonstrated an average flexural strength of 61,500 lb/in² which is about fifty times the average strength of high quality commercial alumina tested under the same conditions.

The replacement of a portion of the zirconia with a stabilizing agent is made to avoid the monoclinic to tetragonal inversion of the zirconia normally occurring at 1800°F and to stabilize the zirconia substantially in a cubic crystal structure. With this substitution the eutectic occurs at 54.5, 38.7 and 6.8 weight percent, respectively of alumina, zirconia and alumina. In the ternary system all compositions existing along the eutectic trough provide the desired ordered microstructure and the trough in the alumina-zirconia-yttria system is oriented such that as the stabilizer amount is changed the alumina-zirconia ratio remains essentially constant.

ALUMINA-CHROMIUM REFRACTORIES

Alumina-Chrome Refractory

E.L. Manigault; U.S. Patent 3,948,670; April 6, 1976; assigned to N L Industries, Inc. and 3,862,845; January 28, 1975; assigned to The Charles Taylor's Sons Co. describes an alumina-chrome refractory composition having increased strength at ambient and high temperatures comprising the following ingredients:

	Percent
Tabular alumina	70 - 96
Calcined alumina –325 mesh	0 - 10
Milled zircon –325 mesh	1 - 10
Chromic oxide refractory grains ground to –10 mesh with at least 70% +200 mesh	3 - 10

(continued)

	Percent
Bentonite	0 - 4
Water	0 - 10
Phosphate compound	0 - 18

The mixture is then formed and fired at 350° to 1750°C for 2 to 8 hours if desired. If phosphates are not employed then the firing temperature should be 1400° to 1750°C. If a ramming or plastic mixture is to be produced, the mixture itself is used directly without forming and firing for the customers use. In the case of preparing a plastic mixture, 1 to 4% bentonite is also added to the mixture.

The tabular alumina should be employed in various sizes. All of the -¼ inch to +325 mesh alumina should be employed in the tabular form and in addition, with respect to the -325 mesh alumina, from ¼ to all of the -325 mesh alumina should be present in the tabular form. If calcined alumina is used, it should be -325 mesh in size. 65 to 95% of the alumina should be -¼ inch to +325 mesh in size, while the remainder of the alumina should be ground to -325 mesh.

The chromic oxide refractory grains should be ground to -10 mesh in size with at least 70% +200 mesh, and should be obtained from either previously fused or sintered chromic oxide. It is preferred that zircon be employed, from 1 to 10%, preferably 3 to 7%. The size of the zircon should be -325 mesh.

Example 1: The following ingredients were added to form a refractory mixture:

Tabular alumina, -8+65 mesh	44 kg
Tabular alumina, -65+325 mesh	22 kg
Tabular alumina, -325 mesh	19 kg
Calcined alumina, -325 mesh	5 kg
Chromic oxide refractory, grains -10 mesh with at least 70% +200 mesh	7 kg
Zircon, -325 mesh	3 kg

The mixture was then tempered with lignin liquor and formed into bricks 9" x 4½" x 3" and the bricks were then fired at 1560°C for 5 hours. After cooling to room temperature, the bricks were analyzed to determine their strengths which are listed as follows: modulus of rupture—ambient temperature, 8777; high temperature (1480°C), 5122.

Examples 2 and 3: The procedure of Example 1 was repeated except that various amounts of zircon were added to the mixture in Examples 2 and 3. Bricks having properties similar to those obtained in Example 1 were produced. The operational details and the results obtained are recorded in Table 1 on the following page, along with those of Example 1. For comparison, a series of control runs were made with various compositions.

Controls A through C: In these runs no chromic oxide was employed. Run A contained no zircon. Runs B and C contained zircon. The results of these runs indicated that modulus of rupture at both ambient temperature and at 1480°C were unsatisfactory. The results are recorded in Table 2 also on the following page.

TABLE 1

Material	Mesh	Examples		
		1	2	3
Tabular alumina, kg	–8+65	44	44	44
Tabular alumina, kg	–65+325	22	22	22
Tabular alumina, kg	–325	19	17	19
Calcined alumina, kg	–325	5	3	5
Milled zircon, kg	–325	3	7	3
Aluminum phosphate, kg	–	0	0	0
Phosphoric acid, kg	–	0	0	0
Lignin liquor, kg	–	2.5	2.5	2.5
Water, kg	–	2	2	2
Chromic oxide refractory grains sintered	–10 with 80%+200	7	7	0
Chromic oxide refractory grains fused	–10 with 88%+200	–	–	7
Chromic oxide (pigment grade)	–325	–	–	–
Silica	–325	–	–	–
Modulus of rupture, ambient	–	8777	8668	8144
Modulus of rupture, 1480°C	–	5122	5011	3349
Porosity, %	–	16.0	–	14.3
Bulk density, g/cc	–	3.22	–	3.23

TABLE 2

Material	Mesh	Controls		
		A	B	C
Tabular alumina, kg	–8+65	48	48	48
Tabular alumina, kg	–65+325	24	24	24
Tabular alumina, kg	–325	20	20	18
Calcined alumina, kg	–325	8	5	3
Milled zircon, kg	–325	–	3	7
Lignin liquor, kg	–	2.5	2.5	2.5
Water, kg	–	2	2	2
Chromic oxide refractory grains sintered	–10 with 80%+200	–	–	–
Chromic oxide refractory grains fused	–10 with 88%+200	–	–	–
Chromic oxide (pigment grade)	–325	–	–	–
Modulus of rupture, ambient	–	5395	6906	6085
Modulus of rupture, 1480°C	–	644	1835	–

High-Strength Alumina-Chromite-Phosphate

E.L. Manigault; U.S. Patent 3,945,839; March 23, 1976; assigned to N L Industries Incorporated describes a refractory material having a low porosity, high bulk density and modulus of rupture. The composition may be produced from alumina, an iron-chromite ore and a phosphate compound.

Iron chromite ore usually contains 35 to 50% Cr_2O_3, 15 to 35% FeO and 25 to 40% oxides of silicon, magnesium, calcium, aluminum and other metal oxides. The phosphate compound is phosphoric acid and/or aluminum phosphate. The phosphate compound is added in amounts from 1 to 10% P_2O_5 by weight. During the firing of this composition a portion of the phosphate content may be lost by vaporization.

This composition is either fired to produce fired refractory products, or the mixture without firing is used either as a ramming mixture, or as a plastic mixture. When it is desired to produce a plastic mixture, small amounts of bentonite

(1 to 4%) and water (1 to 12%), if needed, are added to the mixture. The presence of the bentonite increases the plasticity in the mixture. The plastic mixture contains the following ingredients:

	Percent
Alumina	72 - 94
Iron chromite ore	3 - 20
Phosphate compound (calc. as P_2O_5)	1 - 10
Bentonite	1 - 4
Water	0 - 12

The presence of the bentonite in the mixture is not necessary when the mixture is used as a ramming mix, or when fired to produce a refractory composition. In preparing the refractory composition the iron-chromite ore should be ground to –325 mesh. 65 to 94% of the alumina should be –¼ inch to +325 mesh in size while the remainder should be ground to –325 mesh. Tabular alumina may be used exclusively, but it is preferred to have a mixture of tabular and calcined alumina.

The tabular alumina when used should be employed in various sizes. All of the –¼ inch+325 mesh alumina should be employed in the tabular form and in addition, with respect to the –325 mesh alumina, from ¼ to all of the –325 mesh alumina should be present in the tabular form. If calcined alumina is used, it should be –325 mesh in size.

In preparing the composition, the alumina, the iron-chromite ore and the phosphate compound are mixed together sufficiently to temper the mix properly. Formed shapes are fired at 350° to 1650°C for 2 to 8 hours to form the refractory composition.

Example 1: The iron-chromite ore used was ground to –325 mesh and had the following chemical analysis:

	Percent
Cr_2O_3	45.66
Fe	19.98
Al_2O_3	15.26
SiO_2	1.47
O_2	7.26
MgO	10.24
CaO	0.13

Both tabular and calcined alumina were employed. The –¼ inch+325 mesh tubular alumina had the following size distribution: –¼ inch+60 mesh 66% and –60 mesh+325 mesh 34.0%. In preparing the refractory composition the following ingredients were thoroughly blended to form a homogeneous mixture:

74 kg of +325 mesh tabular alumina
6 kg of –325 mesh tabular alumina
15 kg of –325 mesh calcined alumina
5 kg of –325 mesh iron-chromite ore
5.2 kg of 80% phosphoric acid

Bricks 9" x 4½" x 3" were formed and fired at 1560°C for 5 hours. The fired bricks were examined and found to possess the following properties: modulus of rupture, 7,024 psi; porosity, 13.9 percent; and bulk density, 3.25 g/cc.

Examples 2 through 5: The procedure of Example 1 was repeated using various amounts of calcined alumina, iron-chromite ore, and a phosphate compound. Two control runs were also carried out in which no phosphates were used (control A) and in which chromic oxide was used in place of iron chromite ore and no phosphates were employed (control B). The results of these examples along with those of Example 1 and the control runs are presented in the table below.

Material	Example Number					Control	
	1	2	3	4	5	A	B
Tabular alumina, +325 mesh, kg	74	74	74	66.5	64.1	64.1	90
Tabular alumina, -325 mesh, kg	6	6	6	19.4	12.4	12.4	0.0
Calcined alumina, -325 mesh, kg	15	10	5	4.7	6.1	6.1	0.0
Iron chromite, -325 mesh, kg	5	10	15	10.0	17.4	17.4	0.0
Phosphoric acid (80%), kg	5.2	0.0	0.0	5.3	6.0	0.0	0.0
Aluminum phosphate, kg	0.0	4.8	5.0	0.0	0.0	0.0	0.0
Lignin liquor, kg	0	0	0	0.0	0.0	2.3	3.8
Chromic oxide, kg	0.0	0.0	0.0	–	–	0.0	10

Properties	Example Number					Control	
	1	2	3	4	5	A	B
Modulus of rupture, psi	7,024	8,840	6,543	7,540	7,728	5,775	4,500
Apparent porosity, %	13.9	13.3	14.1	13.9	15.6	17.1	16.5
Bulk density, g/cc	3.25	3.25	2.03	3.25	3.24	3.17	3.16

Mullite Chromite Refractory

E.L. Manigault; U.S. Patent 3,773,532; November 20, 1973; assigned to The Charles Taylor's Sons Company has shown that when an iron-chromite ore is added to a mullite refractory composition, a refractory composition is produced which is less porous and is more resistant to slag and alkali attack than either a mullite body or a chromic oxide-mullite body.

The refractory composition comprises 56 to 80% mullite, 6 to 13% kyanite, 1 to 9% silica and 4.0 to 25% iron-chromite ore, all of the percentages expressed on a weight basis. This refractory material is prepared by admixing the above ingredients, forming the desired shapes to be produced and then firing the shapes at 1500° to 1650°C.

Example 1: 70.8 kg of mullite and 4.8 kg of iron-chromite ore were mixed with 10.1 kg calcined kyanite, 6.6 kg of alumina and 7.7 kg of silica. The mullite contained a size range of from -¼ inch through -325 mesh. The calcined kyanite was ground to -100 mesh while the iron-chromite, the alumina and the silica were all ground to -325 mesh. The iron-chromite ore used had the following typical analysis.

Ingredient	Percent
Chromic oxide	45.66
Iron oxides	27.24
Silica	1.47

(continued)

Ingredient	Percent
Calcium oxide	0.13
Magnesium oxide	10.24
Alumina	15.26
	100.00

The mixture was dry blended for 3 minutes. 3.1 kg of a 30% solution of lignin liquor and water were added to the mixture and it was blended for 10 minutes. Bricks 9" x 4½" x 3" were prepared and fired at 1560°C for 5 hours. The final bricks were examined and possessed the following properties: 15% porosity and 5.8% absorption. The bricks were also subjected to slag and alkali attack tests.

Examples 2 and 3: The procedure of Example 1 was repeated except that various amounts of alumina, iron-chromite ore and silica were employed. In Example 2 the amounts of alumina and silica were decreased while in Example 3 the amount of iron-chromite ore was increased five-fold. Again in all of these examples, improved results were obtained. The results along with the controls are presented in the following table.

	Example Number			Chromic-Oxide Mullite Control	Mullite Control
	1	2	3		
Mullite, kg	70.8	74.3	67.6	68.6	68.6
Kyanite, kg	10.1	10.6	10.6	8.6	8.6
Alumina, kg	6.6	4.6	–	3.6	8.6
Silica, kg	7.7	5.4	1.5	–	–
Iron-chromite, kg	4.8	5.0	22.8	–	–
Chromic-oxide, kg	–	–	–	5.0	–
Clay, kg	–	–	–	14.2	14.2
Porosity, %	15.0	15.3	15.3	16.1	17.3
Absorption, %	5.8	5.9	5.5	6.2	6.8
Slag attack test, cc	9.9	–	–	10.8	15.8
Alkali attack test, cycles	8	–	5	2	1

From the above table it is clearly shown that when iron-chromite ore is used in a mullite composition that the absorption and the porosity of the fired bricks are decreased over a mullite composition or a chromic-oxide-mullite composition. In addition these bricks are attacked less by both molten slag and alkali than the bricks produced by the prior art.

Hardening Dressed and Dimension Ceramics

N.H. Stradley and H.C. Dunegan; U.S. Patent 3,717,497; February 20, 1973; assigned to American Lava Corporation describe a method for hardening dressed and dimensioned ceramics which does not require further dimensioning after hardening.

The method consists of repeated impregnation and curing of a chromium solution in porous alumina ceramic at a temperature sufficient to effect deposition of insoluble reactive chromia in pores, spaces or cavities followed by further heat treatment at a higher temperature sufficient to bring about phase changes and particularly formation of solid solutions.

The pore size and pore volume should be such that the porous structure can be filled after a reasonable number of impregnation and cure cycles, usually at 5 to 15, preferably 8 to 12. It is preferred that the size of the interstitial or pore opening be 0.05 to 1 micron and the pore volume 0.05 to 0.25 cc/g.

BASIC REFRACTORIES

MAGNESIA REFRACTORIES

Alkyd Resin Binder for Periclase

W.T. Bakker; U.S. Patent 3,943,216; March 9, 1976; assigned to General Refractories Company describes the preparation of periclase refractories by using an alkyd resin binder for tempering and pressing periclase grains into refractory shapes such as bricks. The pressed shapes are dried and fired.

After firing, the refractory shapes are impregnated under vacuum with molten pitch and can then be used as lining in basic oxygen steelmaking furnaces. The fired refractory bricks have a greatly improved modulus of rupture at 1482°C of 3,000 to 4,000 psi. The alkyd resin is generally incorporated in amounts of 2.5 to 4.5 weight percent.

The green strength of the alkyd resin bonded bricks can be improved by adding a strong organic acid to the alkyd resin prior to mixing it with periclase grains. The green strength of bricks produced from such a mixture is 30 and 40 psi.

Refractory articles useful in basic oxygen process furnaces desirably should have a low porosity and maximum bulk density, and the sizes of the refractory aggregates should be selected to achieve these results. A typical mixture of coarse, intermediate and fine grain fractions suitable to achieve high bulk density and low porosity, using Tyler standard screen sizes is as follows:

> 30 to 35% passing 4 mesh and retained on 10 mesh
> 30 to 40% passing 6 mesh and retained on 28 mesh
> 30 to 35% ball mill fines (less than 100 mesh)

Preferably, the chemical composition of the ball mill fines is adjusted to provide a CaO/SiO_2 ratio for the ball mill fines of 1.4 to 2.0 preferably 1.6. This ratio can be obtained by adding siliceous periclase fines or finely divided silica together with $CaCO_3$ or $Ca(OH)_2$ powder.

The various grain-size fractions are blended dry in a suitable mixer, commonly used in the refractory industry such as a muller mixer. Generally, one minute dry mixing is sufficient to provide a homogeneous mixture.

The alkyd resins that are preferably used are the resinous reaction products obtained by polymerizing a polyhydric alcohol, a dibasic acid, and a monobasic fatty acid, and are well known in the art for their use in oil-base paint. The monobasic fatty acid is commonly supplied in the form of a triglyceride or oil, and the alkyd resins obtained thereby are commonly referred to as oil-modified alkyds.

Polybasic acids that can be used to prepare the alkyd resins include phthalic anhydride, isophthalic acid, maleic anhydride, fumaric acid, azelaic acid, succinic acid, adipic acid and sebacic acid. Preferably, phthalic anhydride is used.

Polyhydric alcohols that can be used to prepare the alkyds include glycerol, pentaerythritol, dipentaerythritol, trimethylolethane, sorbitol, ethylene glycol, propylene glycol, dipropylene glycol, trimethylolpropane, neopentylene glycol, etc. Glycerol is the preferred polyhydric alcohol.

The monobasic fatty acid that can be used to prepare the alkyd resins are obtained from oils including tung, linseed, soybean, peanut, dehydrated castor, fish, safflower, oiticica, cottonseed, and coconut oil. Soybean and linseed oils are preferred.

Alkyd resins which have proven especially useful contain 70 to 80% resin, 30 to 20% solvent, and have an alkyd number of 15 to 40. These materials are known in the art as medium to long alkyd resins. Short alkyd resins are also useful but they are difficult to disperse in the dry ingredients of the refractory batch because of their high viscosity.

Preferably, a strong acid is used along with the alkyd resin when the dry batch of refractory grains is tempered. The green strength of refractory shapes prepared from a tempered batch of refractory grain and alkyd resin is relatively low.

A dry refractory batch which had been tempered with alkyd resin and a strong acid will provide unfired bricks with greatly increased green strength. Depending on the process conditions, the green strength achieved generally will be 30 to 40 psi, and in some instances can be as high as 50 psi. These higher green strengths are sufficient to permit handling the brick in the green state without significant amounts of breakage.

The acid used is preferably a strong organic acid, although strong inorganic acids can also be used. Exemplary of suitable strong organic acids that can be used are toluenesulfonic acid, xylenesulfonic acid, and mixtures of sulfonic acids. Typical inorganic acids that can be used include H_2SO_4, HCl, H_3PO_4 and the like.

Example: Periclase refractories are prepared in a laboratory from 96% MgO periclase grains. The grains are crushed and sized in accordance with size distribution shown on the following page. Two refractory batches are prepared from the above-described material. In one batch, 3% of a 50% lignosulfonate solution in water and 2% additional water are used as binders. In the other batch, 4.0%

of a long alkyd resin is used as a binder. The alkyd resin is prepared mainly from phthalic anhydride, glycerol, soybean oil, and contains 70% resin and 30% of mineral spirits solvent. The alkyd resin used comprises a mixture of various long alkyds.

Tyler Screen Size	Percent of Refractory Composition
-4+10 mesh	40.0
-10+20 mesh	10.0
-20+48	8.0
-48 mesh	7.0
Ball mill fines	
Ground to 60% -325 mesh	10.0
Ground to 95% -325 mesh	25.0

Each brick batch is processed into 9" x 4½" x 3" brick by first dry blending the sized grain in a muller mixer for 1 minute and then adding the alkyd resin to the sized grains. The mixture of resin and grain is blended for 10 minutes to wet all of the grains with resin and form a shapeable mixture. After blending the mix is pressed at 15,000 psi using a hydraulic press.

The green brick is dried at 230°F (110°C) for 18 hours. A first set of brick from each batch is fired at 1677°C and a second set is fired at 1719°C in a commercial tunnel kiln. After burning, brick with the following properties are obtained:

	Batch 1	Batch 2
Binder	Lignosulfonate + Water (Conventional Binder)	Alkyd Resin (Improved Binder)
Properties after 1677°C Burn:		
Density, g/cc	2.94	3.06
Modulus of Rupture, at 1482°C, p.s.i.	1760	3330
Hot Crushing Strength, at 1538°C, p.s.i.	3530	4860+*
Properties after 1719°C Burn:		
Density, g/cc	2.97	3.06
Modulus of Rupture at 1482°C, p.s.i.	2330	3870

*Samples did not break during test.

These results clearly show that an increase in density and hot strength is obtained by using an alkyd resin binder instead of the conventional lignosulfonate-water binder system.

Zinc Lignosulfonate Binder for Magnesia

F.C. Morman; U.S. Patent 3,923,532; December 2, 1975; assigned to American Can Company describes refractory shapes containing magnesium oxide pressed from a refractory forming mixture which consists of particulate refractory material containing magnesium oxide and a binder solution of zinc lignosulfonate in water. The use of zinc lignosulfonate in solution as a binder stabilizes the refractory forming mixture, to substantially prevent the loss of density of refractory shapes as the refractory forming mixture ages.

Zinc may alternatively be provided in the binder solution by adding a zinc compound to a binder solution composed of water and lignosulfonate salt, the zinc compound dissociating in solution to provide zinc ions. A refractory forming mixture is prepared by mixing zinc lignosulfonate, water, and particulate refractory material containing magnesium oxide, or by mixing a lignosulfonate salt, water, a zinc compound which dissociates in solution to provide zinc ions, and particulate refractory material containing magnesium oxide.

Basic Refractory Bonded with Calcia and Phosphates

J. Damiano; U.S. Patent 3,920,464; November 18, 1975; assigned to Quigley Company Incorporated has shown that the introduction of preformed dicalcium phosphate or tricalcium phosphate into quick-setting refractory compositions affords unexpected advantages over the results obtained with in situ formation of dicalcium phosphate from monosodium phosphate and lime.

Accordingly, the composition comprises at least 85 weight percent particulate basic refractory, 0.5 to 3 weight percent finely divided active calcium oxide, 2.5 to 5 weight percent glassy polyphosphate and 1.5 to 4 weight percent finely divided dibasic calcium phosphate or tricalcium phosphate.

The refractory compositions, since they possess initial set, excellent hot strength and good resistance to erosion and slag attack, are of particular value in casting and gunning operations associated with basic steel-making furnaces.

Examples 1 through 9: Quick-setting refractories of the compositions indicated in Runs 1 through 9 of the table on the following page, were prepared as follows: The listed ingredients for each run, equivalent to 5 kg of refractory composition, were hand blended by mixing with a trowel for one minute. Water equivalent to 11 weight percent of the dry blend was then added and the mixing was continued for an additional minute to give a mixture of good casting consistency. The wet blend was cast into bars using 2" x 2" x 9" (5.1 cm x 5.1 cm x 22.9 cm) molds. The bars were allowed to set for 5 hours and were then removed from the molds and air dried at 220°F (105°C) overnight.

Inspection of the tabulated results demonstrates the advantage realized by substitution of preformed dibasic calcium phosphate for its stoichiometric equivalent in the form of lime and monosodium phosphate. The introduction of preformed dibasic calcium phosphate is exemplified by the "A" runs, with the respective stoichiometric counterparts represented by each corresponding "B" run.

Aside from the case of Runs 1A and 1B, which fall below the calcium phosphate levels indicated each comparison demonstrates the clear superiority resulting from the use of preformed calcium phosphate. Comparable advantages are obtained if dead-burned dolomite is substituted for the 5 to 7 mesh, 7 to 16 mesh and -16 mesh portions of the magnesia or Glass H is substituted for glassy sodium polyphosphate in the refractory composition.

	Run																	
	1A	1B	2A	2B	3A	3B	4A	4B	5A	5B	6A	6B	7A	7B	8A	8B	9A	9B
Refractory (wt %)																		
Magnesia																		
−5+7 mesh	15.0	15.0	15.0	15.0	15.0	15.0	15.0	15.0	15.0	15.0	15.0	15.0	15.0	15.0	15.0	15.0	15.0	15.0
−7+16 mesh	48.0	48.0	48.0	48.0	48.0	48.0	48.0	48.0	48.0	48.0	48.0	48.0	48.0	48.0	48.0	48.0	48.0	48.0
+16 mesh	4.0	4.0	4.0	4.0	4.0	4.0	4.0	4.0	4.0	4.0	4.0	4.0	4.0	4.0	4.0	4.0	4.0	4.0
Pulverized	28.6	27.9	27.1	25.3	25.6	22.6	27.6	26.9	26.1	24.3	24.6	21.6	26.1	25.4	24.6	22.8	23.1	20.1
Hexaphos*	2.5	2.5	2.5	2.5	2.5	2.5	3.5	3.5	3.5	3.5	3.5	3.5	5.0	5.0	5.0	5.0	5.0	5.0
CaO (+16 mesh)	0.9	1.2	0.9	1.7	0.9	2.2	0.9	1.2	0.9	1.7	0.9	2.2	0.9	1.2	0.9	1.7	0.9	2.2
CaHPO$_4$·2H$_2$O (+100 mesh)	1.0	0	2.5	0	4.0	0	1.0	0	2.5	0	4.0	0	1.0	0	2.5	0	4.0	0
NaH$_2$PO$_4$	0	1.4	0	3.5	0	5.7	0	1.4	0	3.5	0	5.7	0	1.4	0	3.5	0	5.7
Average M/R																		
psi at 2300°F	69	135	171	101	165	56	104	77	183	123	179	104	131	87	163	53	130	45
kg/cm^2 at 1260°C	4.9	9.5	12.0	7.1	11.6	3.9	7.3	5.4	12.9	8.6	12.6	7.3	9.2	6.1	11.5	3.7	9.1	3.2

*Glassy sodium phosphate.

Porous Magnesia

According to *K. Ogasawara, G. Kokubo and T. Hojo; U.S. Patent 3,893,867; July 8, 1975; assigned to Nihon Kaisuikako Company Limited, Japan* a magnesia-base sintered article having a desired porosity, which can be selected over a relatively wide range, can be prepared by the following procedure.

That is, (1) finely divided magnesia or magnesia-base oxides, preferably dead burned magnesia or dolomite, is blended with (2) finely divided magnesium hydroxide or a substantially homogeneous mixture of magnesium hydroxide with calcium hydroxide, in a well blended state, preferably 20 to 90% by weight, of the former oxides (1), based on the total weight of the resultant blended mixture.

The resultant blended mixture is formed into shapes and then is calcined to sinter, whereby a relatively high compressive strength of the magnesia-base sintered article is obtained. The sintered product is suitable for use in aggregates of basic refractories.

The particle size of these magnesia-base oxides can be varied to control the desired properties of porosity, compression strength, dimensions, etc., of the sintered magnesia-base product. The desired particle size can be achieved by means of crushing, pulverizing, or the like, accompanied with optional size separation by screening.

The porosity and the size of the pores of the product increase with an increase of the average particle size of the starting magnesia-base oxides, whereas the compressive strength exhibits an inverse tendency.

As the magnesium hydroxide to be blended with the starting magnesia-base oxides, there can be used magnesium hydroxide separated from seawater, naturally occurring brucite or the like. The magnesium hydroxide is usually employed in a finely divided state having a particle size of less than one millimeter. When the size of the intended porous magnesia-base sintered article is small such as of less than 20 mm in diameter, e.g., a briquette, it is desirable to employ more finely divided magnesium hydroxide, e.g., to pulverize the magnesium hydroxide into a powder of less than 0.5 mm particle size.

However, when the article has a larger dimension, such as a brick, the use of the magnesium hydroxide having particle size of less than 1 mm is less preferable. When a homogenized mixture of magnesium hydroxide with calcium hydroxide is employed, the calcium hydroxide content must not exceed 30 weight percent based on the weight of the mixture. The particle size of the calcium hydroxide is substantially the same as that of the magnesium hydroxide.

A binder agent can be added to the mixture so that the mixture can be molded more easily in the succeeding step. When the mixture contains substantially no calcium oxide, water is conveniently employed as the binder agent. Of course, any of the conventional binders such as clay, an inorganic magnesium salt, for example, magnesium chloride or sulfate, carboxymethylcellulose and the like can be additionally incorporated. When the mixture contains a substantial amount of calcium oxide, a nonaqueous binder agent such as liquid paraffin can be employed in order to avoid slaking of the calcium oxide.

Calcining the molded article can be carried out at 1400° to 2000°C, preferably 1600° to 2000°C until the mixture is sintered. The magnesia-base sintered article, e.g., briquette, has a remarkably uniform pore distribution and a compressive strength of more than 100 kg/cm². The purity of the sintered magnesia briquette can be controlled by choosing raw materials of desired purity. The porosity and the compressive strength can be adjusted to a substantially desired value over a broad range by controlling the particle size distribution and the blending ratio of the starting materials.

The magnesia-base sintered article is useful as an aggregate for base refractory materials. It is approximately 20% lower in bulk density as compared with a conventional aggregate consisting of sintered magnesia, and it is also lower in heat conductivity, while the compressive strength thereof is generally similar to that of the conventional aggregate. The magnesia-base sintered article contains many pores which are effective for holding a chemical reaction catalyst, so that it is useful as a catalyst carrier, especially as a basic carrier.

Example: A dead burned magnesia clinker having the chemical composition and the properties shown in Table 1, produced by calcining impure magnesium hydroxide at 1800°C to sinter (the magnesium hydroxide having been prepared from seawater) was crushed, pulverized and sieved to obtain a powder of magnesia clinker passing through a 100 mesh screen.

TABLE 1

	MgO (%)	Mg(OH)$_2$ (%)	CaO (%)	Fe$_2$O$_3$ (%)	Al$_2$O$_3$ (%)	SiO$_2$ (%)	Bulk Density (g/cm³)	Apparent Porosity (%)
Magnesia clinker	95.44	–	1.10	0.13	0.12	3.01	3.26	2.0
Magnesium hydroxide	–	96.99	0.79	0.09	0.11	2.08	–	–

By further sieving, magnesia powder having a particle size distribution consisting of 30% of particles of 100 to 200 mesh size and the remaining 70% of particles of less than 200 mesh size, was prepared. The magnesia powder thus prepared was blended, in the ratios shown in Table 1, with magnesium hydroxide powder having a particle size distribution consisting of 20% of particles of magnesium hydroxide of from 100 to 200 mesh size and 80% of magnesium hydroxide particles of less than 200 mesh size.

To the thus-prepared blended mixture, there was added 5% of water based on the total weight of the mixture, and the resultant mixture was molded in cylindrical shapes of 40 mm in diameter and 40 mm in height under a molding pressure of 500 kg/cm² and then was calcined at 1800°C to sinter.

The apparent porosity, bulk density and compressive strength of the thus-produced magnesia-base sintered articles are shown in Table 2, on the following page, in relation to the blending ratios of the magnesia powder and the magnesium hydroxide employed.

TABLE 2: PROPERTIES OF CALCINED PRODUCTS

					Blending Ratio				
Magnesia powder		20	30	40	50	60	70	80	90
Magnesium hydroxide powder		80	70	60	50	40	30	20	10
Properties:									
Porosity (%)		23	32	33	35	32	28	25	22
Bulk density (g/cm³)		2.71	2.42	2.38	2.32	2.40	2.56	2.61	2.70
Compressive strength (kg/cm²)		165	130	140	180	150	130	190	330

It was observed that the pores of the magnesia-base sintered articles were less than 0.5 mm in diameter and they were uniformly dispersed.

Burned Periclase Brick from Calcined Magnesite

According to *W.S. Treffner and F.P. Filer; U.S. Patent 3,844,802; October 29, 1974; assigned to General Refractories Company* burned periclase brick is prepared from a size-graded batch of calcined magnesite in which the relatively coarse fraction (+48 mesh) has a $CaO:SiO_2$ weight ratio of from above 2:1 to 5:1 but the relatively fine fraction (−48 mesh) has a $CaO:SiO_2$ weight ratio of 1.5:1 to below 2:1.

It has been found that unexpected additional benefits of dicalcium silicate bonding can be obtained when at least a portion of the dicalcium silicate is formed in situ during burning the brick. This is preferably done at the boundaries between the relatively coarse and the relatively fine particles in the brick by changing the chemistry, and consequently the composition, of the accessory mineral phases of the relatively fine particles of the brick thus creating a chemical composition gradient between the relatively coarse and the relatively fine fractions of the brick.

The constituents of the more siliceous fine particles will react with the constituents of the less siliceous coarse particles forming, during the firing of the brick, refractory dicalcium silicates thus forming a bond of a high hot strength between the relatively coarse and the relatively fine particles of the brick. The reaction also forms some secondary periclase and it is believed also to promote direct periclase-to-periclase bonding thus further contributing to increased high-temperature strength.

Calcined magnesite is the basic component of the bricks prepared and will generally have an MgO content of at least 90% and preferably at least 95%, the balance being small and varying amounts of lime (CaO) and silica (SiO_2) and miscellaneous oxides of aluminum, iron, manganese, etc.

Example: Bricks are prepared from a batch mix of 65 parts by weight of coarse calcined magnesite and 35 parts by weight of fine calcined magnesite. The coarse fraction had a particle size distribution shown on the following page.

An oxide analysis of the coarse fraction immediately follows the particle size distribution of the coarse fraction.

Mesh	Percent	Mesh	Percent
+4	0.3	+14	80.3
+6	14.6	+20	89.9
+8	39.5	+28	95.6
+10	65.8	-28	4.4

	Percent
MgO	96.30
CaO	2.46%
SiO_2	0.99
Oxides of Fe, Al, Mn, etc.	0.25

The fine fraction had a particle size distribution as follows:

Mesh	Percent
+48	2.4
+65	8.5
+100	20.7
+200	38.8
+325	46.7
-325	53.3

The fine fraction differed, as set forth in the table below, in terms of $CaO:SiO_2$ ratio. Where the $CaO:SiO_2$ was adjusted upwardly, fine calcium carbonate powder was added until the desired $CaO:SiO_2$ ratio was provided; and where the $CaO:SiO_2$ ratio was adjusted downwardly a fine calcined magnesite having a low $CaO:SiO_2$ ratio was added until the desired ratio was reached.

The bricks were prepared by mixing the grain mix with 1% of a 50% aqueous solution of lignosulfonate and 1.1% H_2SO_4 (as dilute sulfuric acid) and pressing the damp mixture at 12,000 psi. The bricks were then fired at 3000°F in a tunnel kiln following which physical properties were measured as set forth in the table below. The results are tabulated as follows:

Periclase Brick
Variation in Lime to Silica Ratio of Fines

Lime to Silica Ratio of Fines:	1.61				1.83			
Ratio of Fines Adjusted From:	2.65	2.15	1.83	*	2.65	2.15	*	1.61
Bulk density, oz/in³	1.71	1.71	1.72	1.70	1.70	1.71	1.72	1.70
g/cm³	2.96	2.96	2.98	2.94	2.94	2.96	2.98	2.94
Modulus of rupture, psi	1,840	1,755	2,120	1,760	1,970	1,855	1,960	2,030
Cold crushing strength, psi	6,110	4,720	5,070	6,775	4,280	4,250	5,155	5,160
Modulus of rupture at 2700°F, psi	1,285	1,325	1,610	1,660	1,290	1,200	1,195	1,125
Compressive strength at 2800°F, psi	2,785	2,095	3,305	2,620	2,830	2,435	2,865	2,425
Porosity, open, %	1.52	15.4	14.8	16.7	15.7	15.0	14.7	16.4
Lime to Silica Ratio of Fines:	2.15				2.65			
Ratio of Fines Adjusted From:	2.65	*	1.83	1.61	*	2.15	1.83	1.61
Bulk density, oz/in³	1.70	1.70	1.71	1.70	1.70	1.71	1.70	1.69
g/cm³	2.94	2.94	2.96	2.94	2.94	2.96	2.94	2.92
Modulus of rupture, psi	1,770	1,340	1,770	1,835	1,805	1,810	1,780	1,715
Cold crushing strength, psi	3,980	3,720	3,980	5,155	3,805	4,365	3,700	5,075
Modulus of rupture at 2700°F, psi	1,245	995	1,025	1,000	1,130	825	1,240	1,045
Compressive strength at 2800°F, psi	2,230	1,780	2,705	1,920	2,305	1,935	1,935	1,790
Porosity, open, %	15.6	15.5	15.6	16.7	15.9	15.3	14.9	16.3

*Natural ball mill fines (no adjustment).

Hot Strength Magnesia Having High Boron Content

According to *R. Staut and G.L. Mortl; U.S. Patent 3,833,390; September 3, 1974; assigned to General Refractories Company* a magnesia refractory shape possessing good strength properties at elevated temperatures is prepared from a magnesia grain containing 90 to 99% MgO, and over 0.1% by weight B_2O_3 and which can have a Ca-Si mol ratio of less than 2.0.

Fine particles of an alkali metal-containing compound are added to the magnesia grain to help volatilize the boron during firing. Fine particles of calcium and silicon-containing material can be admixed, if needed, with the magnesia grain to increase the Ca-Si mol ratio of the fine (–100 mesh) particles of the admixture to above 2:1 and to increase the total weight percent of CaO and SiO_2 present in the fine particle component of the composition to between 3 to 7% by weight.

A brick or other refractory product produced from a high-boron-content magnesia grain in the process is almost indistinguishable from a similar refractory product produced from a more expensive, low-boron-content magnesia grain. After firing, there is almost no trace of the alkali metal compound, and the boron content can generally be reduced to a level comparable to that of the usual commercially available so-called low boron magnesia grains. The strength properties of the fired refractory products are excellent, and do not appear to be adversely affected by the volatilization of the alkali metal and boron during the firing step.

The elimination of the need for a low B_2O_3 starting material permits using a less expensive starting material, results in production cost savings with respect to inventory requirements, and also eliminates contamination in magnesia grinding systems, and the cleaning cost involved in changing over from one magnesia grain to another.

In accordance with the process an alkali-metal containing compound is added to the magnesia starting material to help volatilize the boron present in the magnesia grain. The alkali metal containing compound should be an oxide, or a compound which is readily converted to the oxide form upon firing, such as nitrate, a carbonate, or a hydroxide. Typical alkali metal-containing compounds which can be used in the process include KOH, NaOH, K_2CO_3, Na_2CO_3, KNO_3, $NaNO_3$, K_2O, Na_2O, and the corresponding lithium compounds.

The amount of alkali metal-containing compound which is admixed with the starting magnesia grain should be sufficient to reduce the B_2O_3 content of the fired composition to below 0.05% by weight. Usually the alkali metal compound is added in amounts to provide 0.2 to 1% and preferably 0.4 to 0.6% by weight of alkali metal on an oxide basis in the admixed composition. Preferably, the alkali metal compound is in finely divided form because uniform distribution of the alkali metal throughout the composition helps promote uniform volatilization of boron from the composition during firing.

Ammonium Salt Binder for Basic Refractories

According to *A.P. Dreyling; U.S. Patent 3,833,391; September 3, 1974; assigned to Quigley Company, Inc.* basic refractory compositions are made with improved binders which after serving their binding function pass off by sublimation or by decomposition into volatile constituents upon firing at 1500° to 3300°F to leave no residual ingredients that reduce refractoriness of the compositions or the end products made therefrom. Bonding properties of binders in this class are obtained either by chemical action, physical bonding, or both, or by chemical breakdown into components which react individually with the predominantly basic refractory grains. These binder reactions may occur at elevated temperatures with considerable speed or slower at normal room temperature to set up the refractory ingredient into a hard dense mass.

The improved binder in particulate form is intimately mixed with a basic refractory material in grain or particle form of various sizes. Water is added to the mixture in an amount dependent upon the method to be used in imparting the desired ultimate shape to the refractory composition and the composition is fired while maintaining its desired shape. By way of example, the desired shape may be cast brick, a cavity in a furnace wall which is filled by a ramming or gunning mix, or a coating sprayed on a furnace wall.

The preferred basic refractory material is magnesia. The purer the refractory material the better, but absolute purity is not expected and if the refractory has a purity of 87% or better it will suffice. The remainder of such basic refractory material will consist usually of magnesium compounds of iron, calcium, alumina, silicon, chrome oxide, clay and combinations thereof.

Preferably 70% of the basic refractory material is in various grain or particle sizes, some capable of retention on a 16 mesh screen while passing through a 7 mesh screen, some on a 7 mesh screen while passing through a 5 mesh screen and some on a 5 mesh screen while passing through a one-quarter inch screen, the various quantities of such sizes, however, not being critical.

The basic refractory material preferably also includes 20% or more of the basic refractory grains or particles capable of passing a 200 mesh screen. These relatively smaller size grains act to fill the interstices between the grains of larger size and impart improved density and strength of the refractory product with a minimum of shrinkage.

The preferred binder compounds, having served their binding function, should be capable of passing off either by sublimation, i.e., directly from a solid to a gaseous state, or by decomposition into constituents which volatilize at the temperatures used, while enabling the refractory material after firing to set up and form a hard dense mass.

Binding compounds that can be used are selected from ammonium chloride, ammonium fluoride, ammonium fluosilicate, ammonium fluosulfonate, ammonium bromide, ammonium iodide, ammonium bisulfate and ammonium sulfate, all of which are crystalline salts that sublime or whose constituents upon decomposition volatilize under the heat of firing. Characteristically the binders contain the ammonium radical and pass off by completion of the firing stage either by sublimation or by decomposition and volatilization of the component parts.

A preferred commercially available salt having the desired property of sublimation is ammonium sulfate. Another preferred salt is ammonium chloride. Binder levels of 0.5 to 10.0 weight percent are employed. In preparing the basic refractory composition the ammonium salt used may preferably have a particle size capable of passing a 200 mesh screen or finer.

If desired, a portion of the basic refractory material, up to 10% of the dry weight of the basic refractory mixture, may be ball milled together with the binder to the preferred grain size of the latter.

Example:

Refractory	Percent by Weight Total Composition
95% Magnesia	
Through ¼" on 5 mesh screen	15.00
Through 5 mesh on 7 mesh screen	31.00
Through 7 mesh on 16 mesh screen	23.00
Pulverized through 200 mesh screen	28.00
Binder	
Ammonium chloride	3.00
	100.00

Ball milled with a portion of the
95% magnesia to pass through 200
mesh screen

The ingredients are mixed thoroughly and intimately with water in amount equal to 8% of the dry weight of the composition. Standard size test bars (2" x 2" x 9") are formed by trowelling into suitable molds. The test bars are air dried for 24 hours, oven dried at 220°F for another 24 hours and then fired at 2950°F for 5 hours. The test bars are removed from the mold and cooled to room temperature.

The product was tested against a test sample similarly made from corresponding percentages and sizes of the same ingredients except for substitution for ammonium chloride of 3% of the commonly used refractory binder ammonium pentaborate (which does not sublime but deposits a stable borate salt), with results as follows:

Physical Properties After Firing at 2950°F

	Refractory Mixture with	
	Ammonium Pentaborate	Ammonium Chloride
Shrinkage	2.2 %	1.0 %
Modulus of rupture	2165 psi	3098 psi
Compressive strength at cold crush	4055 psi	6300 psi
Character of transverse failure	100% binder failure	80% matrix failure 20% binder failure (estimated)

It is apparent from the foregoing that shrinkage with ammonium chloride is less than half that with an ammonium pentaborate type binder, that the modulus of rupture and the compressive strength at cold crush are considerably higher and that the characteristics of transverse failure are different.

One preferred method of making the improved refractory is by forming the ammonium salt directly on the basic refractory grains. The refractory grains are thoroughly dampened with ammonium hydroxide by intimate mixing throughout the entire mass.

The acid corresponding to the salt to be formed (hydrochloric acid where the salt is ammonium chloride) is added and the mass again thoroughly mixed to complete the reaction.

Stoichiometric proportions of ammonium hydroxide and acid are used. Refractory compositions prepared according to this method rupture by cleavage of the refractory grains per se in substantial percentages in preference to failure of the bond between the refractory grains.

Analyses of the resulting matrices have shown residue from the salt to be extremely low or entirely absent. Such a binder, after firing, leaves little or no residue to contaminate the composition or reduce its refractoriness.

Dead-Burned Magnesia

According to *D.R.F. Spencer and D.C. Wooldridge; U.S. Patents 3,808,014; April 30, 1974; and 3,770,467; November 6, 1973; both assigned to Steetley Manufacturing Limited, England* a refractory dead-burned magnesia comprises lime, less than 0.2% by weight silica based on the weight of the magnesia, less than 0.5% by weight R_2O_3 based on the weight of the magnesia and 0.05% to 5% by weight zirconia based on the weight of the magnesia.

The amount of lime in the magnesia is 0.2 to 1.5% by weight, based on the weight of the magnesia, in excess of the weight of lime required to ensure a $CaO:ZrO_2$ mol ratio of 1:1. The magnesia has utility in the production of high hot strength refractory products. R_2O_3 refers to Al_2O_3, Fe_2O_3, Cr_2O_3, B_2O_3 and Mn_2O_3.

Examples 1 through 15: The following Examples 1 through 9 are given to illustrate the process and Examples 10 through 15 are included for the purposes of comparison. In each of these examples, grain has been prepared by compacting a mixture of fine magnesia of the appropriate chemical quality and zirconia into prismatic test bars and firing at 1650°C for 1 hour to densities in the 3.20 to 3.35 g/ml range. After firing, the prismatic grain test bars were surface ground to 6.4 x 6.4 x 40.6 mm and tested on a modulus of rupture apparatus at up to 1600°C.

The results are given in the table on the following page, from which it is observed that the test bars of Examples 1 through 9 produced from magnesias exhibit superior moduli of rupture values at 1500° to 1600°C as compared with the strengths of the test bars of Examples 10 through 15.

EXAMPLE No.	CHEMICAL ANALYSIS (Wt. %)									MODULUS OF RUPTURE GRAIN STRENGTH (MN/m²) AT	
	CaO	ZrO₂	SiO₂	Al₂O₃	Fe₂O₃	Cr₂O₃	B₂O₃	MgO	Excess CaO	1500°C	1600°C
1	0.8	0.10	0.09	0.06	0.06	0.05	0.012	98.83	0.75	90	64
2	0.53	0.50	0.11	0.07	0.08	0.03	0.01	98.67	0.3	86	37
3	0.83	0.50	0.12	0.03	0.02	0.03	0.01	98.46	0.6	101	68
4	1.13	0.50	0.07	0.04	0.02	0.03	0.01	98.20	0.9	97	65
5	0.76	1.00	0.11	0.03	0.05	0.02	0.01	98.22	0.3	58	23
6	1.06	1.00	0.13	0.04	0.03	0.02	0.01	97.71	0.6	72	65
7	1.36	1.00	0.09	0.02	0.05	0.02	0.01	97.45	0.9	67	61
8	1.51	2.00	0.10	0.02	0.08	0.02	0.01	96.26	0.6	78	68
9	2.88	5.00	0.09	0.03	0.04	0.02	0.01	91.93	0.6	77	65
10	0.23	0.50	0.10	0.03	0.04	0.04	0.01	99.05	0.0	11	9
11	0.46	1.00	0.09	0.02	0.03	0.03	0.01	98.36	0.0	12	12
12	0.68	1.50	0.11	0.04	0.02	0.03	0.01	97.61	0.0	12	12
13	0.91	2.00	0.11	0.05	0.02	0.02	0.01	96.88	0.0	12	11
14	1.37	3.00	0.13	0.03	0.02	0.02	0.01	95.42	0.0	12	10
15	2.28	5.00	0.12	0.03	0.02	0.02	0.01	92.52	0.0	13	9

*Excess CaO is that above the requirement to form a 1:1 CaO/ZrO₂ molecular ratio.

Heat Shock Resistant Heat Fused MgO

Y. Fukatsu; U.S. Patent 3,772,044; November 13, 1973; assigned to Asahi Glass Company, Limited, Japan describes a heat fused refractory product having (1) a base structure consisting of melted and resolidified periclase crystals in which needle or tabular ribbon type $CaO \cdot Cr_2O_3$ (calcium chromite) crystals are dispersed, and (2) coarse MgO phase particles. Chemically, this composition comprises more than 50% by weight of MgO, except the coarse particles of MgO phase, more than 5% by weight CaO and more than 3% by weight of Cr_2O_3, which is more than one-sixth of the amount of CaO, and less than 5% by weight of Al_2O_3, when the amount of CaO is more than 3 times that of the Cr_2O_3.

The heated fused refractory product can be prepared by admixing the desirable refractory material components, completely heat melting the mixture in an electric arc furnace, pouring the molten refractory material in a mold supplying MgO clinker to the flow of the molten refractory material, and then cooling and solidifying the molten refractory material in the mold. It can be shown that a crystalline structure is formed by cooling and resolidifying the molten components when coarse particles of MgO are admixed therewith prior to solidification.

The coarse particles are largely incorporated into the structure at the crystal interfaces where a portion of the Cr_2O_3 is diffused into the coarse particulate MgO phase. It can be shown by photomicrograph that the refractory product is made of periclase single crystals with tabular type or needle type $CaO \cdot Cr_2O_3$ crystals dispersed between the base crystals, and a border portion between the periclase crystals.

The heat fused refractory product therefore is characterized by $CaO \cdot Cr_2O_3$ crystals and coarse MgO-containing particles separately dispersed through the molten refractory material. The base crystal structure consists of periclase crystals, where the periclase crystals have an average length of 100 microns, such as 50 to 150 microns.

The actual structure of the resulting product will largely be dependent upon the particular refractory materials used. It is possible, for instance, to obtain a structure having the matrix and/or $CaO \cdot Cr_2O_3$ crystals at the border of all the periclase crystals.

The $CaO \cdot Cr_2O_3$ crystals will usually either partially or wholly cross the periclase crystals and will be crystallized in the form of a needle or tabular ribbon. The actual amount of $CaO \cdot Cr_2O_3$ crystals present in the refractory will depend upon the following conditions, and should be present in sufficient quantities and in sufficient structural form so as to provide sufficient spalling resistance of the periclase refractory product. If the quantity of CaO is too great, single crystals of CaO will be formed with the periclase crystals.

The molten refractory material for forming the base crystal structure is resolidified to form the $CaO \cdot Cr_2O_3$ containing periclase crystals. The quantities of ratios of the components of the refractory, excluding the MgO phase, are as follows on the next page.

Component	Preferable Ratio (wt %)	Most Preferable Ratio (wt %)
MgO	>50	60 - 80
CaO	5 - 40	5 - 25
Cr_2O_3	>3	10 - 25
Al_2O_3 (in $CaO>3Cr_2O_3$)	<5	<5
MgO + CaO		>60
Cr_2O_3	>⅙ CaO	>⅙ CaO
SiO_2		<4
Fe_2O_3 (in $CaO>3Cr_2O_3$)*		<2

*Although Fe oxide can exist in the form of FeO in the refractory product of this process, the quantity of Fe oxide is calculated as Fe_2O_3 for convenience.

High Purity Periclase

According to *R.L. Coatney; U.S. Patent 3,754,951; August 21, 1973; assigned to Kaiser Aluminum & Chemical Corporation* a periclase grain of enhanced high temperature strength is obtained when the subsidiary constituents of the grain are controlled so that substantially all the MgO in the grain is in the form of periclase, substantially all the alumina (Al_2O_3) is present in the form of brownmillerite ($4CaO \cdot Al_2O_3 \cdot Fe_2O_3$ or C_4AF), and substantially all the iron oxide (Fe_2O_3) is present either as brownmillerite or dicalcium ferrite ($2CaO \cdot Fe_2O_3$ or C_2F).

Under these circumstances, all the silica (SiO_2) will be present as either dicalcium silicate ($2CaO \cdot SiO$ or C_2S) or tricalcium silicate ($3CaO \cdot SiO$ or C_3S), and all the lime (CaO) will be present in the preceding compounds or as free lime.

More particularly, the grain of this process is a high purity periclase grain containing less than 0.1% by weight B_2O_3 and consisting of at least 85% by weight MgO and, as subsidiary components, CaO, SiO_2, Al_2O_3 and Fe_2O_3, where the relative amounts of the subsidiary components are controlled so that there are at least 2 mols of CaO present for each mol of SiO_2 plus Al_2O_3 plus Fe_2O_3, and so that there is at least one mol of Fe_2O_3 for each mol of Al_2O_3, whereby substantially all the MgO in the grain is present as periclase, substantially all the SiO_2 is present as a calcium silicate, substantially all the Al_2O_3 is present as brownmillerite, and substantially all the Fe_2O_3 is present as either brownmillerite or dicalcium ferrite, and where the amounts of the subsidiary components are controlled so that the weight ratio of calcium silicates to iron-containing compounds is 1:1 to 20:1.

Example: The compositions of this process are free of spinel materials such as magnesium aluminate ($MgO \cdot Al_2O_3$ or MA) and magnesium ferrite ($MgO \cdot Fe_2O_3$ or MF). Compositions were made by taking the amount of chemically pure (99.9% MgO) magnesium oxide shown in Table 1 and admixing with it, by blending it in an acetone slurry, the requisite amounts of silica, calcium carbonate, alumina, and iron oxide or chromium oxide to form the subsidiary compounds shown in Table 1.

All materials used were of a fineness such that substantially all the material passed a 325 mesh screen. The various admixtures were isostatically pressed into cylinders 4 inches long by 1 inch in diameter. These were fired at 1600°C for 6 hours, producing cylinders about 2½ inches long by ¾ inch diameter. From these fired cylinders, specimens 0.65 inch long by 0.40 inch diameter were machined. These machined specimens were tested for compressive strength at 1500°C, with the results shown in Table 1. All the specimens had, after firing, a B_2O_3 content of less than 0.1% by weight.

TABLE 1

Specimen	MgO	C_2S	C_3S	C_4AF	C_2F	Strength (psi)	Porosity (%)	Crystal Size (μ)
1	100	–	–	–	–	6,200	5.9	40
2	98	2.00	–	–	–	17,200	7.6	20
3	98	1.83	–	0.17	–	21,700	11.2	10
4	98	1.67	–	0.33	–	23,400	11.2	10
5	98	1.50	–	0.50	–	24,800	7.8	20
6	98	1.50	–	0.50	–	28,200	6.8	10
7	98	1.33	–	0.67	–	22,800	10.5	10
8	98	1.00	–	1.00	–	10,600	3.7	35
9	98	0.50	–	1.50	–	2,400	4.1	45
10	98	–	–	2.00	–	2,800	5.0	100
11	98	1.83	–	–	0.17	20,800	10.0	10
12	98	1.67	–	–	0.33	21,900	10.2	10
13	98	1.50	–	–	0.50	25,300	7.6	20
14	98	1.33	–	–	0.67	22,700	9.9	15
15	98	1.00	–	–	1.00	18,200	9.3	15
16	98	0.50	–	–	1.50	5,500	3.3	130
17	98	–	–	–	2.00	6,000	3.3	>150
18	98	–	2.00	–	–	19,100	7.0	25
19	98	–	1.50	0.50	–	19,900	7.0	20
20	98	–	1.50	–	0.50	19,100	6.7	20

From the data in Table 1, it can be seen that the addition of 2% dicalcium silicate (C_2S) increases the high temperature strength of periclase from 6,000 psi to over 17,000 psi (Specimen 2). Similarly the addition of 2% tricalcium silicate (C_3S) to the periclase increases its strength to over 19,000 psi (Specimen 18).

Examination of Specimens 10 and 17 indicates that the addition of 2% dicalcium ferrite (C_2F) has substantially no effect on the strength of pure periclase, whereas the addition of 2% brownmillerite (C_4AF) decreases the strength by more than half. Accordingly, as shown by specimens 3 to 9 and 11 to 16, the addition of either brownmillerite or dicalcium ferrite together with dicalcium silicate to the MgO results in even greater enhancement of the high temperature strength, namely to a value more than four times that found for the pure MgO and about 50% higher than that found for MgO with the addition of dicalcium silicate alone.

Specimens 19 and 20 demonstrate that, while the addition of brownmillerite or dicalcium ferrite to tricalcium silicate does not result in such a dramatic increase in strength compared to specimens with the tricalcium silicate addition alone, these additions do not, as might be expected from their effect on periclase alone, impair the high temperature strength of periclase containing tricalcium silicate.

The data on the average crystallite size in each specimen given in Table 1 might suggest that some of the low strength of specimens 10 and 17 was due to unusually large crystallites in these specimens. However, this effect would be offset by the

fact that specimens 10 and 17 were among those with the lowest porosity. In
any case, the highest strength specimens, 6 and 13, had average crystallite size
and were of higher porosity, suggesting that even higher strengths might be obtained with lower porosity.

Specimens A to K, the compositions of which are indicated in Table 2, were
prepared in the same way as the preceding specimens, using pigment grade Cr_2O_3
as the source of chrome. These specimens, which contain dicalcium silicate and
either dicalcium chromate ($2CaO \cdot Cr_2O_3$ or C_2K) or tricalcium chromate ($3CaO \cdot Cr_2O_3$
or C_3K), indicate that chromium can be substituted for iron.

TABLE 2

Specimen	MgO	C_3S	C_2K	C_3K	Strength (psi)	Porosity (%)	Crystal Size (μ)
A	98	1.83	0.17	–	21,300	10.3	10
B	98	1.67	0.33	–	21,800	12.1	10
C	98	1.33	0.67	–	17,300	16.1	10
D	98	1.00	1.00	–	17,200	17.6	15
E	98	–	2.00	–	3,000	3.3	25
F	98	1.83	–	0.17	21,100	11.2	6
G	98	1.67	–	0.33	20,900	12.5	8
H	98	1.33	–	0.67	19,800	14.8	10
J	98	1.00	–	1.00	15,500	12.8	15
K	98	–	–	2.00	2,000	3.3	20

Fired Periclase Brick from Magnesia, Lime and Silica

*H.L. Hieb; U.S. Patent 3,715,222; February 6, 1973; assigned to Kaiser Aluminum
& Chemical Corporation* has found that a superior refractory is made if, instead
of adding lime and silica separately to a periclase brickmaking batch, these materials are included in the matrix portion of the batch as components of a prereacted periclase grain containing larger amounts of lime and silica than in the
overall batch composition.

More specifically, it has been shown that a superior fire basic refractory can be
made from a batch consisting of (a) 60 to 80% by weight relatively coarse refractory aggregate, at least 90% of which is retained on a 100 mesh screen, the
aggregate containing at least 97.5% MgO; and (b) 20 to 40% by weight relatively
fine matrix material, at least 95% of which passes a 100 mesh screen, the matrix
material containing 80 to 93%, by weight MgO, 5 to 10% CaO, and 2 to 5%
SiO_2, there being at least 1.7 parts by weight CaO for each part by weight SiO_2.
The proportions and amounts of lime and silica in the matrix grain are chosen
so that the predominant secondary phase in this grain is dicalcium silicate.

In forming refractories, the refractory batch, including any bonding and sintering agents desired, will be mixed with a tempering amount of water and formed
into shapes, for example by pressing. Any conventional pressing means may be
used, for example a deairing press. Finally, the shapes will be fired at normal
firing temperatures for high periclase compositions, for example to 1700°C.

An advantage of incorporating the lime and silica in the matrix by the methods of prereaction is that it avoids the problems of correctly measuring and adequately dispersing small amounts of these materials in the refractory batch. In addition, the fire bricks have higher densities, and better volume stability on heating, than comparable refractories made by adding lime and silica to the brick batch.

Example 1: A refractory batch was made from 70.2 parts by weight coarse periclase grain, 26.6 parts by weight fine grain, 1.25 parts sodium nitrate, 1.5 parts lignin sulfonate binder, 0.3 part dispersing agent, and 2.2 parts water. This batch, after thorough mixing, was pressed in a conventional brick press and the shapes thus formed were fired at 1600°C.

The coarse grain, all of which passed a 4 mesh screen and 96% of which was retained on a 100 mesh screen, showed the following chemical analysis: 0.36% SiO_2, 0.23% Fe_2O_3, 0.05% Al_2O_3, 0.09% Cr_2O_3, 1.12% CaO, 0.19% B_2O_3, and (by difference) 97.96% MgO.

The prereacted fine grain, 98% of which passed a 100 mesh screen and 67% of which passes a 325 mesh screen, and which had a specific surface of about 3,000 cm^2/g showed the following chemical analysis on the fired basis: 3.47% SiO_2, 0.34% Fe_2O_3, 0.34% Al_2O_3, 0.22% Cr_2O_3, 6.55% CaO, 0.27% B_2O_3 and (by difference) 88.81% MgO.

Bricks made from the batch of Example 1 had a density of 187 lb/ft^3 (pcf) after pressing, 185 pcf after drying at 150°C and 182 pcf after firing. The average volume change of four specimens upon heating from 150° to 1600°C was a shrinkage of 1.0%. The average modulus of rupture at 1482°C of six specimens was 1,934 psi.

Example 2: A batch of 73 parts by weight coarse refractory grain, 25 parts fine refractory grain, 1 part lignin sulfonate binder and 1.25 parts sodium nitrate was mixed with 2.67 parts water and formed into shapes at a pressure of 10,000 psi. These shapes were dried at 150°C and fired at 1700°C.

The coarse refractory grain, all of which passed a 4 mesh screen and 92% of which was retained on a 100 mesh screen, had the following chemical analysis: 0.4% SiO_2, 0.2% Fe_2O_3, 0.1% Al_2O_3, 1.0% CaO, 0.2% B_2O_3, 0.1% Cr_2O_3, and (by difference) 98% MgO.

The fine refractory grain, 98% of which passed a 100 mesh screen, 67% being –325 mesh material, had the following chemical analysis: 4.0% SiO_2, 0.5% Fe_2O_3, 0.4% Al_2O_3, 7.0% CaO, 0.2% B_2O_3, 0.3% Cr_2O_3, and (by difference) 87.6% MgO.

After drying at 150°C, shapes made from this batch had a density of 184 pcf, the density being substantially the same after firing at 1700°C. The volumetric shrinkage upon heating to 1700°C was 0.2%, and the modulus of rupture at 1260°C was 2,516 psi and at 1482°C, 2,110 psi.

From the preceding examples it can be seen that, preferably, the relatively fine matrix grain material contains 87.5 to 89% by weight MgO, 6.5 to 7% by weight CaO, and 3.5 to 4% by weight SiO_2.

High MgO Brick with Improved Hot Tensile Strength

E.S. Wright; U.S. Patent 3,713,855; January 30, 1973; assigned to Eltra Corporation describes a method for producing a high MgO refractory, having lime and silica as the major impurities. The method involves preparing a batch of size-graded periclase particles, tempering the batch, forming the tempered batch into a shape, drying the shape and firing the shape.

At least 0.1% by weight of a calcium compound, calculated as CaO and based on the dry weight of the batch, is dissolved in the tempering liquid, the weight ratio of lime to silica is controlled to from 1.5:1 to 2.6:1, and the firing is controlled so that the overall refractoriness of the accessory phase is increased. Preferably, the CaO/SiO_2 ratio and firing are controlled so that dicalcium silicate is formed as the principal accessory phase. The addition of silica in addition to the dissolved calcium compound can also be used to control the CaO/SiO_2 ratio.

Example 1: A refractory batch having the following composition was prepared in a muller-type mixer by dry-mixing magnesite, silica and lignin sulfonate and tempering the dry mix with a solution of $Ca(NO_3)_2$:

	Percent
Magnesite	95.1
Silica	0.3
Lignin sulfonate	1.0
$Ca(NO_3)_2$ (added as a 50% aqueous solution)	3.6

The dead burned, natural magnesite used had the following composition:

TABLE 1

	Percent
SiO_2	1.51
CaO	2.99
Al_2O_3	0.07
Fe_2O_3	0.50
Cr_2O_3	0.03
B_2O_3	0.012
MgO (by difference)	94.89

The sizing of the magnesite grain was as follows:

	Percent
−3+6 mesh	27
−6+14 mesh	20
−14+48 mesh	15
−48 mesh	7
−100 mesh	31

The batch ingredients were then formed into bricks in a dry press under 14,000 pounds per square inch, cured at 400°F for 12 hours, and burned at 2900°F with a soak time of 5 hours. The above-described bricks were tested for green bulk density before curing; and after burning they were tested for bulk density, apparent porosity, modulus of rupture at room temperature, at 2700°F and at 2900°F, and hot crushing at 2800°F. All physical test results contained herein represent an average of at least 4 samples. Test results for the brick of Example 1 are tabulated in Table 2.

TABLE 2

Predominant Accessory Phases	Dicalcium Silicate
Relative proportions:	
CaO	3.56
SiO$_2$	1.75
GaO/SiO$_2$	2.03
Green: Bulk density (pcf)	191.9
Burned: Bulk density (pcf)	187.0
Apparent porosity (percent)	15.10
Modulus of rupture (psi):	
Room temperature	2,017
2700°F	2,176
2900°F	1,605
Hot crushing (psi): 2800°F	3,340

Example 2: For purposes of comparison, refractory bricks were prepared and fired by the same procedure detailed in Example 1, except that the CaO/SiO$_2$ ratio was not adjusted by the addition of a soluble calcium compound or silica. The magnesite used had the chemical composition and sizing shown in Table 1. Test results for the following brick composition are given in Table 3.

	Percent
Magnesite	96.2
Lignin sulfonate	1.0
MgSO$_4$ (added as a 22% aqueous solution)	2.8

TABLE 3

Predominant Accessory Phases	Dicalcium Silicate and Merwinite
Relative proportions:	
CaO	2.99
SiO$_2$	1.51
CaO/SiO$_2$	1.98
Green: Bulk density (pcf)	188.1
Burned: Bulk density (pcf)	184.0
Apparent porosity (percent)	16.10
Modulus of rupture (psi):	
Room temperature	1,459
2700°F	1,304
2900°F	806
Hot crushing (psi) 2800°F	1,678

A comparison of the test results of Example 1, with those set forth above for brick demonstrates that a slight increase in the amount of dicalcium silicate from the conversion of merwinite by controlling the adjustment of the CaO/SiO$_2$ ratio by the addition of a soluble calcium compound produces a large increase in the hot tensile strength of the burned magnesite brick at 2700°F, and an increase of almost 100% in the hot tensile strength at 2900°F. A further comparison shows an increase of almost 100% over the prior art sample in hot crushing strength at 2800°F, a substantial increase in room temperature modulus of rupture and bulk density, and a significant decrease in apparent porosity. The increase in the relative green bulk density figure in Example 1 is believed to indicate that the solution of the soluble calcium compound acts as a pressing aid when the solution is used as the tempering fluid.

High Density Grain from Magnesite

B. Davies and D.S. Whittemore; U.S. Patent 3,712,599; January 23, 1973; assigned to Dresser Industries, Incorporated describe a method for obtaining a high density refractory grain, suitable for refractory brick manufacture.

A crude magnesite material, i.e., magnesium carbonate, is crushed to a fine particle size. The material is caustic calcined (light burned). The material is then crushed to a fine particle size of less than 2 microns Fisher Average Diameter. The calcine is formed into the desired shape and deadburned at elevated temperatures.

Preferably, the initial comminution results in particles having a Fisher Average Diameter of less than 10 microns. The material is calcined at 1500° to 1800°F. The material may be formed into shapes, such as, briquettes, cylinders or brick prior to calcining if desired. After comminuting the calcine, the Fisher Average Diameter is less than 1 micron. Subsequent to shaping, the calcine is deadburned above 3000°F to produce a grain having a bulk specific gravity of at least 3.3.

In general, the crude magnesite was comprised of massive, polycrystalline aggregates of magnesite ($MgCO_3$) ranging in crystalline size from several millimeters (1/8 inch or more) down to a few microns. Accessory minerals primarily in the form of talc ($3MgO \cdot 4SiO_2 \cdot H_2O$) and other intermediate weathering products of an enstatite-type mineral ($MgO \cdot SiO_2$), massive, coarse to finely crystalline quartz (SiO_2) and complex vermiculite-chlorite mixed layer clays, were distributed throughout most of the samples. Talc-like accumulations and vermiculites occured generally as intercrystalline or enveloping "stringers" or "pods" surrounding the magnesite rather than being dispersed in the magnesite. Quartz was usually present in a more massive form. A typical chemical analysis of the magnesite samples studied showed the following ranges.

	Percent
Silica (SiO_2)	0.91 - 1.10
Alumina (Al_2O_3)	0.08 - 0.11
Iron oxide (Fe_2O_3)	0.62 - 1.71
Lime (CaO)	0.38 - 1.88
Boric oxide (B_2O_3)	0.0008 - 0.014
Magnesia (MgO)	95.3 - 97.9
Loss on ignition	51.4 - 51.7

MAGNESIA-CALCIA REFRACTORIES

Dolomite Grain Bonded with Magnesite

D.L. Guile and R.K. Smith; U.S. Patents 3,948,671; April 6, 1976; and 3,930,874;

January 6, 1976; both assigned to Corning Glass Works describe basic refractories suitable for use in lining high wear areas of the basic oxygen steelmaking furnaces (BOF). The proportioned, size-graded refractory raw material batch comprises at least one fused or dead-burned dolomitic grain and a fine grained dead-burned magnesite.

The dolomitic grain makes up three particle size fractions, all of which collectively fall substantially in the range –4+65 Tyler mesh and the MgO makes up the single fine fraction which is substantially all –100 Tyler mesh and preferably at least 75% –325 Tyler mesh. The dolomitic grain consists of a fused or dead-burned mixture of, on the oxide basis, 50 to 67% MgO and 33 to 50% CaO and preferably at least 98% MgO plus CaO.

A fused mixture of 70% dolomite and 30% MgO, resulting in an analytical oxide composition (excluding minor impurities) of 59% MgO and 41% CaO, is preferred. Oxides of magnesium and calcium may be substituted for the dolomite but this negates a cost saving resulting from the use of dolomite as a raw material.

Specifically, the basic brick batch is divided into four size fractions separated by three discrete gaps in the distribution of particles. The minimum size of each gap, in terms of the Tyler sieve series, is equivalent to the size of the gap created by removing from the batch the raw material contained on at least one primary Tyler sieve after screening the raw materials through a complete nest of primary Tyler sieves. By primary sieves, is meant those sieves of the Tyler Standard 2 sequence listed in the table below, each of whose openings is related to the next finer screen opening by a factor of two. It is not necessary that the gap occur exactly between identified screens.

Tyler Screen Series (Primary Sieves)	
Mesh Per Inch	Opening Size (Inches)
4	0.1852
6	0.1312
8	0.0928
10	0.0656
14	0.0464
20	0.0328
28	0.0232
35	0.0164
48	0.0116
65	0.0082
100	0.0058
150	0.0041
200	0.0029

In terms of particle size, the size of the finest particle allowed in any fraction must be at least two times the size of the coarsest particle allowed in the next finer fraction.

A preferred particle size distribution is –4+10 (coarse dolomitic grain), –14+20 (coarse-intermediate dolomitic grain), –35+65 (fine-intermediate dolomitic grain) and –100 (fine magnesite). The opening on the 10 mesh screen (0.0656 inch) is the $\sqrt{2}$ times the opening on the 14 mesh screen (0.0464 inch), the opening on the 20 mesh screen (0.0328 inch) is 2 times the opening on the 35 mesh screen (0.0164 inch) and the opening on the 65 mesh screen (0.0082 inch) is the square root of 2 times the opening of the 100 mesh screen (0.0058 inch).

In practice, screening would be completed on the raw materials using the Tyler sieve series of the table on the preceding page (or equivalent) and each of the three gaps in the size distribution would be created by expelling from the brick batch the raw material retained on at least one sieve.

In preparing a batch for fabrication of refractory bricks, the size graded and proportioned raw materials are blended with a nonaqueous (preferably carbonaceous or waxy) binder and pressed to a green density of 170 to 190 pounds per cubic foot. Storage containers for both the raw materials and the pressed brick must be dry and sealed from the atmosphere to avoid hydration of the materials. Following brick fabrication and storage, the bricks are fired to 1600°C to bring about the ceramic bond.

Example: A number of bricks suitable for use in the crash pad area of a basic oxygen furnace were pressed from a batch of:

Constituent	Size (Tyler Mesh)	Amount (Percent of the batch)
Fused dolomitic grain A	−4+10	40
Fused dolomitic grain A	−14+20	20
Fused dolomitic grain A	−35+65	15
Dead-burned Magmaster magnesite	−100	25

Included in the −100 mesh MgO fraction is 20% based on the whole batch, of −325 mesh MgO. The size of the tapered brick varied from 12" x (6 - 5") x 3" to 15" x (4½ - 3¾") x 3".

The proportioned raw materials (fused grain and magnesite) were mixed with 7% (by weight) slurried 40% paraffin wax/60% toluol binder at room temperature for about 10 minutes. (This paraffin binder may optionally be 3% by weight at 60°C). Bricks of the previously mentioned sizes were pressed such that final densities ranged 180 to 184 lb/cu ft.

Firing of the pressed ware was accomplished at 1600°C for 16 hours at peak temperature. A portion of the bricks were tar impregnated using common techniques. Results of hot strength and corrosion resistance of the rebonded grain refractory are compared in the table on the following page, with a commercial tar impregnated dead-burned magnesite refractory.

The slag test herein is a rather severe procedure devised to determine the relative corrosion resistance of various samples. In general, the particular slag under consideration contacts, for a specified period of time, a rotating laboratory furnace cavity which is constructed from the test refractories. An electric arc

maintains the slag at a normal temperature of 1750°C. The resistance is determined by measuring the depth of the slag cut into the refractory and the volume
of refractory eroded (determined by filling the eroded volume with a measured
volume of sand).

	Sized and Rebonded Grain Refractory	Commercial Refractory
Hot Strength (Modulus of Rupture)		
1340°C	1,545 psi	1,280 psi
Electric Slag Test		
Maximum Cut	0.34 inch	0.67 inch
% of Commercial Refractory	51 %	
Cut Volume	0.60 in^3	3.5 in^3
% of Commercial Refractory	17 %	

A normalized value of the slag cut for the refractories is generally reported as
a percentage of the cut of a standard brick used in the same slag test run. This
partially avoids problems in comparing refractories from different test runs. In
this particular test the slag had a composition of 20% FeO, 53.5% CaO, 21.5%
SiO_2 and 5.0% Al_2O_3, giving a lime-silica ratio of 2.5.

Rebonded Dolomite-Magnesite

*R.C. Doman; U.S. Patent 3,901,721; August 26, 1975; assigned to Corning
Glass Works* describes rebonded basic refractory consisting of 65 to 75% –4+65
mesh coarse grain bonded with –100 mesh (preferably at least 65% thereof being
–325 mesh) magnesite fines. The coarse fraction in the batch is a mechanical
mixture of (based on the whole batch) 10 to 55% dead-burned magnesite and
15 to 60% fused dolomite (or equivalent mixed oxides).

Preferably the two coarse materials are divided into each of two discrete particle
size ranges, –4+12 mesh and –20+40 mesh, the former making up 45% of the
batch and the latter 25% of the batch, with the fine dead-burned magnesite being the remaining 30%. The ratio of fused grain to MgO in each of these two
coarse fractions is preferably the same as the ratio of fused grain to coarse MgO
in the whole batch. The burned refractories show superior corrosion resistance
to basic oxygen steelmaking slags.

In preparing the burned refractories, the raw batch materials are proportioned
and then blended with a temporary organic binder such as paraffin wax or a
commercial tar pitch. The refractory bodies are compacted, dried and then fired
to at least 1600°C to bring about the ceramic bond. Fired bodies may also be
tar or wax impregnated according to normal practice in the art.

Example: A variety of commercial furnace bricks 4½" x 3" x 2¼" were mechanically pressed with an impact press using 14 strokes at 10,000 psi per stroke.
After pressing, the bricks were placed in a gas-oxygen furnace and fired to
1600°C. The batches used in making the bricks and the slag corrosion results
are shown in the table on the following page. In addition to the dry ingredients
a 3% binder of paraffin wax was mulled for 10 minutes with the batch at 60°C
prior to compacting the brick.

| Example No. | - - - - - - - Composition (% Total Batch)- - - - - - - | | | Fired Bulk Density (lb/ft³) | - - -Slag Data- - - | |
| | -4+12 and -20+40 Fractions | | | | Max. Cut (in) | Volume Removed (in³) |
	% Fused Dolomite	% Dead- Burned MgO	-100 Mesh Dead- Burned MgO (%)			
1	14	56	30	178	0.35	0.92
2	56	14	30	180	0.39	0.41
3	28	42	30	179	0.34	0.57
4	42	28	30	179	0.36	0.52
5	7	63	30	178	0.48	1.74
6	63	7	30	179	0.44	0.47

The fused grain was made from an electrically melted and resolidified batch of calcined dolomite analyzing 57.8% CaO, 41.2% MgO, 0.5% SiO_2, 0.2% Fe_2O_3, 0.15% Al_2O_3, with 0.15% LOI. The melt was resolidified by casting on a graphite slab and the grain was made by crushing and sizing.

In the examples, the -4+12 mesh particle fraction made up 45% of the whole batch and the -20+40 mesh particle fraction made up 25% of the whole batch. The quantities of fused dolomite and coarse dead-burned MgO were divided into each of the two coarse fractions in the same ratio of fused grain to coarse MgO as in the whole batch. For example, the table below shows the division of materials for one preferred batch of the table above.

Particle Distribution for the Coarse Fraction in Sample 3

Constituent	% of Batch*	-4+12 Mesh Fraction (% of batch)	-20+40 Mesh Fraction (% of batch)
Coarse MgO	42	27	15
Fused dolomite	28	18	10
Total	70	45	25

*Ratio of coarse MgO to fused dolomite in the batch is 1.5.

Typical analysis of the magnesite used was 98.5% MgO, 0.55% CaO, 0.38% SiO_2, and 0.44% others (Fe_2O_3, Al_2O_3, B_2O_3) with 0.13% LOI. Bulk density of the fine MgO fraction should be at least 2.5 g/cc for good corrosion resistance, especially if the particle size is only -100 mesh. The bulk density is not as important in the fine fraction if the -100 mesh MgO is reduced (by milling, etc.) to at least 65% -325 mesh.

The slag test herein is a rather severe procedure devised to determine the corrosion resistance of various samples. In general, the particular slag under consideration contacts, for a specified time period, a rotating laboratory furnace cavity which is constructed from the sample refractories.

An electric arc maintains the slag at a normal temperature of 1750°C. Following the test run, the resistance of the samples is determined by measuring the depth of the slag cut into the refractory and the volume of refractory eroded (determined by filling the eroded cavity with a measured volume of sand). The relative slag resistance is then observed by comparing each refractory result with the results of the other samples of the test run. The corrosion results on each refractory in a run may also be normalized to reduce the effect of dissimilar test runs when comparing refractories from different runs, by recording results as a percent of the corrosion of a control refractory inserted into the test cavity on each run. The slag composition was 20% FeO, 53.5% CaO, 21.5% SiO_2 and 5.0% Al_2O_3 giving a lime-silica ratio of 2.5.

Hydration Resistant Dolomite

R. Staut; U.S. Patent 3,753,747; August 21, 1973 and R. Staut and J.L. Stein; U.S. Patent 3,736,161; May 29, 1973; both assigned to General Refractories Company provide for producing hydration resistant, low flux dolomitic refractory material in which a dolomite containing at least 98% by weight of CaO and MgO is calcined to produce quick-lime which is slaked to at least partially hydrate the oxides, and the slaked oxides are formed into a refractory shape and then sintered.

The improvement comprises incorporating in the slaked oxides at least 0.01% by weight on an oxide basis of a finely divided stabilizing agent such as Al_2O_3, V_2O_5, aluminum fluoride, MgF_2 and CaF_2, and mixtures thereof to increase the total content of the agents to 0.01 to 1.0% by weight on an oxide basis.

It is surprising that the addition of such small amounts of stabilizing agents will produce significant changes in the hydration resistance of high purity CaO·MgO refractory grains. Besides increasing the hydration resistance of low flux CaO·MgO grain, small additions of the stabilizing agent also tend to greatly improve the appearance of the product.

Samples of dolomite grain (CaO·MgO) ordinarily appear nonhomogeneous and display a mottled beige color. A small addition of Al_2O_3, for example, which still yields a material greater than 99% CaO·MgO produces a very homogeneous almost white sample with no detectable second phase formation.

The addition of small amounts, less than 1% by weight, of the stabilizing agents does not significantly lower the refractory properties of the resulting grain, in contrast to the results obtained when higher percentages of traditional fluxing agents are added to dolomite compositions.

The incorporation of stabilizing agents can be used to improve the hydration resistance of natural dolomitic ores, and can also be used to improve the hydration resistance of admixtures of a dolomite ore with a magnesium compound that will decompose to form MgO, or light burned MgO. Thus, the dolomite grain produced can contain 40 to 98% MgO, and 1 to 58% CaO. The density of the material produced is very high, above 3.3 g/cc after sintering. The porosity of the material is very low.

Examples 1 through 7: In these examples, the hydration resistance of a CaO·MgO refractory grain is improved by addition of magnesium fluoride and vanadium pentoxide. Samples are prepared containing 0, 0.25, 0.50, and 1.0 weight percent MgF_2 additions to a hydroxide base composition containing 60% MgO and 40% CaO on an oxide basis.

The sample containing no additive serves as a control. A second series of samples are prepared containing 0.01, 0.05, 0.10 and 0.50 by weight percent V_2O_5

additions to the same base composition. All of the batches are pressed at 15,000 psi and fired to a density greater than 3.3 g/cc with zero water absorption at 1650°C for two hours.

The fired samples are then placed in a closed vessel over water at 72°F for hydration studies. The control sample with no addition of stabilizing agent exhibited slight cracking after 20 days in the closed vessel. The sample containing 1.0% MgF_2 cracked after 20 days in the humidity chamber. After 30 days in the chamber, the control sample is still slightly cracked, as are the samples containing 0.01, 0.05, and 0.10% V_2O_5 and 0.125 and 0.25% MgF_2. No evidence of dusting is apparent in any of the samples containing the added stabilizing agent.

After 75 days in the chamber: (a) the control sample is destroyed, (b) the sample containing 0.125% MgF_2 is only slightly cracked and still in one piece and (c) the samples containing 0.01, 0.05, and 0.10% V_2O_5 are split in several sections, but show no evidence of dusting.

The samples with V_2O_5 additions of 0.01 and 0.05% produced better hydration resistance compared to samples with higher V_2O_5 additions. Also, the sample with 0.125% MgF_2 exhibited better hydration resistance than samples containing larger amounts of MgF_2. Each of the samples containing an additive showed significantly improved hydration resistance compared to the control containing no additive.

Refractory from Dolomite and Hydration Retarding Agent

L. Andersson; U.S. Patent 3,743,524; July 3, 1973; assigned to Strabruken AB, Sweden describes a refractory ceramic mass which is mixable with water, consisting of a graded mixture of grains of one or more refractory materials. At least a portion of the finest fractions having a maximum grain size of 0.1 mm consists of burnt dolomite and there is included in the mixture a water-soluble hydration retarding agent which prevents or sufficiently retards a reaction between the finest grains of burnt dolomite and the added water.

Particularly suitable retarding agents are sodium fluoride, potassium fluoride or ammonium fluoride. Also complex fluorides, such as sodium silico fluoride and magnesium silico fluoride, have been found to be useful although they are less active than the simple fluorides. The amount of the hydration retarding additions can be 1 to 2%. The plastic mass obtained when mixing the dry refractory mass with water must contain in known manner a water-soluble binding agent, e.g., silicate of potassium.

Preferably, the binding agent has been admixed in the dry ceramic mass in powder form. In the selection of binding agents and hydration retarding agents regard must be taken to the effect of the agents on the burnt dolomite at high temperatures.

Example: A mass which is particularly suitable for repairing the lining of e.g., an electric furnace for the production of steel, may be produced in the following manner. The burnt dolomite is first crushed in a hammer crusher or cone crusher and divided into grades or fractions. 40 to 50 parts of a fraction having a grain size of 0 to 0.25 mm are mixed with 50 to 60 parts of a fraction having a grain size of 0.25 to 2.5 mm. During the mixing operation are added 3 to 5 parts of powdered silicate of sodium as a binding agent, and 0.5 to 2 parts of powdered sodium silico fluoride as a hydration preventing or retarding agent.

The repair of the lining may be carried out during the short time interval between the tapping of the furnace and the next charging. The dry refractory mass is introduced into the furnace in the desired position by means of compressed air through a long tubular nozzle and simultaneously water which mixes with the refractory material is blown onto the surface in an amount which gives the material such a consistency that a compact layer is formed at the desired position in the furnace.

The binding agent and the hydration retarding agent rapidly dissolve in the water. Due to the high temperature of the furnace the layer of moist refractory material dries up very rapidly, and thereafter the binding agent starts to bind chemically. The chemical binding gives the applied material sufficient mechanical strength until later on the material is sintered onto the furnace wall at the operating temperature of the furnace. The hydration retarding agent prevents chemical reaction between the fine grains in the refractory mass and the water mixed thereinto during the short period of time elapsing until the mass has dried.

A sprayable mass having the above composition, in which not only the finest fractions but also the coarser fractions consist of burnt dolomite, may be used to particular advantage when it is desired to repair a furnace lining consisting solely of burnt dolomite, in that in such a case the sprayed mass will have the same physical and chemical properties as the backing, whereby the adherence of the sprayed-on mass is improved and the strength is increased.

MAGNESIA-CHROME REFRACTORIES

Direct Bonded Magnesia-Chrome Ore Brick

G.M. Farrington, Jr. and R. Staut; U.S. Patent 3,864,136; February 4, 1975; assigned to General Refractories Company describe a direct bonded refractory brick comprised of 30 to 80% by weight magnesia, 70 to 20% by weight chrome ore and having an overall lime to silica ratio less than 0.30 and a silica content of less than 2.5% by weight on an oxide basis.

By maintaining the overall lime to silica ratio below 0.30 and the silica content at 2.5% or less, a direct bonded brick is obtained having a high number of direct bonds, a high hot strength, and a low open porosity. The microstructure of a hot fracture surface of the brick is characterized by the presence of secondary spinels, by the presence of nonwetting silicates, by the presence of broken grains, and by a high crushing strength at elevated temperature, properties which result from the desired degree of direct bonding.

Preferably, the overall lime to silica ratio and silica content of the brick are controlled by adding a hydrated silicate compound to the magnesia-chrome ore mixture. The hydrated silicate compound is preferably a magnesium silicate such as talc. The overall lime to silica ratio is preferably maintained at less than 0.30.

Direct bonded refractory brick has a low open porosity and a low silica content, and can be produced in conventional forming and firing equipment. An additional advantage of the direct bonded refractory brick is the good slag resistance produced as a result of direct bonding and a low level of isolated silicates.

Example: A brick composition is prepared by blending 60 parts of a fine high purity magnesia containing only ball mill fines, 60% of which is less than 325 mesh and all of which is less than 200 mesh, with 40 parts of a fine chrome ore containing only ball mill fines, 60% of which is less than 325 mesh and all of which is less than 200 mesh. The chrome ore and magnesia employed have the following analysis given in weight percent.

Chrome Ore

LOI (Loss on Ignition)	0.11
SiO_2	2.50
Al_2O_3	29.70
Fe_2O_3	14.60
CaO	0.35
MgO	13.80
Cr_2O_3	36.00
B_2O_3	–

Magnesia

LOI	0.12
SiO_2	0.52
CaO	0.61
Fe_2O_3	0.30
Al_2O_3	0.29
MgO	98.03
Cr_2O_3	–
B_2O_3	0.13

The composition initially has an overall silica content of 1.35% but this content is adjusted to 2.0% by addition of 0.7% Supersil, a finely ground silica containing 100% of –325 mesh particles. As a result of the addition of silica, the composition has a lime to silica ratio of 0.29.

The composition is pressed into a brick at 15,000 psi and fired in a furnace at 3200°F for 6 hours under normal firing conditions. Upon completion of firing and after cooling of the brick to room temperature, it is examined for density, porosity, and hot strength. The results of the examination are set below.

Fired density (g/cc)	3.10
Open porosity (%)	17.3
Hot modulus of rupture (psi) at 2700°F	2300
Hot crush strength (psi) at 2800°F	2265

The brick is fractured at 2700°F and its microstructure is then examined. The microstructure exhibits areas of isolated, nonwetting silicates which indicate that the brick has true direct bonds rather than silicate bonds. The microstructure also exhibits areas of secondary spinels which shows that the silicate phase does not dissolve the secondary spinels of this composition at high temperature.

A dissolving of secondary spinels at this temperature would increase the overall amount of liquid present at the test temperature which would result in decreased strength. The microstructure also shows that the periclase grains are broken and thus indicates that the fracture occurred through the grains. A fracture through the grains indicates that direct bonds are present rather than silicate bonds because silicate bonds would soften at high temperatures showing only fracture through the matrix rather than through both grain and matrix.

Burned Phosphate Bonded Basic Refractory

W.S. Treffner and A.H. Foessel; U.S. Patent 3,839,057; October 1, 1974; assigned to General Refractories Company describe phosphate-bonded basic refractory compositions containing the following compounds in predetermined ratios $CaO:SiO_2$; $P_2O_5:SiO_2$; $P_2O_5:CaO$ and $CaO:(P_2O_5 + SiO_2)$. They are fired at 2500° to 3200°F to provide improvements in dimensional stability on firing, improved hot compressive and transverse load resistance, improved compressive creep behavior, or improved thermal spalling resistance.

In addition, low silica chrome ore is also included in the phosphate-bonded basic refractory composition which is subjected to firing to improve one or more of the above described properties.

The $CaO:SiO_2$ ratio must be at least 4.5:1. While there appears to be no critical upper limit to this ratio, particularly when the SiO_2 content is very low, in practice it generally does not exceed 20:1. The $P_2O_5:SiO_2$ ratio will also be relatively high, that is 3:1 to 12:1. The $P_2O_5:CaO$ ratio will be, as stated, 0.6:1 to 1.2:1, and the $CaO:(P_2O_5 + SiO_2)$ ratio is 0.8:1 to 1.3:1. These ratios are on a weight basis.

Neither the sodium polyphosphate nor the added calcium compound generally exceeds 10% by weight, based on the weight of the dry composition, and each may be present in 1 to 8%, preferably 2 to 8% by weight, based on the weight of the composition. Optionally, up to 4% by weight of a low silica chrome ore, basis combined weight of chrome ore and calcium magnesite, can be included in the composition.

Examples 1 through 3: Dead-burned magnesite of conventional grain sizing (60% -6+35 mesh and 40% -48 mesh) is mixed with a sodium polyphosphate, a calcium-containing additive and a tempering liquid. The mixture is molded into standard sized brick under 12,000 psi pressure and dried. The magnesite has the following composition: SiO_2, 0.3%; CaO, 2.7%; oxides of Fe, Al, etc., 5.0%; MgO, 92.0%.

The properties of this chemically bonded brick are determined by standard methods used in the industry with the results listed on the following page, under the heading, "Example 1" and serve as a control. Brick of the same batch formulation are burned in a commercial tunnel kiln to 2800°F and their properties are also determined. The results are listed on the following page, under the heading "Example 2."

More brick of the same composition are burned in a high temperature commercial tunnel kiln at 3150°F. The properties of these bricks are determined and listed on the following page, under the heading "Example 3."

	Ex. 1	Ex. 2	Ex. 3
Firing change, vol %	NA	-1.1	-1.2
Bulk density, g/cc	2.94	2.98	2.94
Open pores, %	17.0*	16.0	15.3
Hot modulus of rupture at 2700°F, psi	2,840	2,330	1,265
Hot load deformation, % (5 hr at 3100°F)	3.0	1.8	0.7
Reheat change, vol % (3100°F)	-1.0	-1.1	ND

*After 2000°F.

The results show that the brick exhibit significantly improved hot load deformation. While the hot modulus of rupture is lowered by firing, nevertheless the resultant hot modulus is still greater than that of commercial high fired "direct bonded" bricks.

Single Pass Prereacted Grain from MgO and Chrome Ore

W.R. Alder; U.S. Patent 3,817,765; June 18, 1974; assigned to Kaiser Aluminum & Chemical Corporation describes a sintered prereacted refractory grain containing less than 2% SiO_2 which is made from MgO-yielding material and chrome ore by intimately admixing 20 to 80% of MgO-yielding material, calculated as MgO, with 80 to 20% by weight of chrome ore, at least 95% of which passes a 325 mesh screen (i.e., is finer than 44 microns) and 50% of which is finer than 10 microns.

Substantially all of the MgO-yielding material is less than 44 microns in size and has a median particle size of less than 4 microns. The MgO-yielding material contains less than 0.5% SiO_2 on the ignited basis. The chrome ore contains less than 3% SiO_2. The admixture is heated without prior compaction, to at least 1975°C for at least 10 minutes.

The resulting product is a fully recrystallized, uniform refractory grain of periclase solid solution crystallites with exsolved chromium-containing spinels within the periclase crystallites and containing less than 2% SiO_2. The grain is substantially free of any of the original chrome ore in unreacted form, and, in a preferred form, has a total porosity of less than 10 volume percent.

The MgO-yielding material may be any such material, for example brucite or magnesite, but a preferred material is magnesium hydroxide, most preferably magnesium hydroxide produced by reacting CaO-containing material such as calcined limestone or calcined dolomite with brine, for example seawater, containing magnesium values in solution. The chrome ore can be any such material, but will generally be of the so-called "refractory grade." Such chrome ores are found in the Masinloc region of the Philippines, in the Transvaal and other regions of Africa, in Turkey and elsewhere.

A particularly preferred method is to admix the MgO-yielding material and the chrome ore and mill them together until both have the requisite fine particle size and are thoroughly interdispersed. It is an advantage that magnesium hydroxide can be much more intimately interdispersed with the chrome ore than can the calcined magnesia used in the doubleburn process.

The admixed raw materials are then charged, without any compaction, into a kiln, for example a rotary kiln, where they are exposed to a peak temperature of at least 1975°C and preferably at least 2000°C. It is necessary that the material be exposed to temperatures of at least 1975°C for at least 10 minutes. It has been found that shorter exposure times result in insufficiently reacted and nonuniform grain.

During firing, the spinel constituents of the chrome ore are dissolved in the MgO, which is extensively recrystallized. At the peak firing temperatures, substantially all the chromium spinel material is dissolved in MgO. However, upon cooling of the refractory material below the peak firing temperature, chromium-containing spinels exsolve from the MgO, appearing as dendrites in the MgO crystallites. The silicate materials originally present in the chrome ore appear as a minor intergranular phase between the MgO crystallites. A very few intergranular spinel crystallites are also present.

Example: Magnesium hydroxide was produced by reacting calcined dolomite with seawater containing dissolved magnesium sulfate and magnesium chloride. This magnesium hydroxide shows, on the ignited basis, the following typical chemical analysis: 1.2% CaO, 0.3% SiO_2, 0.1% Al_2O_3, 0.2% Fe_2O_3, and (by difference) 98.2% MgO. All the particles of $Mg(OH)_2$ were smaller than 44 microns and the median particle size was 4 microns.

Masinloc chrome ore concentrates were charged to a steel ball mill with steel balls and dry milled to a median particle size of 9.3 microns. 95% of the milled product was finer than 44 microns. The chrome concentrates showed the following typical chemical analysis: 18.7% MgO, 0.4% CaO, 2.6% SiO_2, 29.4% Al_2O_3, 13.9% FeO, and 35.0% Cr_2O_3.

The milled chrome concentrates and damp magnesium hydroxide filter cake were charged to a pug mill in the proportion of 75 parts by weight, dry basis, magnesium hydroxide to 48 parts by weight chrome concentrates. The damp $Mg(OH)_2$ filter cake contained equal parts by weight water and $Mg(OH)_2$, i.e., each 75 parts by weight $Mg(OH)_2$ brought an equal weight of water to the mixture.

From the preceding analyses, it can be calculated that, after ignition, this admixture will contain very close to 60% by weight MgO. After leaving the pug mill, the wet admixture was conveyed by means of a screw conveyor, where further mixing took place, to a rotary kiln. The material was fired to a peak temperature of 2000°C, the residence time of the material in the kiln at temperatures above 1975°C being 15 minutes.

The resulting refractory grain showed the following chemical analysis: 1.5% SiO_2, 7.1% Fe_2O_3, 13.6% Al_2O_3, 15.2% Cr_2O_3, 0.8% CaO, and (by difference) 61.8% MgO. Petrographic examination of the grain showed it to be a thoroughly uniform periclase solid solution-spinel with minor silicate composition, displaying a strong periclase-chrome reaction.

The periclase solid solution crystal size ranged from 60 to 120 microns and averaged 80 microns. The very few irregularly distributed intergranular spinel crystals average 20 microns in size. The inert and well reacted character of the grain is illustrated by the difficulty experienced in putting the material into solution for purposes of running the wet chemical analysis. The porosity of the grain was 9 volume percent.

The grain of the preceding example is to be compared with grain made in similar fashion except that only 75% of the chrome ore concentrates were finer than 44 microns (i.e., passed a 325 mesh screen). This comparison grain had a similar chemical composition to that of the example, but its porosity was 15%.

Microstructural examination showed it to be much less well reacted than the grain of the example. The crystallite size of the comparison grain ranged from 30 to 150 microns, with an average of 60 microns. The greater variability of this comparison grain is thus evident. Although the comparison grain contained periclase crystallites with exsolution chrome, it also contained considerable unreacted chrome material. In general, the comparison grain was not as well bonded as the grain of the example.

The grain of the example can also be compared with a double pass grain made from the same raw materials. The magnesium hydroxide was calcined to 950°C to produce an active magnesia, and admixed with the chrome concentrates ground to an average particle size of 10 microns, over 95% of the chrome concentrates being less than 44 microns in size.

The dry admixture was compacted in a roll-type press at an equivalent pressure of 35 tons per square inch. The compacts so formed were charged to a rotary kiln where they were exposed to a peak temperature of 1975°C. Microscopic examination showed the resulting grain to be made up of very uniform periclase solid solution-spinel crystallites with minor silicate, and having an average crystallite size of 70 microns. The porosity was about 10%. In other words, even the double burn grain made by the more expensive process is not as well reacted as the grain made according to this process.

Refractory Shape from Periclase, Chrome Ore and Prereacted Grain

According to *J.E. Neely and S.F. Brown; U.S. Patent 3,798,042; March 19, 1974; assigned to Kaiser Aluminum & Chemical Corporation* a fired refractory shape is made from a batch consisting of 20 to 80% prereacted grain, 0 to 55% unreacted periclase and 0 to 55% unreacted chrome ore. The prereacted grain is a sintered admixture of 20 to 80% magnesia and 80 to 20% chrome ore, both materials substantially entirely passing a 100 mesh screen, the sintered grain being substantially free of any unreacted chrome particles.

The prereacted grain is made by admixing and firing fine magnesia and fine chrome ore. Both materials substantially entirely pass a 100 mesh screen, and preferably at least 90% of each of them passes a 325 mesh screen. While a dead-burned magnesia such as the periclase used in conjunction with the prereacted grain, can be used, it is generally more economical to use a more lightly burned magnesia.

The chrome ore can be any such material and, where it is desired to limit the SiO_2 content of the refractory, the beneficiated type of chrome ore sometimes referred to as chrome concentrates can be used. The magnesia and chrome ore can be mixed as a slurry or wet sludge and charged directly to a rotary kiln for firing, 1900° to 2000°C or higher. Alternatively, the materials can be compacted prior to firing.

The prereacted grain, after firing, shows complete reaction between the periclase and chrome ore. In other words, the prereacted grain is substantially free of any of the original chrome particles, these all having reacted with the periclase at high temperatures to form a solid solution from which chrome spinels are exsolved during cooling.

Examples 1 through 13: The table on the following page sets forth the compositions and properties of a variety of compositions illustrating the use of different types, as well as amounts, of prereacted grains, unreacted periclase and unreacted chrome ore.

Prereacted grain SPM was made admixing magnesium hydroxide produced from seawater with finely divided (90% –325 mesh) Masinloc chrome ore concentrates (Type MC chrome ore) in amounts to yield about 60% MgO in the fired product, the wet admixture being fed to a rotary kiln where it was fired to 2000°C. The grain showed the following typical chemical analysis: 62.2% MgO, 0.7% CaO, 1.6% SiO_2, 12.8% Al_2O_3, 7.1% Fe_2O_3 and 15.6% Cr_2O_3.

Prereacted grain DPM was made from the same raw materials as grain SPM and in the same proportion. However, the magnesium hydroxide was calcined at 1000°C prior to mixing with the chrome, and the admixture was compacted prior to firing.

Prereacted grain DPT was made from calcined magnesia and finely divided Transvaal chrome concentrates in proportions to yield 80% MgO in the reacted grain. The admixture was compacted before being fired at 1800°C. The reacted grain showed the following chemical analysis: 1.3% SiO_2, 6.1% Fe_2O_3, 5.6% Al_2O_3, 10.4% Cr_2O_3, 0.9% CaO, and (by difference) 75.7% MgO. Type G periclase was a seawater periclase of the following typical chemical analysis: 1.1% CaO, 2.1% SiO_2, 0.3% Al_2O_3, 0.4% Fe_2O_3, 0.3% Cr_2O_3 and (by difference) 95.8% MgO. It was made by firing magnesium hydroxide in a rotary kiln.

Type K periclase was a high purity periclase made by the so-called double burn process with briquetting between the calcination and firing steps. It had a typical chemical analysis as follows: 1.0% CaO, 0.3% SiO_2, 0.1% Al_2O_3, 0.2% Fe_2O_3, 0.2% Cr_2O_3 and (by difference) 98.2% MgO.

Type MC chrome ore was a beneficiated Masinloc chrome ore having the following typical chemical analysis: 18.7% MgO, 0.4% CaO, 2.6% SiO_2, 29.4% Al_2O_3, 13.9% FeO and 35.0% Cr_2O_3. Type TC chrome ore was a beneficiated Transvaal chrome ore of the following typical chemical analysis: 11.0% MgO, 0.1% CaO, 0.6% SiO_2, 14.7% Al_2O_3, 28.5% Fe_2O_3 and 45.1% Cr_2O_3.

The weight percentage given for each of the grain materials in the brick batch is based on the total weight of grain in the batch, and does not take into consideration 2% of a temporary lignosulfonate binder used in all batches and 0.5% ZrO_2 used in batches 6, 7, 8 and 10 as a sintering aid. Each of the batches was pressed into bricks 9" x 4½" x 3" under a pressure of 10,000 psi, dried at 150°C and fired to 1750°C. The firing shrinkage of Specimens 8 and 9 can be compared with that of brick of the same overall composition but made entirely of prereacted grain (and Al_2O_3). This latter brick had 1.5% linear and 4.1% volume shrinkage.

	Specimen Number												
	1	2	3	4	5	6	7	8	9	10	11	12	13
Nominal percent MgO	40	50	50	50	60	60	60	60	60	60	70	80	80
Prereacted grain:													
Type	SPM	DPM	DPM	SPM	SPM	SPM	DPM	SPM	DPM	SPM	DPM	DPT	SPM
Amount (wt %)	55	81	24	40.5	70	60	50	42	42	30	70	70	43
Sizing:													
+28	54	64	100	72	77	75	99	75	100	99.7	87	87	76
28 x 100	2	7	-	4	22	9	1	1	-	0.3	13	13	-
-100	44	29	-	24	1	16	-	24	-	-	-	-	24
Periclase:													
Type	-	-	G	G	G	K	K	G	G	G	K	K	G
Amount (wt %)	-	-	31.5	20.5	18	23	30	33.5	34	41	30	20	57
Sizing:													
+28	-	-	-	-	-	-	-	37	-	17	-	-	40
28 x 100	-	-	2	2	1	1	2	2	3	0.5	2	-	24
-100	-	-	98	98	99	99	98	61	97	82.5	98	100	36
Chrome ore:													
Type	MC	MC	MC	MC	TC	MC	TC	MC	MC	MC	-	MC	-
Amount (wt %)	45	19	44.5	39	12	17	20	24.5	24	29	-	10	--
Sizing:													
+28	47	23	47	46	-	99	3	35	34	35	-	-	-
28 x 100	49	68	48	49	-	1	87	58	59	61	-	-	-
-100	4	9	5	5	100	-	10	7	7	4	-	100	-
Density (pcf):													
Pressed	208	201	204	201	200	199	204	201	202	200	194	196	191
Dried	204	197	200	200	196	194	200	197	199	197	191	191	187
Fired	202	198	199	196	196	192	199	193	194	194	191	195	191
Porosity (vol %)	16.3	16.8	16.4	17.0	15.9	17.2	16.7	16.7	16.4	16.6	16.7	15.4	15.9
Shrinkage (%):													
Linear	0.6	0.9	0.8	0.6	0.9	0.7	0.6	0.6	0.4	0.8	0.9	1.2	1.5
Volume	0.7	1.9	1.2	0.6	1.9	1.1	0.5	0.5	0.3	1.1	1.8	3.1	3.6
MOR (psi):													
1260°C	2431	1942	2682	-	2741	2249	2498	2498	2608	3225	2463	245*	1657
1482°C	766	610	876	800	868	778	1226	1226	1178	1685	552	130**	286

*At 1400°C.
**At 1500°C.

From the example, it will be seen that in most cases at least 75% of the pre-reacted grain is of +28 mesh sizing, that at least 90% of the unreacted periclase is –100 mesh, and that the 28 x 100 mesh fraction of the unreacted chrome ore constitutes at least 45% of the chrome ore and is present in greater amount than the +28 or –100 chrome fractions. Specimens 3, 7, and 9 illustrate the preferred embodiment where all three of the preceding sizing conditions on the three grains are met.

In this regard the superior strengths, and lower firing shrinkage of Specimen 3, containing all three grains may be compared with those of Specimen 2, also containing an overall MgO content of 50%, but containing no unreacted periclase. This comparison indicates the advantage of having both chrome ore grain and magnesia grain together with the prereacted grain.

It has been found that the commercially most useful compositions contain 50 to 70% total MgO, for example as illustrated by Specimens 2 through 11, particularly 60% total MgO, as exemplified by Specimens 5 through 10. As can be seen from the examples, it is preferred that the compositions contain at least 15% unreacted periclase, at least 10% unreacted chrome and up to 75% prereacted grain.

High Chrome Fused Cast Refractories

W.A. Miller and H.R. Weiler, Jr.; U.S. Patent 3,759,728; September 18, 1973; assigned to The Carborundum Company describe fused cast refractory products having the following composition:

	Percent
Cr_2O_3	75 - 85
MgO	5 - 10
FeO	5 - 8
SiO_2	1 - 4
Al_2O_3	Less than 6

In addition, the mol ratio of Cr_2O_3 to Al_2O_3 is greater than 10 and the R_2O_3:RO mol ratio is 1.5 to 2.3. The compositions have a high specific gravity, typically over 4.1, and excellent resistance to molten glasses, particularly alkaline-earth borosilicate glasses such as "E" glass. They are preferably produced from chrome ore, Turkish and Transvaal ores being particularly suitable, with added chromium oxide and magnesium oxide to produce compositions within the above-stated range. Preferably, also, there is further added a small amount of an arc-stabilizer.

A number of these are known. The amounts of arc-stabilizer employed are very small, 1% or less being effective, and consequently the composition of the products is not materially affected. In fact, in most cases it is extremely difficult if not impossible to ascertain from the product what, if any, stabilizer was used in its production.

The refractory products are very dense, with an average specific gravity greater than 4.1. In general, the products are characterized by a structure consisting of grains of a magnesia-chrome oxide spinel having lath- or needle-like inclusions of an R_2O_3 phase, and a quite small amount of glass occurring intergranularly.

The R_2O_3 phase is a solid solution of indefinite and varying composition that is predominantly Cr_2O_3 but which may contain, in various proportions Al_2O_3 and Fe_2O_3. It is to be noted that magnesium oxide as periclase is absent.

Burned Basic Refractory from Size-Graded Batch

M.A. Nelson, R.F. Patrick and T.M. Wehrenberg; U.S. Patents 3,751,273; August 7, 1973 and 3,726,699; April 10, 1973; both assigned to Corhart Refractories Company describe a burned basic refractory, and a batch from which to make it, that exhibits further improved hot strength in combination with excellent resistance to thermal shock and chemical attack by molten metal and slag and vapors or fumes thereof. Such refractory body, when burned at 1600°C or higher is characterized by a modulus of rupture at 1340°C above 1,750 psi and, in certain cases, of at least 2,000 psi or higher.

Accordingly, the size-graded particulate batch, and a compacted and burned refractory body consist of:

(1)   a mixture of chemical grade chromite grain containing less than 4% SiO_2, periclase or dead-burned magnesite grain, and fused grain composed of a melted and resolidified mixture of chemical grade chromite and dead-burned magnesite or magnesia, which fused grain analyzes 40 to 58% (desirably 53 to 58%) MgO on an oxide basis, and ZrO_2-yielding material,

(2)   10 to 35% (preferably at least 25%) being a chromite fraction whose particles are substantially all −10+150 mesh, with less than 12% thereof being −100 mesh, and composed of the chromite grain,

(3)   25 to 60% (preferably 30 to 45%) being a coarse fraction whose particles are substantially all −4+20 mesh and composed of, with percentages based on whole batch, 5 to 60% the fused grain and 0 to 40% the periclase grain,

(4)   0 to 30% (preferably 5 to 15%) being an intermediate fraction whose particles are substantially all −10+35 mesh and composed of the fused grain and/or the periclase grain,

(5)   15 to 60% (preferably not more than 30%) being a fine fraction whose particles are −35 mesh and composed of, based on whole batch, 10 to 40% the periclase grain and 0 to 50% the fused grain,

(6)   4 to <50% (preferably 15%) being +28 mesh particles of the chromite grain and/or the periclase grain, and

(7) ZrO_2-yielding material being −65 mesh particles and in amount sufficient to provide 0.5 to 7% (preferably 0.7 to 2.5%) ZrO_2 based on whole batch.

In an especially beneficial form, the batch fractions are more particularly characterized as follows:

(a)   the chromite fraction is 0 to 45% +28 mesh particles and 60 to 95% +65 mesh particles,

(b)   the fused grain of the coarse fraction is 73 to 89% +10 mesh particles.

(c) the periclase grain of the coarse fraction is 69 to 85% +10 mesh particles,

(d) the intermediate fraction is 70 to 86% +28 mesh particles,

(e) the fine fraction is 0 to 10% +65 mesh particles, 8 to 23% +150 mesh particles, and 27 to 43% +325 mesh particles,

(f) the ZrO_2-yielding material is at least about 90% -325 mesh particles.

Example 1: A variety of commercial size bricks are mechanically pressed from a tempered batch made from the following constituents: chromite fraction—32% Transvaal chromite grain; coarse fraction—10% fused grain, 25% dead-burned magnesite; intermediate fraction—5% fused grain, 3% dead-burned magnesite; fine fraction—5% fused grain, 18% dead-burned magnesite; ZrO_2-yielding material—2% milled zircon flour.

Each of these fractions of grain particles is size-graded so as to have the following cumulative percentages (maximum, minimum and typical) of the whole fraction left on each of the several mesh size screens:

(a) Chromite Fraction

Mesh	Maximum	Minimum	Typical
10	0	0	0
28	45	0	15
35	70	20	35
48	85	40	53
65	95	60	78
150	100	80	96

(b) Fused Grain of the Coarse Fraction

Mesh	Maximum	Minimum	Typical
4	0	0	0
6	21	1	10
8	57	39	48
10	89	73	78
12	100	85	92
20	100	94	97

(c) Magnesite of the Coarse Fraction

Mesh	Maximum	Minimum	Typical
6	0	0	0
8	40	24	32
10	85	69	77
12	97	87	92
20	100	95	97

(d) Intermediate Fraction

Mesh	Maximum	Minimum	Typical
10	0	0	0
14	21	3	9
20	61	43	52
28	86	70	78
35	100	85	92

(e) Fine Fraction

Mesh	Maximum	Minimum	Typical
35	0	0	0
65	10	0	5
100	15	0	8
150	23	8	15
200	30	15	22
250	33	18	26
325	43	27	35

(f) ZrO_2-Yielding Material

Mesh	Maximum	Minimum	Typical
325	100	90	99*

*Average particle diameter: 20 to 25 microns

Typical analyses of the four materials from which these constituents are formed are as follows: Transvaal chromite—46.0% Cr_2O_3, 24.9% $FeO + Fe_2O_3$, 15.2% Al_2O_3, 11.3% MgO, 0.3% CaO and 2.3% SiO_2; fused grain—55 to 56% MgO, 20% Cr_2O_3, 8% Al_2O_3, 11% FeO, 2.5% SiO_2, 0.5% CaO, 0.3% F and 1.5% TiO_2; dead-burned magnesite—95.8% MgO, 1.0% CaO, 1.8% SiO_2, 0.4% Al_2O_3, 0.6% Fe_2O_3 and 0.05% Cr_2O_3; and milled zircon flour—99.0% $ZrSiO_4$ (65.8% ZrO_2), 0.25% SiO_2 (free), 0.18% TiO_2, 0.25% Al_2O_3, 0.05% Fe_2O_3 and 0.01% P_2O_5.

After the pressed grain bricks are dried, they are burned at about 1675°C for at least 9 hours, whereupon extensive direct bonding was developed between the refractory crystals of periclase and chrome spinel. Typical property data for these bricks are set forth in the table below.

Example No.	1	2	3
% ZrO_2 (calculated)	1.3	3.3	6.6
% Apparent porosity	15.5	17.4	18.8
Bulk density (lb/ft³)	199.4	197.1	196
Modulus of rupture at 1340°C (psi)	2000	1750	1750

Example 2: Commercial furnace size bricks are made using the same materials, proportions and size-grading thereof, and procedures as in the preceding example, except that the zircon flour is made 5% of the batch and the dead-burned magnesite of the fine fraction is made 15% of the batch. Typical property data of such bricks are set forth in the table above.

Example 3: Additional commercial furnace size bricks are made using the same material, proportioning and size-grading thereof, and procedures as in Example 1, except that the zircon flour is made 10% of the batch and the dead-burned magnesite of the fine fraction is made 10% of the batch. Typical property data for such bricks are set forth in the above table.

Basic Magnesite-Chrome Refractories

E.S. Wright; U.S. Patent 3,715,223; February 6, 1973; assigned to Eltra Corporation describes the production of magnesite chrome and chrome magnesite refractories. The principal batch ingredients are 40 to 80 parts by weight of dead-burned magnesite or synthetic periclase and 60 to 20 parts by weight of chrome ore. The magnesite or periclase and the chrome ore are suitably sized, mixed with minor batch ingredients, e.g., lime hydrate and lignin sulfate as a temporary binder, and tempered.

The desired shape is then pressed from the tempered batch, and the shape is fired, e.g., to 2500°F or higher. Calcium nitrate is dissolved in the tempering liquid, prior to pressing of the desired shape. As a consequence of the calcium nitrate dissolved in the tempering liquid it has been found that the porosity of the final refractory is substantially reduced and as a consequence, it is believed, of the reduction in porosity, the final refractory has outstanding slag-resistance. In the following example all mesh sizes refer to the Tyler Sieve Series.

Example: Magnesite chrome refractories were produced from a batch consisting of 34 parts of –6+14 mesh magnesite, 26 parts of –150 mesh dead-burned magnesite, and 40 parts of chrome ore. The batch ingredients were dry mulled with 1½ parts of lignin sulfate to produce a substantially uniform mixture, and this mixture was then tempered with 3.6 parts of a 50% aqueous solution of calcium nitrate.

The tempered batch was then dry pressed into bricks, under a total pressure of about 14,000 pounds per square inch. The bricks were dried at 425°F and were stacked on kiln cars and fired at 3100°F in a periodic kiln. The total time at temperature was 10 hours. The average porosity of the bricks was found to be 13.45% by volume. The modulus of rupture was found to be 1,676 pounds per square inch at 2350°F and 366 pounds per square inch at 2700°F. Brick produced have been found to have excellent resistance to spalling and to erosion by liquid slags.

For purposes of comparison, brick were also produced except that 0.76 part by weight of hydrated lime was mixed with the dry batch ingredients and these ingredients were tempered with 3.2 parts of a solution containing 22% of magnesium sulfate; these brick had a modulus of rupture at 2350°F of 1,346 pounds per square inch and a total porosity of 14.7% by volume.

Of the two samples on which a determination of modulus of rupture at 2700°F was made, one gave a modulus of 299 pounds per square inch and the other gave a modulus of 50 pounds per square inch, but the latter broke off center, probably indicating a defect in the brick.

The dead-burned magnesite used in making brick as described was produced by burning natural magnesite of Euboean origin in a rotary kiln. A typical analysis on an ignited basis, is 95.24% of MgO, 2.72% of calcium oxide, 1.46% of silica, 0.01% of Cr_2O_3, 0.53% of iron oxide (Fe_2O_3), 0.03% of aluminum oxide and 0.012% of B_2O_3.

The chrome ore used in preparing brick as described was beneficiated Transvaal ore. Typical analysis, on an ignited basis, and sizing, presented as cumulative percent retained on various mesh screens are set forth in the following tables:

SiO_2	0.62
Cr_2O_3	45.07

(continued)

CaO	0.04
MgO	9.74
Fe_2O_3	29.08
Al_2O_3	15.45

Cumulative Percent Retained	Mesh
4	14
14	20
31	28
48	35
67	48
81	65
90	100
95	150
98	200

REFRACTORY FIBERS

CARBON FIBERS

Carbon Fiber Reinforced Carbon Bodies

According to *F.C. Cowland and H. Appleby; U.S. Patent 3,790,393; February 5, 1974; assigned to Beckwith Carbon Corporation* chemically resistant bodies of high mechanical strength are obtained by first producing a composite body in which partly fired carbon fiber material, made by low temperature firing of natural or synthetic organic fiber material in an oxidizing atmosphere while restraining the fibers from shrinking is incorporated in a matrix of carbonizable thermosetting resin and then heat treating this body to carbonize the material of the matrix and at the same time complete the conversion of the organic precursor fibers into carbon fibers.

The heat treatment may be carried on to a temperature sufficient to convert the resin of the matrix into vitreous carbon, but greater mechanical strength through somewhat reduced chemical resistance may be obtained by use of carbonization temperatures which though well above the curing point of the resin are below the temperatures necessary for the production of vitreous carbon. The use of a resin for the matrix whose shrinkage during the carbonizing heat treatment is less than that of unrestrained partly fired carbon fibers results in the production of a carbon fiber reinforced carbon body in which the fibers are under tensile stress while the material of the matrix is under compressive stress.

Example 1: In order to produce carbon fiber reinforced vitreous carbon, a layer of aligned partly fired high modulus carbon fibers is laid out in the form of a mat in a tray coated with polypropylene sheet and this mat is soaked with a solution of 40% formaldehyde. Excess of the solution is poured off and a solution of novolak resin with a phenol:formaldehyde ratio of 1:1 is added to cover the layer of carbon fibers.

A proportion of methanol may be added to the original solution of novolak resin for the purpose of diluting the resin to ensure wetting of the fibers, and it is believed that this also serves to improve bonding of the resin to particular points on the carbon molecule. Additional layers of fibers are positioned on top of this first layer, and each of these layers is covered with a layer of the same novolak resin. The tray is then heated to 90°C for 24 hours, during which time the fibers are allowed to settle under the influence of gravity, whereafter excess resin is poured off, leaving just sufficient resin to completely cover the fiber layers. The resin is then cured by heating on a schedule to 160°C for 7 days. If the fibers are randomly oriented, a loading of 40% is likely to be achieved and if all fibers are oriented in the same direction, the loading can be increased to approximately 50% of the total weight. The fiber composite is then heated to selected temperatures above the normal curing temperature of the resin to form vitreous carbon.

The following heating cycle, which has been found to be particularly suitable for carbonizing resin bodies is conducted in a furnace in two stages. For the first stage a furnace atmosphere of an inert gas is employed, and in the second stage the articles in the furnace are placed under vacuum.

First Stage — Inert Gas: The heating schedule of the furnace is as follows.

- - - - - - - - - - - -Furnace Temperature- - - - - - - - - - - - -

100°-160°C in 24 hours	420°-440°C in 24 hours
160°-190°C in 24 hours	440°-510°C in 24 hours
190°-280°C in 24 hours	510°-600°C in 24 hours
280°-340°C in 24 hours	600°-750°C in 24 hours
340°-380°C in 24 hours	750°-890°C in 24 hours
380°-410°C in 24 hours	890°-300°C in 24 hours
410°-420°C in 24 hours	300°-40°C in 24 hours

This 14-day schedule may be modified to 7 or 28 days by halving or doubling the time intervals, according to the thickness of material to be furnaced.

Second Stage — Vacuum: Heating from ambient temperature to 1800°C is effected in 3 days and is followed by cooling. This schedule remains the same for 7, 14, or 28 days first-stage schedules.

A vitreous carbon body which was prepared in this manner with a 50% loading of partly fired high modulus carbon fibers had a transverse strength of 50,000 psi compared with only 30,000 psi for a body similarly prepared with the use of fully fired carbon fibers and with 15,000 to 25,000 psi for a body made from unloaded resin, and the matrix of the body had sufficient elasticity to withstand 0.7% strain before fracture. The body was capable of being used in air up to 600°C without appreciable loss of weight.

Example 2: A composite body formed of phenol-formaldehyde resin with a loading of 50% by weight of partly fired oriented high modulus carbon fibers and having an ultimate tension strength of 2 to 3 x 10^5 psi was prepared in the manner described in Example 1 and heat treated by following the first stage of the heating cycle described in Example 1, i.e., up to the indicated temperature of 890°C, no second stage being employed.

The composite carbon body obtained according to this example had a crossbreaking strength of 65,000 psi three point test according to BSI 598/1949, compared with a strength of 50,000 psi for the carbon fiber reinforced prestressed vitreous carbon body obtained by the method of Example 1, and with respective breaking strengths of 38,000 and 30,000 psi for carbon bodies in which the heat treatment of Example 2 was applied respectively to a resin body having a 50% loading of fully fired carbon fibers and to an unloaded resin body. The body, while less resistant to oxidation than vitreous carbon, could be used at 300°C compared with a maximum of 150°C for normal carbon fiber reinforced epoxy resin.

Low Thermal Expansion Composites

J.O. Gibson, R.L. Schumacher and K.L. Myler; U.S. Patents 3,766,000; October 16, 1973; and 3,736,159; May 29, 1973; both assigned to McDonnell Douglas Corporation provide graphite composites by treating fibers with a mixture of a refractory metal compound or mixture of such compounds of extremely small particle size, e.g., NbC of 0.02 micron size, and sufficient binder to form a slurry, and forming a coating on the fibers. The coated and impregnated fibers are then heat treated at very high temperatures, e.g., 3000°C for graphite fibers coated and impregnated with NbC. If desired, prior to heat treating the coated fibers can be formed to the desired shape. By hot pressing, the coated fibers can be formed and heat treated in one operation. There is thus produced a composite having graphite fibers dispersed in a matrix of refractory metal compound, e.g., NbC.

Briefly, the process for producing a composite of low thermal expansion comprises treating fibers selected from the group consisting of amorphous carbon fibers, graphite fibers, pyrolytic graphite fibers, natural organic fibers, synthetic organic polymer fibers, inorganic fibers, and mixtures thereof, with a coating composition comprising (a) a refractory metal compound or refractory metal compound mixture, in the form of particles of a size not larger than about 0.6 micron; and (b) a binder for bonding the particles of refractory metal compound to the fibers, sufficient binder being employed to form a slurry; forming a coating on the fibers, and heating the coated fibers at a temperature of 40 to 98% of the melting point of the refractory metal compound or refractory metal compound mixture.

Composites produced by the method exhibit lower coefficients of expansion, greater resistance to thermal shock, low compressive creep at operating temperature and times when used for structural parts, and have of the order of 2½ times the flexural strength of prior art composites. At the same time the composites exhibit good resistance to hydrogen and other chemical corrosion.

Examples of binders which can be used are: (1) protein or protein derivatives; (2) starch, sugars, dextrans, dextrins, cellulose, glycogens, pentosans or gums and their derivatives; (3) thermoplastic synthetic resins; (4) thermosetting synthetic polymers and prepolymers; (5) natural resins and bitumins; and (6) inorganic adhesives. A particularly preferred binder is based on furfuryl alcohol with 1 to 50% by weight maleic anhydride.

Example: A coating composition is first formed by mixing together the following ingredients: 8.2 parts by weight maleic anhydride; 29.5 parts by weight niobium carbide of 0.02 micron particle size; 3.3 parts by weight niobium metal of 5 micron particle size; and 59.0 parts by weight of a binder, chiefly furfuryl alcohol. The above composition is characterized by high viscosity. Graphite yarn is then drawn continuously through the above composition ef   ing a coating thereon. A typical graphite yarn has the following properties:

Plies/yarn	2
Filament/ply	720
Twist-ply/ply	1.6
Yarn diameter	~0.02 in
Yield	5,600 yd/lb
Breaking strength	~8 lb

A typical graphite fiber of which the graphite yarn is composed has the following properties:

Modulus	25×10^6 psi
Specific modulus	490×10^6 in
Strength	180×10^3 psi
Specific strength	3.5×10^6 psi
Density	1.42 g/cm^3
Diameter	7.4 microns

The yarn emerging from the high viscosity slurry is uniformly covered with a coating of the slurry. The coated yarn is then dried by exposure to hot air at 200°C. The hot air causes the coating to thermally set on the yarn forming a strong adherent coating on the yarn and individual fibers of which the yarn is composed.

After hardening of the coating, the coated yarn is then chopped into pieces 0.375 inch long and placed in a graphite die case. The coated fibers are then vacuum hot pressed in the die at 3150°C and at 3,000 psi for 10 minutes to form a billet. The composite billet contains 65 volume percent graphite fiber and 35 volume percent NbC. The billet is found to have the following properties set forth in the table below.

Linear coefficient of thermal expansion 0° to 2500°C	4.6×10^{-6} in/in/°C
Compressive creep at 2500°C, 4,000 psi load, for 1 hour	<0.1%
Flexural strength at 20°C	28,000 psi
Carbide particle size	Fine
Hydrogen corrosion resistance	Satisfactory

Oxidation Resistant, High Strength Composites

According to *J.O. Gibson, R.L. Schumacher and T.W. Gore; U.S. Patent 3,770,487; November 6, 1973; assigned to McDonnell Douglas Corporation* a highly oxidation resistant, thermal shock resistant and high strength composite can be produced by hot pressing a mixture consisting of by weight (a) 40 to 70% hafnium diboride, zirconium diboride, titanium diboride or their mixtures; (b) 16 to 50% tungsten disilicide and/or molybdenum disilicide; and (c) 10 to 16% graphite, amorphous carbon, pyrolytic graphite or their mixtures, the hot pressed mixture or composition being free of silicon carbide. The resulting composite, in addition, has the advantageous properties of high thermal conductivity and high impact resistance, assuring good resistance to oxidation due to the hot pressing procedure employed in producing the composite. Where the substrate is a metal part, such as tantalum, tungsten, nickel and cobalt alloy, or steel, prior to introduction of the substrate or part to be coated in the hot press, the part is first coated with a suitable adhesive to provide a tacky surface, an illustrative type of such adhesive being Duco Cement, an acetone-butyl acetate composition, or Nicrobraze, a lacquer plastic, comprising polymethyl methacrylate in a Freon.

As a result of the hot pressing operation, the composite produced has a high density ranging from 2.5 to 10 g/cc, and is compressed to 95 to 99% of theoretical density. Such densification of the composite improves its impact resistance. The thickness of the composite coatings produced is 0.050 to 3 inches, usually 0.2 to 0.5 inch. The composite forms a glass coating which provides a temperature gradient of 400° to 900°C across a 60 mil thickness of the glass layer. If the composite is operating in an ablation mode, the composite continuously regenerates a protective glass coating which is selfhealing. In other words, the composite functions as a regenerative type oxidation resistant composite.

In addition, the composite of the process has good hardness and machining properties, and the metalography of the composite shows good dispersion of the carbon particles, e.g., graphite fibers, throughout the refractory metal boride and silicide matrix. The composite and the substrate coated therewith have a wide variety of applications, including use in fabrication of turbine blades and aircraft structural material for high speed aircraft, use on the leading edge of high speed aircraft and space vehicles, requiring high strength at elevated and rapidly changing temperatures and high oxidation resistance, coatings to generally protect metal and graphite substrates of the types noted above in oxidizing atmospheres at 2000°C for long periods, and coatings for thermionic diode applications involving combustion heating at temperatures of the order of about 1800°C. The composite has particular applicability to high temperature load bearing materials in oxidizing atmospheres, for protecting isotope fuel blocks and rocket nozzle liners.

Example: A tantalum substrate or member is treated on a surface thereof with Duco Cement to form a tacky surface. The substrate or member is then contacted on the tacky surface thereof with a fine HfB_2 powder to uniformly coat the tacky surface. The tantalum substrate or member having an initial HfB_2 coating on its surface is inserted into the die chamber of a vacuum hot press.

A mixture of 66% HfB_2, 20% WSi_2, both in powder form and having a particle size of 0.1 to 5 microns, and 14% of discontinuous graphite fibers having an average length of 4,000 microns, by weight, are mixed together in a jar mill for 4 hours. The mixed powder is then placed in the die of the vacuum hot press in contact with the HfB_2 coated surface of the tantalum member or plate, and the mixture and the HfB_2 coated tantalum substrate in the die are subjected to hot pressing in the press at 2100°C, under a vacuum corresponding to a pressure of 1 micron, and at a pressure of 4,000 psi for 15 minutes. The billet containing the composite and tantalum member is then removed from the press. The properties of the composite produced are shown in the table below.

Density	7.1 g/cm^3
Thermal conductivity	0.12 cal/cm-sec·°C
Thermal shock properties	Excellent
Flexural strength (20°C)	40,000 psi
Melting point of composite	~2300°C
Oxidation resistance (2200°C), wt loss in 120 minutes	<0.08%

(continued)

Machining properties	Good
Hardness of composite (HfB_2-WSi_2 matrix), Knoop at 75°F	~3,300 kg/mm^2

From the above table, it is seen that the hafnium diboride-tungsten disilicide-carbon composite produced as described above by hot pressing, has high oxidation resistance, high flexural strength, high density, good thermal conductivity and excellent thermal shock resistance.

Carbonized Material from Expanded Graphite and Carbon Fibers

J. Gellon and J.-P. Slonina; U.S. Patent 3,726,738; April 10, 1973 describe a method of manufacturing a carbonized material that is resilient and/or resistant to high temperatures consisting in associating expanded graphite and carbon fibers, the carbon fibers being interengaged with the expanded graphite. Advantageously, the expanded graphite is homogeneously mixed with carbon fibers having an elementary diameter of 5 and 15 microns.

Carbon fibers are first carded in the form of wadding, of apparent diameter 5 to 10 mm, and the expanded graphite and carbon wadding assembly is then suspended in a cold fluidized bed. The carbon wadding is gradually surrounded by the vermicular particles of expanded graphite. After stopping the fluid bed, the assembly is compressed at 20 bars to 50 hectobars, according to the properties required, or it may be laminated.

In addition, carbon fabric or carbon braid regularly wound is interposed between two plates of expanded graphite which have previously been slightly compressed, e.g., at 0.1 to 0.2 bar/cm^2. At the time of compression or final lamination, the fabric or braid is interengaged in the expanded graphite to form a massive assembly whose method of manufacture is described in Example 3.

There are two major and important properties of the material thus obtained: (1) a sealing capacity against liquids and gas (neutral or reducing) at temperatures higher than 600°C; (2) an elasticity of a joint made therefrom when a unidirectional pressure is applied. This elasticity is 15 to 20%, the material retaining its initial value when the stress ceases to be exerted. The material thus obtained may be made in all sizes. However, it may be suitable to make it in the form of thin sheets, i.e., of one or two mm thickness, which are cut up so as to form joints of any desired size, or alternatively to form joints of standard dimensions.

Example 1: In a vertical tube 200 mm in diameter and 1 m in height, 50 grams of expanded graphite is introduced which represents 3 dm^3 in volume, together with an identical quantity in volume of carbon wadding made of short and fine fibers in bulk form, of dull black color, and whose carbon content is 90 to 95% and whose residual ash is 0.6%. A gaseous current fed in at the base of the tubes allows the whole to be mixed in a homogeneous manner. After mixing for 10 minutes, the mixture which is very voluminous, is introduced into a mold made of a plastics material of an inner diameter of 90 mm. The pressure applied is 10 bars and the material is placed in a metal matrix of a slightly greater diameter than that of the mold made of a plastics material. Then a pressure of one hectobar is applied which results in a disc 92 mm in diameter and 2 mm in depth. The material thus produced has been tested: cycles of successive charges and discharges at a pressure of 7.5 bars allow an elasticity of $^4/_{10}$ mm to be exhibited without hysteresis and without loss of elasticity over a series of 20 cycles at ambient temperature.

Example 2: A sheet 2 mm in thickness is prepared by laminating between two supple cardboard sheets a mixture which has been prepared in a fluidized bed such as described in Example 1, and comprising, in volume, 70% of carbon wadding, whose carbon content is between 90 and 95% and which is formed from fine and short elementary fibers whose diameter is 5 to 12 microns, and 30% of expanded graphite. The products obtained according to this example may form a joint 30 mm in diameter. Inserted into an assembly in place of an ordinary rubber toric joint, the joint produced can retain a primary vacuum, i.e., $^1/_{10}$ mm of mercury.

The resilience of the joint which is obtained according to this example may be measured by means as described below. The joint is arranged between two graphite cylinders externally heated by heating collars and arranged between two water-cooled plates. The whole is retained between the two plates of a press. Pressure applied is 10 bars at 600°C. The rebound measure is 17% without hysteresis after 10 successive cycles of charges and discharges.

Example 3: On a sheet of porous cardboard is placed a sheet of expanded graphite 1 mm thick, slightly compressed to a density of 0.1 to 0.2, then a layer of carbon fabric whose characteristics are as follows:

Diameter of threads	0.6-0.7 mm
Thickness	0.4-0.5 mm
Weight per m^2	200-250 g
Electrical resistance of a square measured along one side	1.25 ohm
Carbon content	99-100%
Residual ash	0.1-0.2%
Volatile content	$<$1%

This is covered with another sheet of expanded graphite similar to the first, and then a further sheet of porous cardboard. The whole is then compressed at a pressure of one hectobar. The porous cardboard is removed after the compression operation. The sandwich of expanded graphite/carbon/cloth/expanded graphite obtained, remains supple and has properties of elasticity when hot of 20% at 600°C. As many sheets of expanded graphite and layers of carbon cloth may be superposed as is necessary to obtain the desired thickness, the whole also being compressed between two sheets of porous cardboard.

Carbon Fiber Composites

D.H. Leeds; U.S. Patent 3,713,865; January 30, 1973; assigned to Ducommun Incorporated describes a composite product having a substrate or matrix comprising carbonized fibers in which a metallic salt solution consisting of a soluble metal salt selected from a group consisting of Hf, Zr, Ta, Cb, W, Mo, Ti, V, Cr, Si, B, P and Pb, which form either carbides which are refractory or oxides which melt and are capable of forming a liquid surface (which will slow the diffusion of oxygen to the graphite surface causing $C + \frac{1}{2}O_2 \rightarrow CO$ or $C + O_2 \rightarrow CO_2$, depending upon both the particular pressure and temperature employed in the end use). The metallic salt when dissolved in water, acetone, alcohol, ether or the like, is introduced into the interstices of the substrate where, after drying, it is left in fine particles. The composite is then heated to 1500° to 2500°F with or without hydrogen atmosphere to effect partial reduction of the metallic salt. Almost simultaneously, carbon is infiltrated into the substrate to enclose the fibers and the metal particles. Subsequently, the substrate is raised to 1400° to 2800°C to anneal the pyrocarbon and convert the reduced metallic salt into carbide, thus producing a composition product having the substrate fibers surrounded by dead soft annealed graphite intimately mixed with hard carbide particles of very high melting point. Preferably the substrate is preformed into a matrix of the desired size and shape, e.g., a nozzle for use in rockets.

Example: A 2 x 2 x 4 inch sample of 20 lb/ft^3 was used as the substrate in this test. This sample is referred to as RPG and consists of a fibrous body of rayon in which the fibers had been carbonized and infiltrated with pyrocarbon. This sample was impregnated with $HfCl_4$ and methyl alcohol solution in proportions of one gram to one cm^3 by dipping, drying at 70°C in an oven and redipping and redrying until the weight pick-up diminished below 10 wt %. The sample was machined into a hemispherical cone, then placed in a furnace at 1800°F and 14 mm Hg vacuum and infiltrated with CH_4 for 120 hours, which reduced the metal salt and infiltrated the body with pyrolytic carbon surrounding both the carbon fibers and impregnant particles.

At this point a fine machining removed 0.001 inch from the surface and the sample was again infiltrated with pyrolytic carbon until a density of 2.03 g/cc was achieved. This sample was tested in a typical reentry thermal environment at 70 atmospheres pressure and performed excellently, showing a minimum of erosion. This sample was then remachined and further heat treated at 2500°C to drive some carbon into solid solution with the hafnium. The sample thus treated thereupon became an example of molecularly impregnated metal in a carbon matrix.

ALUMINUM FIBERS

High Strength Aluminum Oxide Fibers

M. Mansmann and L. Schmidt; U.S. Patent 3,865,599; February 11, 1975; assigned to Bayer AG, Germany describe a process for the manufacture of aluminum oxide fibers containing a proportion of silicon dioxide and, optionally, small amounts of modifying additives, by spinning a solution of the fiber forming systems and minor amounts of a dissolved linear polymeric high molecular substance and subsequent heat treatment. The solution containing a neutral or basic aluminum salt of a monobasic lower carboxylic acid and a hydrolyzed silicic acid ester or a hydrolyzed organoalkoxysilane is spun to fibers in the presence of dissolved polyethylene oxide having a degree of polymerization of above 2,000, and the fibers are subsequently subjected to a heat treatment up to 1800°C.

It is possible thereby to obtain, from relatively dilute and therefore stable, easily filtrable and very simply degassed solutions, continuous fibers which display a high tensile strength and a high E-modulus and are outstandingly suitable for reinforcing plastics, glasses, ceramic materials and metals. These fibers comprise the following composition in percent by weight.

	Percent	Preferred %
Al_2O_3	61 - 98	76 - 96
SiO_2	1 - 20	3 - 15
P_2O_5 and/or B_2O_3	0 - 10	0 - 5
MgO	0 - 5	0 - 1
C	0.1 - 4	0.5 - 2.8

Within the indicated analytical composition, the fibers can, depending on their heat treatment, be characterized by their phase content, which can be determined by x-ray methods; it is possible to differentiate four types:

(1) Aluminum oxide fibers which are amorphous to x-rays; they are formed by a heat treatment at 400° to 950°C. Their density is 2.8 to 3.0 g/cm^3 Typical values of their tensile strength lie between 80 and 180 kp/mm^2 and typical values of their E-modulus between 9,000 and 14,000 kp/mm^2.

(2) Polycrystalline γ-Al_2O_3 fibers; they are produced by a heat treatment at 950° to 1150°C. Their SiO_2 content is not detectable in the x-ray diagram. Their density is 2.9 to 3.4 g/cm^3. Typical values for their tensile strength lie between 100 and 270 kp/mm^2 and for their E-modulus between 20,000 and 24,000 kp/mm^2. The crystallite size of the γ-Al_2O_3 phase is about 40 to 70 A.

(3) Polycrystalline γ-Al_2O_3/mullite fibers; they are produced by a heat treatment at 1200° to 1400°C. Their density is between 3.1 and 3.6 g/cm^3. Their strength data corresponds to those of the γ-Al_2O_3 fibers; their E-moduli are between 21,000 and 26,000 kp/mm^2. The crystallite size of the γ-Al_2O_3 is 40 to 60 A and that of the mullite phase 700 to 1400 A. δ-Al_2O_3 is detectable in the x-ray diagram.

(4) Polycrystalline α-Al_2O_3/mullite fibers; they are produced by briefly rapidly heating the other type to above 1400°C. They have a density of 3.4 to 3.9 g/cm^3 depending on the SiO_2 content. The tensile strength and E-modulus are 80 to 170, and 23,000 to 32,000 kp/mm^2 respectively. The crystallite size of the α-Al_2O_3 phase and of the mullite phase are 500 to 1100 A.

In detail, the following procedure can be employed to manufacture the fibers. An aluminum salt solution is produced by dissolving aluminum or aluminum hydroxide in a monobasic lower carboxylic acid such as, for example, formic acid, acetic acid or propionic acid, preferably formic acid, at 40° to 100°C, employing, in general, 1.7 to 3.2 mols of the organic acid per one mol of Al, so that either basic or neutral salts are present in the solution. It is, however, also possible to employ commercially available aluminum formates or aluminum acetates as the starting material.

The amount of water used during dissolving can be so chosen as to give directly a suitable concentration of Al_2O_3 for the subsequent spinning process. It is however preferred to prepare a substantially more dilute solution, for example, with a concentration of 6 to 9% of Al_2O_3, and only to concentrate it to the desired concentration after adding the SiO_2 component. The concentration of Al_2O_3 in the spinning solution is 8 to 18%, preferably 11 to 15%.

The SiO_2 constituent of the aluminum oxide fibers is present in the spinning solution in the form of a hydrolyzed silicic acid ester or of a hydrolyzed organo-alkoxysilane. The hydrolysis of the silicic acid esters, which are the preferred

SiO_2 source, takes place in accordance with the equation below via the intermediate stages of the alkoxypolysiloxanes, up to SiO_2, which remains in colloidal solution.

$$pSi(OR)_4 + pnH_2O \longrightarrow [Si(OR)_{4-2n}O_n]_p + 2pnHOR$$

In the equation p equals degree of condensation and n equals 0 to 2. The degree of the hydrolysis, like the degree of crosslinking, depends on the amount of available water. As is shown by the equation, at least 2 mols of water are required per mol of silane for the complete hydrolysis of a tetraalkoxysilane. Instead of tetraalkoxysilanes, alkoxypolysiloxanes can also be used. Methoxy, ethoxy, propoxy and butoxy groups, individually or as mixtures, are used as alkoxy groups. Methoxy and ethoxy compounds are preferred. Hydrolysis of these compounds is generally accelerated catalytically by strong acids. Hydrochloric acid in extremely low concentrations, e.g., about 0.1% or less by weight, relative to the total solution, is preferably added. Usually, a lower saturated aliphatic alcohol, preferably methanol or ethanol, is used as the solubilizing agent for the hydrolysis reaction.

The hydrolysis of the silicic acid ester or of the organoalkoxysilane can be carried out directly in the aluminum salt solution. After bringing the components together, the reaction takes place, with stirring and gentle warming, within a few minutes, after which the mixture is stirred for a further ½ to 2 hours. The solutions obtained in this way are water-clear and not opalescent. The silicic acid ester can, however, also be hydrolyzed separately and subsequently mixed, while stirring, with the aluminum carboxylate solution.

Before manufacturing the actual spinning solution, yet further additives, with which the properties of the resulting fibers can be modified, can optionally be added to the conjoint dilute solution of the aluminum salt and the silane hydrolysis product. These additives serve to modify the fiber surface with regard to the use of the fibers as a reinforcing material, through the incorporation of basic, neutral or acid constituents, in such a way that depending on the matrix material used a firm bond between the matrix and fiber is formed in each case, which is of great importance for the strength of the composite material.

It is furthermore possible, through incorporation of the additives, to give the fiber a surface character which resembles that of glass fibers, so that the adhesion promoters developed for the latter can also be used directly for the aluminum oxide fibers according to the process. Desirably the additives should not adversely influence the strength of the fibers. Proportions of up to 5% of MgO and of up to 10% P_2O_5 and/or B_2O_3, relative to the total oxide sum, have already proved favorable.

The oxide concentrations of 8 to 18%, preferably 11 to 15%, are made spinnable by adding small amounts of polyethylene oxide, with the resulting viscosities only amounting to a few poises. The higher the degree of polymerization of the polyethylene oxide the lower is the amount required to obtain spinnable solutions. Degrees of polymerization of above 2,000, especially above 5,000 are preferred. The polyethylene oxide content is less than 2%, relative to the spinning solution. At degrees of polymerization of 100,000 and above, additions of less than 0.8%, down to 0.1%, already suffice to give solutions of excellent spinnability. Analytically, the glossy white gel filaments thus manufactured generally contain 35 to 45% of oxide. The difference is attributable to the different degree of drying resulting from different column temperatures and different filament gauges. All constituents present in the spinning solution are still present in the gel filaments. The filaments show no tendency to stick together and possess a strength which easily suffices for their further processing. This further processing consists of a heat treatment which aims at expelling all volatile constituents from the filament and sintering the remaining oxide components together to give a filament of high strength and high E-modulus. For this purpose, the continuous filaments are drawn continuously through a furnace having an appropriately designed temperature profile.

Fibers from Alumina Sol and Hexamethylenetetramine

J.C. Hayes and J.E. Sobel; U.S. Patent 3,814,782; June 4, 1974; assigned to Universal Oil Products Company describe a method of preparing porous, flexible, refractory inorganic oxide fibers which comprises forming a substantially liquid phase reaction mixture consisting of: (1) an acid anion-containing sol of alumina and/or chromia-alumina sols; and (2) a sufficient amount of hexamethylenetetramine to effect 5 to 50% neutralization of the acid anion in the sol, concentrating the reaction mixture to a tacky consistency characterized by a viscosity of 1 to 20,000 poises, fiberizing the reaction mixture in an atmosphere characterized by a relative humidity of less than 80% and at 5° to 90°C, and calcining the resulting fibers in air at 300° to 1000°C.

Chromia-alumina sols are prepared by dissolving chromium trioxide in aqueous hydrochloric acid in an amount of 0.2 to 2.0 mols of hydrochloric acid per mol of chromium trioxide. The solution is admixed with formaldehyde as a reducing agent whereby a chromia sol containing chromium in the +3 valence state is formed. The chromia sol is then commingled with an alumina sol. Alternatively, the chromia-alumina sol can be prepared by digesting aluminum metal in an acidic chromium chloride solution, suitably an aqueous chromium chloride solution.

The alumina sols can be prepared by the hydrolysis of a suitable acid salt of alumina such as aluminum chloride, aluminum sulfate, aluminum nitrate, aluminum acetate, etc., in aqueous solution, and treating the solution at conditions to form an acid anion deficient solution or sol. Reduction in the concentration of the acidic anion formed by the hydrolysis reaction may be accomplished in any conventional manner.

Another method of producing a suitable alumina sol is in the electrolysis of an aluminum salt solution, such as an aqueous aluminum chloride solution, in an electrolytic cell having a porous partition between anode and cathode whereby an anion deficient aluminum salt solution, or sol, is recovered from the cathode compartment. Preferably, the sol is an aluminum chloride sol prepared, for example, by digesting aluminum pellets or slugs in aqueous hydrochloric acid and/or aluminum chloride solution, usually at about reflux temperature. Aluminum chloride sols are preferably prepared to contain aluminum in a 1:1 to 2:1 atomic ratio with the chloride anion.

The inorganic oxide hydrosol is mixed with a soluble organic amine to form a substantially liquid phase reaction mixture at fiberizing conditions. The selected organic amine must be substantially stable at fiberizing conditions, i.e., at 5° to 90°C. Suitable organic amines thus include n-butylamine, n-amylamine, sec-amylamine, tert-amylamine, ethylenediamine, trimethylenediamine, tetramethylene-

diamine, pentamethylenediamine, hexamethylenediamine, etc., preferably hexamethylenetetramine. This amine has the effect of inhibiting crystallite growth
during fiber formation and produces a flexible, resilient, porous fiber.

Example 1: An alumina sol was prepared by digesting an excess of aluminum
metal in aqueous hydrochloric acid under reflux conditions (98° to 115°C). The
aluminum/chloride ratio of the sol was controlled by monitoring the amount of
hydrogen liberated during the reaction. The resulting sol analyzed 12.03% aluminum and 10.38% chloride for an aluminum/chloride ratio of 1.16. The specific gravity of the sol was 1.3455.

To about 3 liters of this sol was added, with stirring, 714 ml of a 28% aqueous
hexamethylenetetramine solution, sufficient to neutralize 50% of the chloride
anion. The resulting mixture was thereafter concentrated by allowing 855 ml of
water to evaporate therefrom at room temperature. The concentrated mixture
analyzed 12.38% aluminum, 10.73% chloride and 50.8% total solids at 105°F.
The specific gravity of the concentrated mixture was 1.383 and the kinematic
viscosity 26.5 cs.

The concentrated mixture was drawn into fibers in a closed chamber and in an
atmosphere characterized by a relative humidity of 45 to 55% and at 20° to
30°C. The fibers were collected and calcined in air at 550°C for 15 minutes.
Optical examination of the calcined fibers indicated a clear, transparent fiber.
The mean diameter of the fibers was 10.6 microns. The fibers were quite flexible and resilient and showed little tendency toward attrition. X-ray examination
of the fibers indicated an overall gamma-alumina pattern but of such fine crystallite size that the material could be considered amorphous. Measurement of surface properties by nitrogen absorption indicates a surface area of 230 m^2/g, a
pore volume of 0.21 cc/g and a pore diameter of 37 A.

Example 2: Alumina fibers were prepared identically as described in Example 1
except that the aqueous hexamethylenetetramine was omitted from the sol formulation. After the calcination treatment, the fibers were opaque and friable and
crumbled readily upon handling. The mean diameter of the fibers was 13.5 microns. X-ray examination indicated that the fibers were of gamma-alumina of
large crystallite size. Thus, the addition of hexamethylenetetramine to the sol
formulation has the effect of retarding crystalline growth and results in a porous
flexible fiber which is not dependent upon high temperature sintering to acquire
strength.

Example 3: A chromia-alumina sol was prepared by digesting aluminum metal
in aqueous chromium chloride solution. The sol analyzed 10.06% aluminum,
2.83% chromium and 10.15% chloride. The specific gravity measured 1.349.
To 64.3 grams of the sol was added 11.0 ml of a 28% aqueous hexamethylenetetramine solution. The mixture was then concentrated and drawn into fibers
substantially as in Example 1 and under the same conditions of temperature and
humidity. After calcination in air at 550°C for 1 hour, the fibers were flexible
and resilient with a mean diameter of 12.1 microns. The fibers were greenish-
gold in color. Nitrogen absorption data indicated the fibers were porous with
a surface area of 300 m^2/g, a pore volume of 0.32 cc/g and a pore diameter of
44 A.

Polycrystalline α-Alumina Fibers

L.E. Seufert; U.S. Patent 3,808,015; April 30, 1974; assigned to E.I. du Pont de Nemours and Company describes polycrystalline ceramic oxide fibers containing at least 60% Al_2O_3 and comprising an α-alumina crystalline phase. They are prepared from a spinnable mix containing solid particles that are at least 80% alumina and an aqueous phase that has dissolved in it at least one precursor of alumina or at least one precursor of an oxide that forms a solid solution with alumina. It has been found that the amount and size of the solid particles in the mix has an effect on the grain size, roughness and orientation of the fibers produced from the mix.

The mix is extruded through a spinneret, and the emerging fiber is wound on a collapsible bobbin, usually continuously, and fired, usually in two stages. The result is a strong polycrystalline alumina fiber. It is believed that the presence of the solid particles minimizes filament-to-filament sticking during the spinning and firing steps. It is also believed that the presence of the solid particles aids in the firing of the fibers by reducing the amount of shrinkage of the fiber. The resulting fibers have the following properties:

(1)   At least one dimension less than 0.01 inch, preferably less than 0.001 inch;

(2)   A crystallinity greater than 85%;

(3)   A porosity of less than 10%, preferably less than 5%;

(4)   A distribution of grain size wherein substantially all the grains are less than 3 microns in diameter, at least 30% of the grains are less than 0.5 micron in diameter, and at least 10% of the grains are larger than 0.04 micron in diameter;

(5)   A microscopic roughness height of 1100 to 7000 A and a microscopic roughness period of 4000 to 15,000 A; and,

(6)   A uniform orientation, as determined by optical birefringence.

Transparent Aluminum Borate and Borosilicate Fibers

H.G. Sowman; U.S. Patent 3,795,524; March 5, 1974; assigned to Minnesota Mining and Manufacturing Company describes aluminum borate and aluminum borosilicate refractory material essentially free of any crystalline alumina. The refractory material is also transparent, has no discernible graininess under binocular (e.g., 48X) microscopic examination, and shaped articles thereof are additionally smooth-surfaced. Fibers of the refractory material can be made which are continuous in length, strong, glossy, have high moduli of elasticity, and they as well as other shaped articles can be deeply internally colored.

The refractory material can be made in an amorphous form and subjected to higher temperature to densify, strengthen, and convert them into a homogeneous mixture of amorphous and crystalline phases, or into an all crystalline form, with retention of desired shape, integrity, and transparency.

The refractory fibers in their green or unfired form can be made by shaping and dehydrative or evaporative gelling, e.g., by extruding, drawing, spinning, or blowing, or combinations thereof, a viscous concentrate of an aqueous mixture comprising alumina and boria precursors, with or without silica precursor, such as colloidal silica, present in the mixture. Alternatively, the aqueous mixture can be gelled by evaporating water from a film thereof or by gelling droplets of such an aqueous mixture in a dehydrating alcohol to form spheres or bubbles, or by spray-drying the aqueous mixture in air to form beads, spheres, bubbles, or fine particles. The shaped green articles are then heated to remove further water and other volatiles, decompose organic material, and burn off carbon, and convert the article into a monolithic refractory comprising aluminum borate or aluminum borosilicate compositions.

Generally, the aluminum and boron compounds are present in the solution or two-phase system in amounts sufficient to provide an equivalent $Al_2O_3:B_2O_3$ mol ratios of 9:2 to 3:1.5, preferably 9:2 to 3:1⅓, and typically 3:1. The silica component of the two-phase system can vary and be as much as 65 weight percent based on the total weight of three metal oxides present, the silica component preferably being present in the amount of 20 to 50 weight percent. Generally, the green articles made from two-phase systems containing high amounts, e.g., above 20 weight percent, of the silica component can be fired more rapidly, and thus more economically, to obtain the desired physical properties including physical strength, integrity, and transparency; such articles also will withstand higher temperatures in use.

Suitable aluminum compounds which can be used include aluminum chloride hexahydrate, aluminum formoacetate, and aluminum nitrate. Suitable boron compounds which can be used are boric acid and boria. Especially useful raw material which can be used to prepare an aqueous solution is a boric acid-stabilized aluminum acetate, $Al(OH)_2(OOCCH_3)\cdot\frac{1}{3}H_3BO_3$. The pH of an aqueous solution of this powder will be inherently on the acid side, e.g., less than 5, and concentrations of 30 to 40 weight percent generally will be useful.

The two-phase system used to prepare the aluminum borosilicate refractories can be prepared by admixing an aqueous colloidal dispersion of silica with the above described aqueous solution of the aluminum and boron compounds, the basic aluminum acetate solution being preferred. The two-phase dispersion can be prepared by adding the dispersion of silica to the aqueous acetate-boria solution with mixing in order to obtain a uniform dispersion without formation of a gel, floc, or precipitate. The pH of the resulting dispersion will be inherently on the acid side, e.g., below 6. The pH is preferably 3 to 5. If desired, a heat fugitive acid, such as hydrochloric, nitric, or acetic acid, can be added to the colloidal silica dispersion to acidify the same prior to use and prevent premature gelling.

A heat fugitive organic agent is incorporated in the two-phase system dispersion to improve shelf-life of the subsequently concentrated system or to increase viscosity to improve the fiberizing nature of the latter. Such organic agents representatively include polyvinylpyrrolidone, polyvinyl alcohol, dimethylformamide, and glucose (e.g., corn syrup), these additives being oxidized and removed during the firing of the green articles produced from such systems.

The silica aquasol or aqueous dispersion of colloidal silica can be used with SiO_2 concentrations of 1 to 50 weight percent, preferably 15 to 35 weight percent. However, the colloidal silica can be used in the form of an organosol, the silica being colloidally dispersed in such water-miscible, polar organic solvents as normal or isopropyl alcohol, ethylene glycol, dimethylformamide, and various Cellosolve glycol ethers. The size of the colloidal silica particles in the aquasols or organosols can vary, preferably from 10 to 16 millimicrons.

Example: Concentrated HCl (4.8 ml) was added to aqueous dispersion of colloidal silica (214 grams) to give a dispersion having a pH of 1. White Karo corn syrup (86 grams) was added to the dispersion to increase viscosity and the resulting mixture was filtered. Basic aluminum acetate powder (300 grams) was dissolved in 400 ml of water to give an aqueous solution having a pH of about 5 and the solution was filtered. The latter solution was added to the syrup-containing aqueous dispersion of colloidal silica. The resulting mixture was concentrated in a rotating flask partially immersed in a water bath, the temperature of the bath gradually being raised to 35°C. The viscosity of the resulting liquid concentrate was 75,000 cp.

Fibers were formed by extruding this concentrate through a spinnerette having 30 holes, each with a 4 mil diameter, at 150 to 180 psi, and drawing the resulting extruded fibers at 100 to 130 ft/min on a rotating 24" diameter drum having a cylindrical polyester film facing. The resulting continuous green fibers were dried by room temperature air as they were pulled 6 feet from the spinnerette vertically downward to the drum.

The fibers were cut and removed from the drum in 6 foot long bundles and placed in an electric air atmosphere oven, heated from room temperature to 950°C, and held at 950°C for one hour. Some of the resulting fired fiber bundles were further heated to 1150°C and some to 1400°C. The 950°C and 1150°C-fired fibers were continuous, transparent, colorless, clear, and strong. The 1400°C-fired fibers ranged from transparent-to-translucent-to opaque and were weak and fragile. The final overall calculated composition of these fired aluminum borosilicate fibers was $3Al_2O_3:1B_2O_3:3SiO_2$.

The density of the fired fibers varied between 2.65 and 2.80 g/cc. The modulus of elasticity for the 950°C-fired fibers varied between 20.4 and 25.8 x 10^6 psi. The 1150°C-fired fibers had a modulus of elasticity of 15 to 18.5 x 10^6 psi. The fibers fired at 1400°C had a modulus of elasticity of 16.8 to 19 x 10^6 psi. X-ray diffraction patterns of the various fired fibers were run and the pertinent, distinctive portions of the patterns were found to be as follows:

Firing Temperature, °C	Relative Intensities of Pertinent Diffraction Lines	
	5.4	3.4
950	100	40
1150	100	60
1400	40	100

The relative intensities of the 5.4 and 3.4 diffraction lines of the 950°C- and 1150°C-fired fibers are significantly different from those of the characteristic mullite pattern. These differences in relative intensities are such that the 950°C- and 1150°C-fired fibers can be characterized as having a reverse mullite x-ray diffraction pattern, whereas the corresponding relative intensities for the 1400°C-fired fibers are consistent with those of mullite. No diffraction line at 4.35 or free alumina were discernible in the x-ray diffraction patterns of the three samples.

SILICA FIBERS

Insulation for Turbine Casings

P.F. Hawthorne; U.S. Patent 3,904,427; September 9, 1975; assigned to Foseco International Limited, England describes a shaped article of refractory heat insulating material which is formed of a dry composition comprising inorganic refractory fiber and a water-soluble inorganic binding agent, and which is in the form of a composition of which the hardness, strength and density do not vary across the bulk material. Usually, though not necessarily, such a composition will be one in which the binding agent is homogeneously distributed.

Preferably, such a shaped article is made by forming a wet or damp mixture of the ingredients to the desired shape and then drying the so-formed shape by means of a drying method which does not cause migration of any of the components of the composition. Microwave drying or dielectric heating may be employed, the first of these being preferred.

The inorganic fibers used may be aluminosilicate, calcium silicate, asbestos, alumina, silica, zirconia or carbon fibers either singly or as a composite mixture of two or more. The binding agent used may be, for example, colloidal silica sol, colloidal alumina sol, sodium silicate, potassium silicate, ethyl silicate, or a metallic phosphate or borate. The refractory heat insulating composition may also include a proportion, e.g, up to 10% by weight of finely divided refractory material, e.g., alumina, silica, calcined rice husks, diatomite, kieselguhr, magnesia, silicon carbide, silicon nitride or fireclay, if desired. One group of compositions of particular value are mixtures of colloidal silica sol and aluminosilicate fiber, the silica content of the dry composition being 55 to 90% by weight.

Specific compositions of value may be made by drying shaped mixtures of aqueous colloidal silica sol and aluminosilicate fibers. The dried shape may have a composition of for example, (by weight) fiber 34%, silica 66%. Calcined rice husks may replace some of the fiber in such material, for example in a composition consisting of silica 66%, calcined rice husks up to 10% and the residue being the inorganic fiber.

The refractory heat insulating materials are of particular value in the protection of turbine engine casings. These are, in use of the engine, subjected to severe conditions: for example, the temperature of gases inside the casing may reach 900° to 1400°C and the pressure changes inside the casing (because the load on or power output of the engine changes) may range from atmospheric up to ten atmospheres or more, the pressure outside the casing remaining atmospheric. As well as the mechanical forces generated by these changes, the casing is subjected to vibrational stresses over a wide range of frequencies during running.

Turbine casings for road vehicles are usually made of spheroidal graphite cast iron, and this material does not stand up well to the conditions of service. At 250°C and above, progressive modification of the internal structure of the metal takes place, and the desirable mechanical properties are lost. Eventually, the turbine casing must be replaced with a new casing. Such replacement is not only uneconomic, but very inconvenient since the majority of other components of a turbine engine have comparatively much longer lives. By affixing in a turbine casing a lining of material, preferably 12 to 75 mm thick and most preferably 25 to 40 mm thick, the protection of such casings may be achieved and their life considerably extended, for example, to 3 years (300,000 miles) or more. During such life, the casing requires little or no maintenance, the layer of heat insulating refractory material giving rise to no maintenance problems. Further, by similar use of materials the life of the turbine casing and exhaust system of an aero jet engine may be materially increased.

In the lining of the casing of aero jet engines compositions are preferred in which a proportion of the fiber (up to 10% by weight of the whole composition) is replaced by calcined rice husks. If a two layer material is used for lining a turbine or aero engine casing, the lower density layer is preferably adjacent to the casing itself and the higher density layer is exposed to the high temperature gases inside the casing.

Example 1: A preformed web of colloidal silica sol bonded aluminum silicate fiber, of thickness 25 mm, was adhered with a sodium silicate based adhesive to the inside of a turbine casing, part of the engine of a gas turbine driven truck, and riveted in place at the edges thereof. The casing was installed in the engine which was then used in a gas turbine truck for 300,000 miles. The engine was dismantled and the lining inspected. Little erosion had taken place and the casing itself was metallurgically unchanged. In contrast, a similar engine casing without the lining or other protection failed (cracked) after only 15,000 miles use.

Example 2: A web of 25 mm thickness was formed of a composition comprising silica sol, calcined rice husks and aluminum silicate fiber, which was dried by microwave drying to give a composition as follows:

	% by Weight
Silica	66
Aluminum silicate fiber	27
Calcined rice husks	7

This web was adhered to the inside of the exhaust of an aerojet engine and riveted in place at its edges. The jet engine was used for 5,000 flying hours and then dismantled and the exhaust inspected. It was found that very little erosion had taken place and the casing itself was metallurgically substantially unchanged.

Glossy, Light Gray Silazane Fibers

G. Winter, W. Verbeek and M. Mansmann; U.S. Patent 3,892,583; July 1, 1975; and W. Verbeek; U.S. Patent 3,853,567; December 10, 1974; both assigned to Bayer AG, Germany describe the production of a shaped article comprising forming a mass consisting of a silazane from ammonia and a halogenosilane into a shaped article, and heating the shaped article to 800° to 2000°C in an inert atmosphere, whereby the silazane decomposes. The resulting shaped article comprises a homogeneous mixture of silicon carbide with silicon nitride and optionally silicon dioxide and/or carbon. The shaped articles which are obtained have the following composition:

	% by Weight
Si	30 to 70
C	5 to 60
N	5 to 35
O	0 to 12

The halogenosilane to be used for the formation of suitable resinous silazanes can have the general formula R_nSiX_{4-n}, where R is hydrogen, alkyl, alkenyl and/or aryl, X is F, Cl, Br or I, and n is 0, 1, 2 or 3.

The silazane compounds are very readily soluble in conventional solvents such as chlorinated hydrocarbons, aromatics and higher alkanes but can, in a dilute or concentrated form, only be converted into fibers with difficulty since, because of their partly three-dimensional crosslinking, the silazane compounds do not display any filament-forming properties. The spinnability of the silazane solution— preferably chlorinated hydrocarbons are used as solvents—can be brought about, by adding to the solution linear polymeric high molecular auxiliaries with degrees of polymerization of above about 2,000, in a concentration of about 0.01 to 2% by weight. Polyethylene oxide, polyisobutylene, polymethyl methacrylate, polyisoprene and polystyrene have proved particularly suitable. The silazane compound can be present in the spinning solution in a concentration of 10 to 90% by weight, preferably 20 to 40% by weight.

A conventional dry spinning process can be used for the spinning of the silazane solutions. For this, the spinning solution is forced from a spinneret, which preferably has orifices of 50 μm in diameter, into a heated spinning column, and the resulting continuous filaments are wound up while being stretched. When the filaments formed pass through the spinning column, the solvent is expelled and condensation, with elimination of ammonia, results in a substantially infusible silazane fiber. The silazane solutions can furthermore be spun by a jet-blowing process or a centrifugal spinning process to give staple fibers. In this way, glossy fibers of light gray color, having a cross-section of 10 to 20 μm, can be spun.

Since the fiber-forming silazane compound already crosslinks during the spinning process, with elimination of ammonia, infusible fibers in most cases result, which can immediately be subjected to pyrolysis under an inert gas. However, if appropriate, the fibers can also be subjected to a heat treatment at 20° to 400°C, under air or inert gas, before the pyrolysis. The silazane fibers thus obtained are insensitive to oxidation up to 400° to 500°C, infusible and resistant to a large number of organic and inorganic solvents and can advantageously be used for heat resistant filter fabrics or textile articles of low inflammability. During the subsequent heat treatment, all remaining volatile constituents are eliminated. For this purpose, the fibers are heated, if appropriate under tension, to 800° to 2000°C in a tubular furnace.

The rate of heating can be varied within wide limits and is, for example, 1° to 100°C/min. The weight loss is approximately 10 to 30%. When heated up to 1400°C under an inert gas such as nitrogen, ammonia, argon or hydrogen, the black, glossy fibers consist of homogeneous mixtures of x-ray amorphous silicon carbide and silicon nitride and possibly silicon dioxide and/or carbon. At higher temperatures, β-SiC and β-Si$_3$N$_4$ are formed. The fibers subjected to the heat treatment are completely oxidation resistant up to 1200°C and possess good mechanical strength. The tenacity of the fibers heat treated up to 1200°C under an inert gas, as measured in a commercially available tensometer 90 to 130 kp/mm^2 and the modulus of elasticity is 8,000 to 10,000 kp/mm^2. The modulus of elasticity can be increased further if the fibers are briefly brought up to 2000°C under argon. Because of their good mechanical properties and oxidation resistance even at high temperatures, the fibers are very suitable for use for the reinforcement of plastics, and especially of glasses, ceramics and metals. Furthermore, the fibers are outstandingly suitable for high temperature insulation and for use as a filter material for hot, corrosive gases and melts.

Example: 35 grams of ammonia are passed into a solution of 50 grams of methyltrichlorosilane and one liter of methylene chloride at 40°C, while stirring. The reaction mixture is subsequently freed of the resulting ammonium chloride by filtration. The silazane solution is concentrated to 93 grams and 29 grams of a solution of 2 grams of polyethylene oxide (degree of polymerization, 100,000), 8 grams of carbon tetrachloride and 90 grams of methylene chloride are added. The mixture is homogenized for one hour while stirring slowly. The spinning solution contains 13.1% of silazane and 0.47% of polyethylene oxide. The spinning solution is spun at a spinning pressure of 0.01 atmosphere gauge, and a spinning column temperature of 30°C (column head) and 170°C (middle of column) through spinnerets of cross-section 400 μm, using a draw-off speed of 210 meters per minute.

The fibers are heated to 1200°C over the course of 3 hours, under nitrogen. The weight loss is 8.8%. Black, glossy, completely oxidation-resistant, x-ray amorphous fibers result, which possess a tenacity of 60 to 115 kp/mm^2 and a modulus of elasticity of 9,000 to 10,000 kp/mm^2. The fibers consist of 50% by weight of silicon carbide and 50% by weight of silicon nitride.

Transparent Thoria-Silica-Metal Oxide Fibers

D.D. Johnson; U.S. Patent 3,909,278; September 30, 1975; assigned to Minnesota Mining and Manufacturing Company describes transparent refractory fibers of at least 50% thoria with silica, alumina, boria and optionally chromia for refractory articles. In the green form they are made by conventional means, followed by dehydrative or evaporative gelling of a viscous concentrate of an aqueous mixture of the above oxides as such or as precursors to give a green fiber. The green fibers are heated to remove further water and other volatiles, decompose organic material and burn off carbon. A fiber comprising thoria, silica and aluminum borate or aluminum borosilicate is obtained. The compositions comprise 55 to 90% ThO_2, 5 to 35% SiO_2 and 5 to 25% R_2O_3 [metal(III) oxide]. The compositions from which the refractories are formed are viscous sols or solutions of Th compounds with Al and B compounds and containing colloidally dispersed silica.

Preferably aqueous sols, solutions or dispersions of water-soluble or water-dispersible compounds of thorium and of aluminum and boron are used which can be calcined to the respective metal oxides, thoria, alumina and boria. In some instances a soluble or dispersed chromium compound which converts to or may be chromia, Cr_2O_3, is also added. Generally, the thorium and silicon compounds are present in aqueous dispersions in amounts sufficient to provide weight ratios of 1.5:1 to 10:1 of ThO_2:SiO_2.

Thoria is conveniently provided as a sol of thorium oxynitrate or oxychloride prepared from thorium hydroxide by dissolution or dispersion in less than the stoichiometric amount of nitric or hydrochloric acid, suitably with the addition of lactic acid in some instances. Generically, these compounds and others formed this way are termed thorium oxysalts whether they actually form solutions or are colloidal dispersions. A convenient and preferred procedure for preparation of thorium oxynitrate is to precipitate thorium hydroxide from a solution of the nitrate in water and after washing the precipitate until wash waters have a pH of 7 the hydroxide is then dispersed or dissolved in dilute nitric (or hydrochloric) acid to give the basic salt, believed represented as $Th(NO_3)_x(OH)_y$, where x and y are about 2, dissolved or dispersed in water.

Silica is conveniently employed in the form of colloidal dispersions in water with SiO_2 concentrations of 1 to 50 weight percent. Preferred concentrations are 15 to 35 weight percent which are commercially available and from which less water has to be removed in raising the viscosity of the mixtures than when using more dilute dispersions. The colloidal silica can also be used in the form of organo-sols, the silica being colloidally dispersed in such water-miscible, polar organic solvents.

An especially useful source of both alumina and boria is the powdered boric acid-stabilized basic aluminum acetate which is represented by the formula $Al(OH)_2(OOCCH_3)\cdot\frac{1}{3}H_3BO_3$. This material contains an equivalent mol ratio of $Al_2O_3:B_2O_3$ of 3:1 and approximately 44.6% by weight of alumina and boria together. For convenience, this will sometimes be referred to as aluminum borate inasmuch as the acetic acid is ultimately lost.

Example: Thorium hydroxide is precipitated from 450 grams of thorium nitrate tetrahydrate dissolved in 1.5 liters of deionized water by addition of 275 ml of 28% ammonium hydroxide at 20° to 25°C. The precipitate is collected on a filter. It is resuspended in deionized water and collected repeatedly until the wash waters have a pH of 7. The thorium hydroxide (not dried) containing 200 grams ThO_2 is suspended in 1.0 to 2.0 liters of water and 50 ml of concentrated hydrochloric acid added. The suspension is stirred and the precipitate gradually dissolves in one hour. This solution of thorium oxychloride is used in compositions alternatively to the oxynitrate described above. Analysis by evaporating to dryness and igniting shows that 533 grams of the solution contains 50 grams ThO_2.

To 533 grams of the solution are added 21 grams of a boric acid-stabilized basic aluminum acetate (corresponding to 7.53 grams Al_2O_3 and 1.71 grams B_2O_3) and the mixture stirred until the solid has dissolved. The pH of the solution is determined and the pH of 48 grams of colloidal silica adjusted to approximately the same value in the range of 3 to 5 using nitric or hydrochloric acid.

The acidified colloidal silica dispersion is added to the thoria-alumina solution gradually with agitation at room temperature. Failure to adjust the acidity may result in some precipitation when the solutions are mixed. The total solution is evaporated at 30°C to a small volume of viscous fluid which if necessary is de-aerated and filtered to remove any suspended solids. The viscous solution is extruded in a small laboratory spinner (30 orifices) at 17.5 kg/cm^2 pressure and gives satisfactory unfired continuous green fibers.

Firing in air, at a maximum of 750°C, by gradual admission to a zone at that temperature, produces strong continuous white transparent refractory fibers containing 68% ThO_2, 20% SiO_2 and 12% Al_2O_3 and B_2O_3 combined in a molar ratio of 3:1. These fibers can be knotted and can be woven to produce cloth. They can be used to reinforce plastics.

SILICA-ALUMINA FIBERS

Thermal Insulation Board from Ceramic Fibers

*E.W. Olewinski and L.J. Pluta; U.S. Patent 3,835,054; September 10, 1974;
assigned to Nalco Chemical Company* describe a method of making thermal insulation board by mixing 2 to 5% of an aqueous slurry of ceramic fibers with
1 to 15% of silica sol based on the weight of the fibers and having a particle
diameter of 4 to 100 mμ, adding 0.1 to 10% alum or sodium aluminate, adjusting the solution pH to 5 to 7, adding 0.01 to 2% based on fiber of an anionic
polymer flocculating agent consisting of an acrylamide/acrylic acid latex, and
drying and forming the board.

A highly preferred ceramic fiber material usually of equal parts of alumina and
silica known as Fiberfrax is composed of 51.3% of Al_2O_3 and 47.2% SiO_2 where
the fibers range in length up to 1½ inches and have a mean diameter of 2 microns.
Such fibers are capable of withstanding continuous use temperatures up to 2300°F
and short-term exposures to higher temperatures. Colloidal silica sol may be prepared and utilized under a preparation using an input of sodium silicate through
a bed of cation-exchange resin in the hydrogen form.

Example 1: 140 grams of Fiberfrax were slurried in 5,600 ml tap water, 51.3
grams silica sol and 40 grams sodium aluminate were added and the slurry was
mixed for 15 minutes. The pH was adjusted to 5.5 with H_2SO_4 and 350 ml of
0.1% solution of acrylamide/acrylic acid copolymer were added. Mixing was continued slowly until the system was flocculated. The slurry was then poured into
a 10½ x 10½ inch box with a screen on the bottom and vacuum applied to draw
off the water. The pad was then dried overnight at 110°C. The dried pad had
a strength of 90 psi and a density of 18.5 lb/ft^3. After firing at 1300°F for one
hour, the strength was 100 psi.

Example 2: 140 grams of Fiberfrax were slurried in 5,600 ml tap water. 57.9
grams silica sol and 54.9 grams alum were added and the slurry was mixed for
15 minutes. The pH was adjusted to 6.5 with NaOH and 130 ml of 0.1% acryl-
amide/acrylic acid copolymer were added. Mixing was continued slowly until
the system was flocculated. The slurry was then poured into a 10½ x 10½ inch
box with a screen on the bottom and vacuum applied to draw off the water.
The pad was dried overnight at 110°C. The dried pad had a density of 17.7 lb/ft^3
and a strength of 102 psi. After firing at 1300°F for one hour, the strength was
107 psi.

Example 3: Following the procedure of Example 1, a 5% ceramic fiber was
slurried in water and the same percentage of colloidal silica and sodium aluminate
were utilized. The slurry was mixed 15 minutes and the pH adjusted to 6 with
sulfuric acid. Then 0.25% of an anionic copolymer identified as an acryl-
amide/acrylic acid was added where the weight was based on the mineral fiber
and the polymer was added as a 0.1% product solution. The mixture was slowly
agitated until flocculation occurred whereupon it was vacuum formed into a fiber
mat and dried for use as thermal insulation.

Coating of Al_2O_3/SiO_2 Fibers Bonded to Al_2O_3

With the intensified interest in eliminating air pollution, considerable effort is being expended towards improving catalysts and supports therefor which might be useful in this area. An important requirement for such materials is that the catalyst structure be such as to permit high space velocity, i.e., the passage of substantial volumes of exhaust gas or fumes per unit of time, per cubic foot of catalyst, without substantial pressure drop or slowdown of the gas. Generally speaking, honeycomb ceramic structures are satisfactory in this respect but the processes by which such structures are made are complicated and time consuming.

Accordingly, *J.B. Hunter; U.S. Patent 3,799,796; March 26, 1974; assigned to Matthey Bishop, Inc.* describes the preparation of a ceramic or refractory support material which comprises a porous outer layer or coating of Al_2O_3/SiO_2 fibers surrounding and integrally bonded to an inner portion which may be entirely or essentially Al_2O_3 or it may include a core of metal, e.g., mild steel, stainless steel, Inconel, aluminum alloy, or other material such as a composite of metal particles in a ceramic matrix.

The process is carried out by applying one or more ceramic cement coatings containing primarily Al_2O_3/SiO_2 fibers to an aluminum structure, e.g., a conventional aluminum honeycomb held together with an epoxy or like binder, drying to form a porous fibrous coating on the honeycomb or other structure and firing the coated product to melt the aluminum and convert it to alumina. To modify the structure for certain applications it may be desirable to also include in the ceramic cement some added high surface area Al_2O_3 powder and/or a metal powder having a melting point above that of the aluminum structure. The metal powder being wetted by the molten aluminum enhances the diffusion of the aluminum and at the same time alloys with it to form a strengthening metal inclusion.

Despite the fact that the aluminum is melted, the fibrous ceramic holds the molten aluminum in place while permitting oxidation or alloying of the aluminum to occur so that the fired product retains the original shape of the starting structure. This product has outstanding properties for use as a catalyst support and because of the porous outer coating of ceramic fibers, it may not be necessary to wash coat before metallizing to provide an adequate catalyst surface. Thus, the fired structure can be directly metallized, using conventional metallizing techniques to form a catalyst although it will be appreciated that a wash coat may be applied before metallizing if this should be desired.

In one way of preparing a product, a desired number of discs or other shapes, of required size, are punched or otherwise cut from conventional aluminum honeycomb slab, thereafter a continuous surface coating is applied to the cut shapes by dipping the latter into an aqueous slurry of a ceramic cement containing Al_2O_3/SiO_2 fibers, followed by air drying and then firing in at least two stages at temperatures and times sufficient to drive off the binder, e.g., epoxy, which is normally used to form an aluminum honeycomb and to melt the aluminum.

In a second way the aqueous slurry of ceramic cement may contain an added amount of high surface area Al_2O_3 to give a substrate of greater catalytic surface area. In a third variation the aqueous slurry may contain an added amount of metal powder to give a substrate of greater physical strength. A fourth and preferred method involves a combination of the second and third ways described above. In this modification the aluminum structure is first dip coated in a slurry of ceramic cement containing a metal powder having a melting point above that of aluminum and capable of alloying with the aluminum. After air drying, the once coated structure is dip coated a second time with a dispersion of a high surface area Al_2O_3 in the ceramic cement. After air drying a second time the twice dipped substrate is fired at a temperature below the melting point of the aluminum to burn out the epoxy binder and then fired again at a temperature above the melting point of the aluminum to cause melting and alloying of the aluminum with the metal powder adjacent to its surface. All aluminum metal not alloyed in this way is subsequently converted to Al_2O_3 by the diffusion and reaction with oxygen in the furnace atmosphere.

The fibrous cement coating effectively envelopes the aluminum and retains the honeycomb structure even though the aluminum is melted, sufficient air passing through the fibrous coating to oxidize the aluminum while the latter though molten is prevented from escaping. It is difficult to determine the degree to which the aluminum is oxidized to alumina during firing. In any event, the molten aluminum is either converted to alumina or alloyed with the metal particles that may lie adjacent to the original metal structure. The fired product, after cooling, and with the same shape as the starting structure, is then ready to be impregnated with catalytic metal to complete the catalyst. The success of the process is due to the use of a ceramic cement containing Al_2O_3/SiO_2 fibers.

The amount of fibrous cement employed will also vary dependent upon other operating factors, however, sufficient cement should be used to entirely cover the aluminum structure with the cement. Preferably, to facilitate dip coating, the cement is made into an aqueous slurry by diluting with water. Upon drying, the cement, because of the ceramic fiber content, gives the coated structure the necessary strength for handling and further processing. The honeycomb or equivalent structure may be coated a number of times with the fibrous ceramic cement provided a continuous coating is finally obtained.

Firing is preferably carried out in at least two stages continuously, one stage below the melting point of the aluminum, e.g., at 900° to 1200°F, for 1 to 3 hours, followed by heating above the melting point of aluminum, for instance, at 1400° to 1600°F for 1 to 3 hours. Additionally, it is important that the first stage of firing be sufficient to drive off the binder, which is used to make the honeycomb or equivalent starting structure. In the second firing stage, which should directly follow the first stage as noted above, the conditions used should be adequate to melt the aluminum and diffuse the same into the adjacent surface of the fibrous cement.

Binder for Inorganic Fibers

V.W. Weidman; U.S. Patents 3,785,838; January 15, 1974; and 3,775,141; November 27, 1973; both assigned to E.I. du Pont de Nemours and Company describes totally inorganic fibrous refractory compositions of 50 to 98 parts by weight fibers such as aluminosilicates, mineral wool, fiber glass, asbestos and quartz, and 2 to 50 parts by weight binder plus flocculent where the weight ratio of binder to flocculent is 3:1 to 1:5 useful for high temperature applications. The binders are positively charged colloidal particles of silica coated with a polyvalent metal-oxygen compound, boehmite alumina, amorphous fumed alumina and basic aluminum chloride. Suitable flocculents are negatively charged clayminerals such as montmorillonite, hectorite and attapulgite. These structures can be made by forming a dilute (1 to 5% solids) aqueous slurry of inorganic fibers and adding the positively charged binder. Usually the binder is added as an aqueous suspension. After the binder is mixed with the fiber slurry the flocculent is added. The slurry with all ingredients added is mixed for an additional 5 to 30 minutes and the refractory objects are vacuum formed on a screen. The moist cake formed can be dried as is or it can be shaped further by molding on forms, wrapping on mandrels or the like, and then dried. The totally inorganic structures may be used at temperatures as high as 2300°F without smoking or significant strength loss or shrinkage.

Example 1: A dilute slurry was prepared containing 2 pounds of an aluminosilicate refractory bulk fiber in 111 pounds water. To this slurry, 0.53 pound of an acidic aqueous dispersion of positively charged colloidal particles consisting of a dense silica core coated with positively charged polymeric alumina, was added with good stirring, which was continued for 5 minutes. Next, 0.2 pound of sodium exchangeable bentonitic clay, a montmorillonite type mineral, was added to the slurry with good stirring, as a 6% (solids) colloidal (negatively charged) suspension in water. After 30 additional minutes of stirring, a test pad (9½" diameter by 1½" thick) was prepared in the slurry by vacuum forming techniques.

A 240 mm porcelain Buchner funnel containing a fitted bronze screen (70/52) was attached by means of a rubber hose to an aspirator pulling a vacuum of 29 inches of water. This suction mold assembly was immersed upside down in the slurry. Within 2 minutes' time, a pad of sufficient thickness ($\approx$1½ inches) had formed on the bronze screen. The mold containing the wet test pad was then withdrawn from the slurry with vacuum still applied. After 15 seconds, excess water had been removed by the suction and the damp pad was recovered from the mold. The pad was transferred to a perforated metal plate support and dried in an oven. After thoroughly drying the pad at 250°F, test bars (1 x 1 x 6 inches) were cut from the pad and weighed. Strength, density, and shrinkage properties of this composite product are given below. (Ratio by weight of ceramic fiber: positively charged binder:negatively charged flocculent = 100/8/10.)

Density	21.6 lb/ft^3
Fired (linear) shrinkage, 2 hours at 2300°F	5.3%
Modulus of rupture, dry	85 psi
Fired 4 hours at 1000°F	112 psi
Fired 2 hours at 2300°F	121 psi

Examples 2 through 11: Example 1 was repeated using varying ratios between the ceramic fiber, the positively charged binder and the negatively charged mineral flocculent.

Example	Wt Ratio of Fiber/Binder/ Flocculent	Density Lbs/Ft^3	Modulus of Rupture, psi			Fired Linear Shrinkage %
			Dry	4 Hr at 1000°F	2 Hr at 2300°F	2 Hr at 2300°F
2	100/4/6	15	31	42	66	4.1
3	100/6/8	17.5	48	65	93	4.4
4	100/8/6	20	61	80	99	4.8
5	100/8/8	22.5	72	87	80	7.0
6	100/8/12	24	96	137	112	4.3
7	100/10/10	20	100	133	121	7.0
8	100/15/18	22.5	101	115	92	5.3
9	100/0/0 (Control)	—	—	—	(Sample too weak to test)	
10	100/0/10 (Control)	11.5	9	13	20	1.5
11	100/10/0 (Control)	—	—	—	(Sample too weak to test)	—

Mineral Wool-Ceramic Fiber Insulation

M.D. Ash; U.S. Patent 3,754,948; August 28, 1973; assigned to Morganite Ceramic Fibres Limited, England describes refractory insulating compositions which comprise mineral wool and ceramic fiber in admixture. Up to 60% by weight of the composition preferably can be mineral wool for low temperature applications; at higher temperatures a higher proportion of ceramic fiber is needed. The compositions may be mixed with carriers, binders and/or stiffeners.

The mineral wool includes slag wool and rock wool. Mineral wools can, for example, be produced by passing a stream of molten rock or slag in front of high pressure steam or air jets. Slag, which can be formed into such wools, is obtained as a by-product from various metallurgical processes and the composition of the slag will depend on the particular process from which it has been obtained.

Rock wools are made from molten rocks or minerals, e.g., limestone, which may contain other rocks, e.g., shales. Clearly the precise constitution of a rock wool will depend on the starting mineral or rock. Wollastonite has been used; as has basalt, which is a dark colored igneous rock which comprises plagioclase, feldspar and augite. Other rocks have been used and these typically contain silica, alumina, lime, magnesia and iron oxide. Irrespective of origin, the main chemical constituents of mineral wools in percentages by weight are usually 25 to 50% SiO_2, 3 to 20% Al_2O_3, 20 to 45% CaO and 3 to 18% MgO.

Ceramic fibers are usually made from fused natural minerals such as kaolins, bauxite, kyanite and certain fireclays or from mixtures of alumina and silica and modifying agents. They can also be made from viscous solutions. It seems that these fibers owe their highly refractory nature to a relatively high alumina content. Thus a typical mineral wool as defined above, contains 3 to 20% alumina, whereas ceramic fibers usually contain not less than 40% alumina.

The refractory compositions can be prepared by mixing the ceramic fibers with the mineral wools. For some applications it may be desirable to chop the fiber tufts before they are mixed. The refractory compositions can be used to prepare by air or water deposition or vacuum forming, fibrous articles such as blankets, felts, blocks, boards, special shapes, paper products; by dry or wet blending, castables, concretes and cements, ramming, tamping and moldable materials; by the use of clay or plastic forming machinery, pressed and injection molded articles; by the use of conventional textile machinery, textiles, ropes, threads and cloths and other woven or unwoven textiles. The fibers may be used in the bulk form in which they are produced and processed into blankets or other desired shapes with no additional ingredients.

Example 1: The following composition was prepared: 69.5% by weight rock wool; 29.5% by weight ceramic fiber; and, 1% by weight of a methyl hydroxypropyl cellulose. The composition was thoroughly mixed and was then mixed with water and formed into blocks and boards. The products were found to be useful at temperatures up to 1150°C and had a shrinkage of less than 3% at this temperature.

Example 2: The following composition was prepared: 59.5% by weight slag wool; 39.5% by weight ceramic fiber; and, 1% by weight of a methyl hydroxypropyl cellulose. The composition was prepared and used in the same way as in

Example 1. Products made from this composition had an even smaller shrinkage.

Example 3: The following composition was prepared: 25% by weight slag wool; 10% by weight ceramic fiber; 30% by weight colloidal silica; 5% by weight micronized alumina; and, 30% by weight water. The composition was thoroughly mixed and then formed into blocks and boards which were allowed to set. The products were found to be useful at temperatures of up to 1150°C having a shrinkage below 5%.

ZIRCONIUM FIBERS

Zirconia-Silica Fibers

According to *H.G. Sowman; U.S. Patents 3,793,041; February 19, 1974; and 3,709,706; January 9, 1973; both assigned to Minnesota Mining and Manufacturing Company* refractory fibers of zirconia and silica mixtures are made by shaping and dehydratively gelling, for example by extruding in air, an aqueous mixture of a zirconium compound, such as zirconium diacetate, and colloidal silica, and heating the resulting gelled fiber in a controlled manner to decompose and volatilize undesired constituents and convert the fiber to a refractory fiber having a desired microstructure.

The refractory fibers can be used for a variety of purposes, particularly where high temperature stability is desired or required. For example, fibers of such refractory material can be fabricated into woven, felted, or knitted textiles and used for heat resistant upholstery or clothing, and for other purposes where thermal stability is desirable or required. Such articles can be fabricated into textiles which are brilliantly internally colored with inorganic colorants and used to make decorative clothing, draperies, wall covering, and the like.

Such articles can also be used as reinforcement for plastic, elastomeric, metal, or ceramic composites and as sound suppression material or as filtering or adsorptive material. The fibers also can be processed into lightweight insulation, paper, batting, and the like and used to insulate furnaces and other heating or high temperature equipment.

Example: 80 grams of an aqueous colloidal dispersion of silica were added with mixing to 225 grams of an aqueous solution of zirconium diacetate to provide a colloidal dispersion of silica in aqueous zirconium diacetate having a pH of 4, an equivalent $ZrO_2:SiO_2$ mol ratio of 1:1 and an equivalent solids content of 48 weight percent. The resulting aqueous dispersion was slightly cloudy on mixing but became clear when concentrated. Such concentration was achieved by placing the dispersion in a rotating flask partially immersed in a 30° to 50°C water bath and rotating the flask while maintaining the contents under a vacuum (ca 29 inches Hg) with a water aspirator, this concentration step being continued until the resulting concentrate became viscous enough (50,000 to 70,000 cp) to enable fibers to be pulled therefrom with a glass rod.

The viscous concentrate was then centrifuged for 15 minutes in a laboratory test tube centrifuge to remove bubbles. The resulting clear viscous concentrate was then extruded (130 psi into ambient air, ca 22°C) through a gold-platinum spinnerette having six round 3 mil orifices. The fibers or filaments so extruded were essentially continuous and straight and were drawn for 3 feet in the air and wound in a parallel fashion on a variable speed take-up drum covered with a sheet of polyester film, the speed of the drum being adjusted to exert a slight pull on the extruded fibers and hold them taut as they were extruded and wound, thereby effecting attenuation of the fibers. The green fibers so spun were essentially dry on their surface, they did not stick together after being wound, and were glossy, looking much like spun glass fiber, and under a binocular microscope appeared water-clear, transparent, smooth, round in cross-section, and straight. As wound on the drum in the form of a coil or winding, the fibers appeared white and glossy.

The coil was removed from the drum and fired in air by placing the same in an electric furnace, raising the temperature from room temperature up to 550°C, and holding at 550°C for 2 hours. Some of the 550°C-fired fibers were further fired in air in the furnace for 1 hour at 775°C, and some of the 550°C-fired fibers were further fired in air at 1000°C overnight (12 hours). Some of the resulting 775°C-fired fibers were fired in air to 1050°C. Some of the fibers fired to 1050°C were fired in air at 1050°C overnight. Some of the latter 1050°C-fired fibers were then fired in air for 48 hours at 1100°C.

The 550°C-fired fibers had diameters of 10 to 17 μ with tensile strengths of 75,000 to 150,000 psi. The 775°C-fired fibers had diameters of 15 μ with tensile strengths of 95,000 to 155,000 psi. The fibers fired to 1050°C had diameters of 15 μ and tensile strength of 90,000 to 160,000 psi. The 550°C-, 775°C-, and 1000°C-fired fibers and those fired to 1050°C all appeared transparent, glossy, smooth, round, and essentially continuous and straight under a binocular microscope. Those fibers which were held overnight at 1050°C had diameters of 15 μ and tensile strengths of 60,000 to 75,000 and exhibited a sharp extinction when examined with a microscope using polarized light, and exhibited some small crystallites or seeds, probably zircon. All fibers, except the 1100°C-fired, were flexible.

X-ray diffraction analysis of the 775°C-fired fibers revealed the presence of exclusively ZrO_2 crystallites, in the cubic form, the SiO_2 component being undetected and being apparently present in the amorphous state. The 1000°C-fired fibers did not break in handling, i.e., they were strong, and x-ray diffraction analysis showed the presence of ZrO_2 predominantly in the tetragonal form (relative intensity 100) and to a small extent in the monoclinic form (relative intensity 25), the SiO_2 component again being apparently amorphous. The 1100°C-fired fibers were found under x-ray diffraction analysis to comprise a mixture of zircon (relative intensity 100), tetragonal ZrO_2 (relative intensity 80) and monoclinic ZrO_2 (relative intensity 40), the balance of the SiO_2 component, not combined in the form of zircon, again being apparently amorphous.

When these 1100°C-fired fibers were examined under an optical microscope, they were found to be composed of opaque crystalline areas some of which were as wide as the diameter of the fibers (15 μ) with lengths up to 5 times the diameter (i.e., 75 μ), these opaque crystalline areas being separated by translucent milky areas. These 1100°C-fired fibers were very fragile and broke in handling.

The above results and evaluations of the various fired fibers show inter alia that the strongest transparent fibers were those obtained by firing to a temperature where substantial amounts of stable tetragonal ZrO_2 are formed, e.g., to 1050°C, and that firing for extended periods at 1050°C and above (e.g., 1100°C) reduced their strength and caused them to become opaque, such fibers no longer having tetragonal ZrO_2 in a dominant amount but containing zircon crystallites and monoclinic ZrO_2 in substantial amounts.

Zircon-Coated Zirconia Fibers

B.H. Hamling; U.S. Patent 3,861,947; January 21, 1975; assigned to Union Carbide Corporation describes the preparation of zircon-coated zirconia fibers by: (a) contacting zirconia fibers with a silicon-containing compound to form a coating on the fibers, and (b) heating the coated fibers to a temperature and for a period of time to form a zircon phase on the fibers by interaction of the silicon-containing compound and the zirconia. These fibers have improved high temperature properties which render them ideally suited as lightweight thermal insulation. The fibers are characterized by improved resistance to shrinkage, sagging and thermal shock up to above 2500°F and particularly 2800° to 3100°F.

In practice, the silica which is applied to the fibrous zirconia can be employed in a variety of forms. For example, the coating can be applied as ethyl orthosilicate. This compound is soluble in alcohol and can be used to coat the fibers. After removal of the solvent by drying, the organic portion of the compound is volatilized by heating at 300° to 1200°F; alternatively, the compound can be hydrolyzed with water leaving a silica coating on the surface of the fiber.

Silicon tetrachloride can also be employed. This compound reacts with moisture to produce a silica solid and gaseous HCl. When contacted with moist fibers, silica is deposited on their surfaces. $SiCl_4$ has a high vapor pressure and may be passed through a fiber bed or structure in the vapor form. The amount of silica to be deposited is proportional to the moisture content of the fibers. Alternately, the compound may be used in its normal liquid form. However, it is preferable to dilute it to 2 to 10% concentration with an inert solvent such as benzene.

Other silicon-containing compounds such as silicone resins, silicon metal, alkali silicates and organo-silanes can be employed in coating the fibers. The organo-silanes upon contact with moist fibers, react with water, condense and form a silicone polymer on the surfaces of the fiber. After removal of reaction products, HCl or alcohol, the polymer is decomposed to form silica by heating in air at 800° to 1000°F.

The improved characteristics are attributed to the formation of a zircon phase by reaction of the silica with the zirconia surfaces. Zircon phase forms on unstabilized fibers (i.e., zirconia fibers, not doped with a stabilizing agent such as Y_2O_3) by heat treatment at 2000° to 2500°F for 3 or more hours to form sufficient zircon for imparting high temperature stability. Fibers fully stabilized in the cubic or tetragonal forms must be heated at 2300° to 2500°F for 3 or more hours to form sufficient zircon for imparting high temperature stability. Preferably heat treatment at 2400°F for 16 hours is used for stabilized fibers. While as little as a few percent of zircon formation imparts improved stability, 20 to 50% is preferred for best properties.

Tetragonal Zirconia Fibers Stabilized with Group III-b Oxides

Zirconia (ZrO_2) has a fusion point of 2677°C, and is therefore a very useful refractory material. However, pure zirconia bodies are rarely used as refractories because of a phase change that occurs near 1000°C. Below 1000°C, pure zirconia exists in the monoclinic form, and when heated above 1000°C, it transforms to the tetragonal form. The phase change at about 1000°C is accompanied by a large volume change which would cause a pure zirconia article to shatter. However, it is known that zirconia can be stabilized by firing with certain other refractory oxides to produce the cubic form which is stable over a wide range of temperatures.

For instance, zirconia is frequently stabilized in the cubic form by firing to 1700°C, or higher with 11.5 to 15 weight percent yttria, by firing to 1500°C or higher with 8 to 15% magnesia, or by firing to 1500°C or higher with 6 to 15 weight percent calcia. Zirconia can also be stabilized in the cubic form by silica, scandia, and oxides of the rare earth metals (i.e., lanthanide metal oxides).

According to a process that is described by *B.H. Hamling; U.S. Patent 3,860,529; January 14, 1975; assigned to Union Carbide Corporation* zirconia fibers and textiles can be stabilized in the tetragonal form by small, carefully controlled amounts of oxides of metals of Group III-b of the Periodic Table. The oxides of metals of Group III-b which are useful as stabilizers include scandia, yttria, oxides of the rare earth metals such as lanthana, ceria, and the like, and oxides of the metals of the actinide series such as uranium oxide. The preferred metal oxide stabilizers are yttria, ceria, and mixed rare earth metal oxides. The stabilizers can be used in mixture, if desired. The stabilized tetragonal zirconia fibers and textiles are produced by the relic process.

Briefly, the relic process for producing stabilized tetragonal zirconia fibers comprises: (a) impregnating an organic polymeric fabric or textile with a mixture of a zirconium compound and a Group III-b metal compound, and (b) heating the impregnated fabric or textile, at least partly in an oxidizing atmosphere, in order to carbonize and volatilize the organic polymer without igniting the polymer, and at the same time to insure conversion of the zirconium and Group III-b metal compounds to their oxides. The resulting stabilized zirconia fiber or textile will have the same form as the original fiber or textile, although the dimensions will be reduced.

Following impregnation of the fiber or textile with the metal compounds from a solvent solution, it is preferred to remove excess solution from the surfaces of the individual fibers in order to prevent accumulation of caked metal compound. This can be done by blotting, pressure rolling, centrifugation, or the like. The impregnated fiber or textile is then dried, for example, by air drying or heating in a stream of warm gas, prior to the heating steps which volatilize the organic polymeric precursor fiber or textile and form the tetragonal zirconia. The heating step for volatilizing the precursor fiber or textile and forming the metal oxide is carried out under controlled conditions so as to avoid ignition of the fiber or textile.

The exact amount of stabilizer oxide that is desired to be used depends, to an extent, upon the end use of the zirconia fiber or textile. In uses wherein the excellent resistance of zirconia to alkali is important, and wherein the use temperature is not excessively high (e.g., from 300° to 500°C), as little as 0.5 mol percent yttria can be employed. An addition level of 2 to 4 mol percent yttria is desired to enable the zirconia fiber or textile to be fired to the 800° to 1000°C range in order to eliminate impurities and permit maximum densification, and to thereby achieve maximum chemical resistance. As much as 10 or 20% of monoclinic zirconia is permissible in such uses. When the zirconia is to be used above 1000°C, 2 to 4 mol percent yttria is normally desired in order to achieve the best physical properties. The stabilized tetragonal zirconia fibers and textiles have enhanced utility in many high temperature and corrosion resistant applications. The stabilized zirconia fibers and textiles can be used as battery separators, fuel cell separators, pipe liners and troughs for transferring molten metal, heat shields, and the like.

Example: A series of zirconia cloths containing varying proportions of yttria were produced by the relic process. The initial cloth substrate was square weave, textile rayon. The rayon was preswollen in 1 N hydrochloric acid, then rinsed in water. The cloth was impregnated by immersing it for 18 hours in 2.5 molar aqueous $ZrOCl_2$ containing varying quantities of YCl_3. The impregnated cloth was centrifuged three times at 4,000 to 4,600 rpm in an 11-inch diameter basket for a total of 20 minutes in order to remove excess salt solution. The cloth was then heated in a forced air oven. The initial temperature was 25°C, and the temperature was gradually increased over 24 hours to 650°C. Thereafter, the cloth was fired as indicated in the table below.

Crystallographic Phases of ZrO_2/Y_2O_3 Cloth

Cloth No.	YCl_3 Content of Solution, g/l	Mol % Y_2O_3 in Cloth	Crystal Structure After: 1 Hr at 1000°C	1 Hr at 1200°C	1 Hr at 1400°C	1 Hr at 1400°C
1	0	0	93 % M 7% T	100% M	100% M	100% M
2	3.6	1.29	90% M 10% T	100% M	100% M	100% M
3	7.2	2.08	9% M 91% T	43% M 57% T	96% M 4% T	100% M
4	13.9	2.73	100% T	100% T	100% T	100% T
5	26.2	5.52	100% C	100% C	100% C	100% C
6	37.4	7.10	100% C	100% C	100% C	100% C
7	47.8	9.57	100% C	100% C	100% C	100% C

The table displays the YCl_3 content of the impregnating solution, the mol percent Y_2O_3 in the ZrO_2 cloth, and the crystallographic phases in the cloth after heat treating in air. The crystallographic phases were determined at room temperature by x-ray diffraction analysis. In the table, M, T and C denote the monoclinic, tetragonal, and cubic phases, respectively, present in the ZrO_2 cloths.

Fibrous Zirconia-Cement Composites

According to *B.H. Hamling; U.S. Patent 3,736,160; May 29, 1973; assigned to Union Carbide Corporation* fibrous zirconia/cement composites can be prepared by impregnating fibrous zirconia with a mixture of a liquid containing a zirconium compound and a refractory powder, followed by subjecting the impregnated fibrous zirconia to a temperature sufficient to convert the zirconium compound to zirconia. It is preferred that the zirconia fibers that are used be stabilized. The preferred stabilizer oxide for use in stabilizing the fibrous zirconia in the tetragonal form is yttria. The zirconia cement comprises a mixture of a liquid containing a zirconium compound and a refractory powder. The liquid containing the zirconium compound is preferably an aqueous solution of the zirconium compound.

The zirconium compound employed can be, for example, zirconyl chloride, basic zirconium chloride (zirconium hydroxychloride), zirconium acetate, zirconium citrate, zirconium oxalate, and the like. Preferably, the liquid containing the zirconium compound also contains a compound of a metal whose oxide will stabilize the zirconia in the tetragonal or cubic form. Such metals whose oxides stabilize the zirconium in the tetragonal or cubic form include yttrium, calcium, magnesium, cerium, and certain other rare earth metals. Preferably, the stabilizer oxide is yttria, therefore, the yttrium compounds such as yttrium acetate and yttrium chloride are preferably employed in the liquid containing the zirconium compound.

The refractory powder which is employed in the cement can be a refractory metal oxide, zircon, barium titanate, strontium titanate, or other refractory powder. The preferred refractory powders for use in preparing the cement include zirconia, hafnia, ceria, thoria, yttria, rare earth oxide, zircon, or mixtures thereof. More preferably, the refractory powder is zirconia, and most preferably, it is zirconia stabilized with a stabilizer metal oxide such as yttria, ceria, or other rare earth oxide. The fibrous zirconia/cement composites are useful as heat shields, flame barriers, corrosion protection barriers, and the like.

Example: Fabrication and Properties of a Rocket Nozzle Throat Insert — A composite structure was fabricated suitable for use as a throat insert in a rocket nozzle where a constant throat area (no erosion) is required under an oxidizing environment. The composite was fabricated with zirconia cloth which contained 8 weight percent yttria and which had the following properties:

Cloth construction	Five-harness satin weave
Thickness	28-31 mils
Weight	22 oz/yd^2
Bulk density	63 lb/ft^3
Porosity (void content)	83%
Breaking strength	6 lb/in width
Elongation at break	8%

The cement used was prepared by mixing together a ratio of 100 ml of liquid binder to 150 grams of zirconia powder. The liquid binder contained 35.6 weight percent ZrO_2 in the form of dissolved ZrOOHCl and 2.75 weight percent Y_2O_3 in the form of dissolved YCl_3 (8 weight percent Y_2O_3 relative to ZrO_2). The ZrO_2 powder also contained 8 weight percent Y_2O_3 stabilizer and was prepared by ashing and wet ball-milling $ZrOCl_2$ + YCl_3-loaded wood pulp. The ZrO_2 cloth was saturated with cement, thoroughly padding cement into both sides of the cloth and removing excess cement from the cloth prior to lamination. The lamination was prepared by cutting round discs 2-inches in diameter from the cement-filled cloth and stacking 100 layers high. The stack was placed in an aluminum die which had been lined with Mylar film. The laminate was pressed at 150 psi, the temperature being raised over three hours to 250°F. After holding for three hours at 250°F, pressure was released and the laminate removed from the die. (At this point the dried laminate was hard and strong. The cement had been dried and formed a strong, green body.) The laminate was next cured in a gas-fired kiln, which was raised to 3000°F over six hours and held at 3000°F for ½ hour. The cured laminate measured approximately 1¾" high by 1¾" diameter. A ⅜" diameter hole was machined in the center of the laminate using diamond-grinding media, such that the cloth layers were normal to the axis of the hole. The finished specimen had a bulk density of 280 lb/ft³ (4.5 g/cc).

In separate tests on zirconia-bonded satin-weave cloth laminates prepared and fired in the manner described above (except that maximum firing temperature was 2800°F), the properties of such laminates are as follows:

Porosity	32%
Flexural strength	6,000-8,000 psi
Modulus of elasticity	3-5 x 10⁶ psi
Coefficient of thermal expansion (25°-1000°C)	105 x 10⁻⁷ in/in/°C
Melting point	4800°F

NICKEL FIBERS

Nickel Oxide Fibers

B.D. Brubaker; U.S. Patent 3,875,296; April 1, 1975; assigned to The Dow Chemical Company has prepared whiskers of nickel oxide and/or solid solutions of nickel and magnesium oxides by: (a) forming a molten film of a mixture of a metal salt with at least one auxiliary salt, the proportion of the metal salt being up to 80 and preferably 0.5 to 80 mol percent of the mixture; (b) reacting the mixture with water vapor at 700° to 1200°C to initiate metal oxide fiber growth; (c) continually supplying a molten film of the mixture over the fibers to sustain growth; and (d) separating the fibrous metal oxide product from any residual mixture, for example, by treating it with an aqueous leaching solution.

The fibrous oxide product so-prepared and obtained comprises, after drying, a loosely packed, bulky, fibrous mass of nickel oxide or solid solutions of nickel and magnesium oxides having a low density, the individual fibers having high temperature resistant properties. Auxiliary salt refers to the alkali metal chlorides and bromides, alkali metal sulfates, calcium, barium or strontium chlorides, or mixtures thereof, employed in a reaction mixture with particular nickel and magnesium inorganic salts. Metal salt includes a nickel salt, such as chlorides and sulfates, and the chlorides, bromides and sulfates of magnesium or mixtures thereof.

In carrying out the process a metal salt is mixed with preferably at least 30 and more preferably at least 40 mol percent of at least one auxiliary salt to form the reaction mixture. A double salt such as, e.g., carnallite ($MgCl_2 \cdot KCl \cdot 6H_2O$) may also be conveniently used in combination with the nickel salt. In such instance, no separate addition of auxiliary salt is necessary. A mixture of the nickel salt together with either a sodium or potassium halide and especially the chlorides of such alkali metals is preferred. When calcium chloride and lithium salts are employed they should constitute less than 60 mol percent and 80 mol percent, respectively, of the auxiliary salt.

Examples 1 through 11: Various reaction mixtures were prepared containing nickel chloride ($NiCl_2 \cdot 6H_2O$), magnesium chloride ($MgCl_2$), and as auxiliary salts potassium chloride (KCl) and calcium chloride ($CaCl_2$), then furnace heated in a crucible (crucible comprising 96% silica) for the time and temperature indicated in the table below.

Upon heating, the liquid mixture wet the crucible walls forming a film. Reaction of the mixture with water vapor in the furnace atmosphere caused formation of solid solution fibers. Liquid mixture was supplied by capillary action. Magnesium oxide-nickel oxide solid solution fibers of varying length were obtained. The reaction condition, and proportions of materials are shown in the table below.

Ex. No.	$NiCl_2 \cdot 6H_2O$	MgCl$_2$	KCl	CaCl$_2$	Time (hr)	Temp (°C)	NiO	MgO	Fiber Length (microns)	Mol Ratio in Starting Composition of Ni:Mg
		Mol Percent					Fiber Composition in Mol Percent		Fiber Length	
1	2.5	22.3	66.3	8.9	2	900	18	82	125	1:9
2	7.4	17.3	66.4	8.9	2	900	45	55	125	3:7
3	12.4	12.4	66.3	8.9	2	900	57	43	300	1:1
4	17.3	7.5	66.3	8.9	2	900	73	27	750	7:3
5	22.3	2.5	66.3	8.9	2	1000	91	9	300	9:1
6	22.3	2.5	66.3	8.9	2	850	–	–	400	9:1
7	22.3	2.5	66.3	8.9	2	950	–	–	400	9:1
8	22.3	2.5	66.3	8.9	2	1000	–	–	500	9:1
9	13.2	1.5	75.2	10.1	2	1000	–	–	1,400	9:1
10	8.4	0.94	79.9	10.7	2	1000	–	–	800	9:1
11	12.9	1.4	73.4	12.3	1	1000	–	–	600	9:1

CARBIDES, BORIDES AND NITRIDES

CARBIDES

Free-Flowing Flame Spraying Powder

R.F. Cheney, D.J. Port and J.R. Spencer; U.S. Patent 3,881,911; May 6, 1975; assigned to GTE Sylvania Incorporated describe a free-flowing flame spray powder consisting of subparticle agglomerates of at least one inorganic material having a melting point above 1800°C, the inorganic material selected from W, Mo, Cr, Ta, Nb, their alloys, and the oxide, boride, carbide and nitride compounds of these refractory metals. The subparticles are held together by diffusion bonds to one another and have diameters up to $\frac{1}{5}$ the diameters of the agglomerates, and at least 80% of the agglomerates have sizes within a range of 30 microns.

Accordingly the binder related deficiencies of contamination, low apparent density and low agglomerate strength can be overcome by heating the agglomerates above the vaporization temperature of the binder to remove the binder and thereafter heating the powders to a still higher temperature causing the subparticles of the agglomerates to sinter or fuse together by diffusion bonding.

This second temperature must not exceed $\frac{1}{2}$ of the melting point of the material since that would result in the agglomerates themselves sintering or fusing together to form an unusable cake or sponge. The sintering of the subparticles restores or improves the strength which the agglomerates had prior to removal of the binder, and also causes a densification of the agglomerates, thus improving the apparent density. By controlling this sintering within the permissible range, the apparent density can be intentionally adjusted over a considerable range.

For successful strengthening of the agglomerates, it is also necessary to control the relative sizes of the subparticles of the agglomerates. In general, these subparticles should be no larger than $\frac{1}{5}$th the agglomerate diameter and generally $\frac{1}{10}$th or less. This is important because if the subparticles are too large, heating under the conditions necessary to avoid appreciable sintering of the agglomerates

134

will result in agglomerate strengths insufficient to permit handling after the sintering operation. Accordingly, the subparticles should preferably be as small as possible, e.g., $1/100$ the agglomerate diameter or less.

For particles within the above size range, sintering results in a product which is a free-flowing binderless powder having sufficient strength to withstand considerable handling, and being useful in flame spraying applications.

The binder should volatilize at 50°C above the highest temperature to which the material is subjected during agglomeration, and at least 50°C below the lowest temperature at which sintering is carried out. Typical binders include polyvinyl alcohol, stearic acid, paraffins, polyethylene glycol, methylcellulose, and various resins.

Example 1: A molybdenum flame spray powder is manufactured as follows. A slurry is made by dissolving 1.5% (wt percent solids basis) polyvinyl alcohol in hot water and adding (FSSS) 3 to 6 μm Mo powder to a 60 to 70%, preferably 70%, solids concentration. The slurry is pumped at low pressure to a two fluid nozzle located at the top of a commercially available spray dryer. The spray is continually agitated throughout the spray drying run. The atomization air pressure to the nozzle is 40 to 60 psi. The inlet air temperature is 370° to 430°C, preferably 400°C, with an outlet temperature of 140° to 150°C, preferably 150°C of the drying air passing through the dryer at 20 to 30 cfm.

The product from the spray dryer first receives a presintering cycle of 4 hours at 600°C in a hydrogen atmosphere to remove binder. The powder is then sintered for 4½ hours at 1100°C in H_2. This sintering gives the powder enough granule strength to withstand classification, handling during shipping and feeding to the flame spray gun. The sintered powder is separated into the desired size fractions for plasma flame spraying by screening. A typical size distribution obtained is given below in U.S. Standard Sieve sizes:

1% max	+170 mesh
15% max	+200 mesh
80% min	−200+325 mesh (44 to 74 microns)
20% max	−325 mesh

The out of size material is deagglomerated by milling and screened 200 mesh. The −200 mesh fraction is used along with virgin powder in subsequent slurries for spray drying while the +200 mesh fraction is returned to the mill.

Example 2: A cemented tungsten carbide flame spray powder is manufactured as follows. The powder is prepared by ball milling a blend of WC powder 0.5 to 4.0 μm FSSS with Co powder 1 to 10 μm FSSS to produce a powder of composition WC-12% Co. The slurry is prepared by combining this powder, paraffin wax and stearic acid in the ratio of 97.6:2:0.4 respectively, with enough trichloroethane to make an 80 to 85% solids concentration. Spray drying is carried out as in Example 1 with the following exceptions.

The drying conditions are 120° to 125°C inlet air temperature, 65° to 70°C outlet temperature. The slurry is pumped to the nozzle at low pressure with 38 psi atomization air pressure in the two fluid nozzle. The spray dried powder is composed of very soft agglomerates of wide size distribution. This material is presintered to remove binder in H_2 for 4½ hours through a temperature gradient of 500° to 700°C. The powder is then sintered in H_2 for 15 minutes at 1240°C resulting in sufficiently hard granules to withstand screening and handling.

The typical size distribution desired of this powder is 100%, –325 mesh +10 μm. The out of size material is deagglomerated by milling to return it to a particle size which could be used as starting material.

Silicon Carbide Hot Pressed with Boron Compound

S. Prochazka; U.S. Patent 3,853,566; December 10, 1974; assigned to General Electric Company describes a method of making a dense silicon carbide ceramic by forming a homogeneous dispersion of a submicron powder of silicon carbide and a boron-containing additive, the amount of the additive being equivalent to 0.5 to 3.0 parts by weight of boron per 100 parts of silicon carbide, and hot pressing the dispersion at 1900° to 2000°C and 5,000 to 10,000 psi. The resulting ceramic has a density of at least 98% of the theoretical density for silicon carbide and is suitable as an engineering material for example in high temperature gas turbine applications. The boron-containing additive may be either elemental boron or boron carbide.

During hot pressing an atmosphere must be used which is inert to the silicon carbide. Thus, oxidizing atmospheres such as air cannot be used since they would tend to oxidize the silicon carbide to silica, interfere with sintering and degrade the high temperature properties. Useful inert atmospheres include argon, helium, and nitrogen.

As revealed by x-ray diffraction, electron diffraction and metallography, the fine grained SiC is composed of β-SiC and a minor, varying amount of β-SiC (6H). There is also a small amount of silica located at grain edges, typically one volume percent. No separate boron-containing phase was detected which suggests that boron formed a solid solution with SiC.

Example 1: To submicron size SiC powder, was added 3.6 parts by weight per 100 parts of SiC of submicron particle size boron carbide. A dispersion was formed of 50 g of the mixture in 100 g water by tumbling in a plastic jar with ¼ inch tungsten carbide balls. The slurry was dried, the powder obtained was sifted through a 40 mesh sieve and charged in a 1 inch bore graphite die.

A 1 inch diameter sample was hot pressed under an atmosphere of 1 mm Hg pressure of argon at 1950°C and 5,000 psi for 30 minutes. Full pressure was applied at 1700°C on heating up and released at 1700°C on cooling. The pressing had a density of 3.18 g/cc (99% of theoretical density), less than 1% residual porosity and an average grain size of 4 microns. Room temperature strength in 3 point bending on a 0.625 span of as machined test bars 100 mil square cross section and 1 inch long was 80,000 psi.

Example 2: To submicron size SiC powder was added 1 part by weight per 100 parts SiC of submicron ball milled amorphous boron and this was mixed with an equal weight of CCl_4 to form a thin slurry. The slurry was milled in a plastic jar with tungsten carbide balls for 5 hours. The liquid was then allowed to dry off and the powder was fed into a graphite die. The die was subsequently inductively heated in an atmosphere of 1 mm argon to 1750°C under

1,000 psi for 1 hour. After this temperature was achieved, the load was increased to obtain 10,000 psi and the temperature was further increased to 1950°C in 30 minutes and held at this temperature for another 30 minutes. The atmosphere was maintained at 1 mm Hg of argon. Thereafter, the temperature was gradually reduced maintaining the full load on the specimen. When 1700°C was reached, both load and power was shut off and the system allowed to cool off.

Test bars having a cross section of 100 mils square and 1 inch long were prepared and ground with a 200 mesh diamond wheel. The product having a density of 3.20 g/cc was characterized as follows:

Properties	Results
Strength (3 point bending)	
RT	80,000 psi
1300°C	80,000 psi
1400°C	80,000 psi
1500°C	64,000 psi
1600°C	44,000 psi
Time to fracture at 1600°C	
25,000 psi	>4,500 min
30,000 psi	>4,500 min
35,000 psi	~40 min
Creep rate at 1620°C/25,000 psi (bending)	1.5×10^{-9} cm/cm sec
Oxidation rate in air at 1600°C (weight gain in 64 hours)	0.5 mg/cm^2

Infiltrating Silicon Carbide with Boron Carbide-Silicon Carbide

S. Prochazka; U.S. Patent 3,852,099; December 3, 1974; assigned to General Electric Company describes a procedure for making high strength silicon carbide articles by an infiltration procedure. This technique involves initially forming a porous silicon carbide compact and then infiltrating the compact with a lower melting alloy. The product obtained is a nonporous silicon carbide ceramic and is at approximately the theoretical density.

The silicon carbide ceramic is prepared by preforming a porous silicon carbide body, infiltrating the pores of the body with a boron carbide-silicon carbide melt at a sufficient temperature, and directionally cooling the filled body to advance the solidification front unidirectionally through the body whereby a dense, substantially nonporous ceramic is formed. The resulting dense, strong, two-phase ceramic is suitable as an engineering material, for example, in high temperature gas turbine applications.

The porous silicon carbide compact or preform may be prepared by pressing or slip casting a particulate silicon carbide mixture.

A preferred mixture of silicon carbide consists of 55 weight percent 150 mesh, 15 weight percent of 600 mesh and 30 weight percent of –10 μ, the mesh being defined as Tyler sieve sizes. The silicon carbide is the α-type and may be commercial abrasive grades. The silicon carbide particles are compacted at ambient temperatures at pressures as low as 10,000 pounds per square inch. Preferably a sufficient amount of a lubricant, such as 1% by weight of aluminum stearate, together with a binder, such as 3% by weight of a phenolic novolak resin, should be used during the compacting step. The preformed porous silicon carbide body is then placed in a graphite crucible.

Example 1: A preformed compact was prepared from a particulate mixture of finely divided silicon carbide having the following Tyler mesh sizes: 144 g of 150 mesh, 66 g of 600 mesh, and 40 g of –10 microns. The mixture was dispersed in a 1% solution of aluminum stearate in carbon tetrachloride to obtain a thick slurry which was then stirred for 3 hours. Some of the solvent was dried off and the resulting slurry was pressed in a steel die into a 1½ inch diameter 3 inches long cylinder at a pressure of 20,000 psi. During pressing the solvent was removed to yield a strong green compact having a density in excess of 80% of the theoretical density.

An infiltration mixture was prepared from commercial grade materials by mixing 2 parts by weight of –325 mesh boron carbide with one part by weight of –325 mesh silicon carbide and then pressing the mixture into circular discs 1½ inches in diameter in a steel die.

The pressed silicon carbide compact was placed on a massive graphite block as a setter and then an excess, i.e., about 22 wt %, of the eutectic infiltration mixture was placed on top of the compact. The sample and the graphite block were placed into a graphite tube furnace in such a manner that the graphite block extended beyond the hot zone of the furnace.

Thus, the graphite block served as a heat sink and created a temperature gradient to be established in the sample during the heating cycle. The temperature difference was measured through a top sighting window at the face of the setter and the body, and then adjusted at 2000°C by moving the carbon block up and down. The furnace was then heated to 2300°C in an argon atmosphere of 10 mm mercury pressure. After holding at this temperature for 5 minutes, the temperature was gradually decreased at a rate of 5°C per minute until the temperature reached 2150°C as read on the top of the sample at which temperature the furnace was shut off and the sample allowed to cool.

The top and bottom face portions of the sample were ground off by a diamond wheel to remove the remainder of the eutectic melt. The resulting product had a density of 3.10 grams per cc and corresponded to 99% of the theoretical density. The product was extremely hard, strong, impervious and abrasion resistant.

Diamond cut test bars annealed in argon at 1800°C to remove surface damage introduced on machining, had an average modulus of rupture of 40,000 psi at room temperature and 22,000 psi at 1500°C in air. The material was exposed to 10 repeated heat-ups with a gas-oxygen torch to 1300°C and cooled with a blast of pressurized air without failure. It also resisted repeated quenching from 700°C into water.

Oxidation at 1400°C in air caused a weight loss 0.53 mg/cm^2 after 24 hours which did not change any more at extended exposure. This may be explained by the fact that a thin protective oxide coating had formed on the surface. At 1600°C in air the weight loss after 27 hours was 2.83 mg/cm^2, was not yet stable but the rate of oxidation was falling off. These data show that the silicon carbide formed bodies were sufficiently oxidation resistant.

Example 2: A soft moldable mass was prepared by dispersing a –40 mesh graphite powder in a 40% by weight solution of a phenolic resin dissolved in ethanol. A paraffin wax cylinder 1 inch in diameter and 2 inches long was embedded into the moldable graphite mass contained in a paper container and was dried at room temperature for 4 days. The drying was continued at 40°C, just below the melting temperature of the paraffin wax, for another 2 days.

Thereafter, the temperature was slowly increased to 60°C at which the paraffin wax core melted and the wax was drained off. The graphite mold which contained a cavity in the shape of the wax cylinder was heated up slowly to 600°C in nitrogen to form a porous strong graphite die.

A silicon carbide casting slip was prepared by dispersing 126 g of 150 mesh silicon carbide, 24 g of 600 mesh silicon carbide and 30 g of –10 micron silicon carbide powders by ball milling in 60 cc of a ½% solution of aluminum stearate in carbon tetrachloride. The first two silicon carbide powders were commercial abrasive quality and the latter silicon carbide was prepared by wet ball milling and acid leaching of the 600 mesh silicon carbide.

The slip was poured into the graphite die and the dispersant drained off until the die cavity was filled with a casting. The procedure was the same as that used in ceramic forming called solid slip casting. Thereafter, the die with the casting was dried at 80°C to evaporate the solvent and was then placed in a carbon resistor furnace on a solid graphite block. A 20 g cylindrical compact made from one part by weight of –325 mesh silicon carbide and two parts by weight of –325 mesh boron carbide powders was placed on top of the casting and the furnace was heated to 2300°C and subjected to a heating and cooling cycle as set forth and described in Example 1. The powder compact melted and the melt infiltrated the casting but did not react with the carbon die.

After infiltration was completed, the sample was removed from the furnace and refired at 800°C in air until the carbon die burned off. The remainder of the silicon carbide-boron carbide infiltrant was ground off from the top of the ceramic sample. The density of the product was 3.05 g/cc and the body was impermeable and had a smooth surface appearance.

Inhibiting Grain Growth in Boron Carbide

D. Stibbs, R. Thompson, and O.W.J. Young; U.S. Patent 3,811,899; May 21, 1974; assigned to United States Borax & Chemical Corporation describe a method for inhibiting grain-growth in ceramic powders by treating the powder with a solution or suspension of the appropriate metal alkoxide, decomposing the alkoxide to produce the corresponding oxide in situ, and hot-pressing the mixture to form a ceramic composite of improved mechanical strength. The improved strength apparently results from the smaller grain size.

The alkoxides can be defined by the formula $M(OR)_n$, wherein M represents a divalent or trivalent metal of the group consisting of aluminum and the alkaline earth metals, such as magnesium, calcium, strontium and barium, R represents an alkyl group containing one to five carbon atoms, and n is a positive integer equal to the valency of the metal M.

Accordingly, the alkoxide is mixed, in the form of a solution or suspension in an organic solvent, with a ceramic powder based on boron. Preferably, the amount of alkoxide is up to 0.375% by weight of the ceramic powder. This ceramic powder may be selected from boron carbide (B_4C), other boron-carbon containing ceramics, or borides of titanium, zirconium, tantalum, etc. The mixture is heated while being stirred at a temperature sufficient to evaporate the solvent and decompose the alkoxide to the corresponding oxide.

The resultant oxide deposits on the surface of the ceramic particles. The mixture is then ball-milled to a uniform particle size and compacted by hot-pressing following known procedures. Although it is preferred to decompose the alkoxide completely before hot-pressing, it is possible to hot-press a ceramic mixture containing some residual alkoxide.

Example 1: Boron carbide (40 g) was added to powdered magnesium oxide (0.4 g) and the mixture ball-milled to an average particle-size of 1 micron or less. The mixture was hot-pressed. Some restriction of grain-growth was observed and the resultant product was found to have a particle-size of 10 to 15 microns.

Example 2: Boron carbide (40 g) was mixed with a 7.5% solution of magnesium methoxide [$Mg(OCH_3)_2$] in 2 ml of methanol. The mixture was heated, with continuous stirring, to 150°C, at which point magnesium oxide deposited on the boron carbide particles and the solvent evaporated. The mixture was ball-milled to an average particle size of 1 micron or less and formed into a composite by pressing at 2.5 tons/in² for 14 minutes to 2000° to 2200°C. The product was found to have an average particle size of 2.5 microns.

Examples 3 through 5: The following mixtures were ball-milled to an average particle size of 1 micron or less:

Example	B_4C, g	Alkoxide
3	40	$Mg(OMe)_2$ in MeOH:7.5% soln (1.0 ml)
4	40	$Al(OPr^i)_3$ in EtOH:10% soln (1.0 ml)
5	40	$Mg(OBu)_2$ in MeOH:7.5% soln (1.0 ml)

The mixtures were then subjected to the process of Example 2. The average particle size of the products was 2.0 to 2.5 microns. Thus, as illustrated by Examples 3 to 5, the addition of up to 0.25% of the alkaline earth metal oxides, added as alcoholic solutions of their alkoxides, results in a greater grain-growth inhibiting effect than the addition of magnesia powder shown in Example 1.

Metal Carbide Dispersions in Carbon

According to *E.M. Wewerka, R.J. Imprescia and R.D. Reiswig; U.S. Patent 3,809,565; May 7, 1974; assigned to the U.S. Atomic Energy Commission* dispersions of microsize, metal-carbide particles are achieved in carbon matrices by dissolving or dispersing organometallic compounds in carbon precursor materials, e.g., thermosetting resins and coal tar pitch, and subjecting the mixture to a high temperature heat treatment.

It is desirable but not essential that the organometallic additives be completely soluble in the liquid carbon precursor material. Complete solubility of the additives ensures uniform blending of the metals throughout the carbon precursor material. However, even if the additive is nonsoluble or only partially soluble in the liquid carbon precursor material, uniform blending may be achieved through well-known blending techniques. The mixture of additive and carbon precursor material is then heat treated to carbonize the precursor material and reduce the organometallic compound to the metal carbide. The resulting carbonaceous matrix which contains a dispersion of fine metal carbide particles may be further heat treated as desired, e.g., to enhance graphitization.

This additional heat treatment may also be used to control to a certain extent the size of the metal carbide particles within the carbon matrix. The organic components of the organometallic additives are not critical except insofar as they aid in making the additives soluble in the liquid carbon precursor material.

Example: The samples listed in the table were made by dispersing or dissolving either 10 or 20% by weight of organometallic compounds of Zr, V, Ta, or Ti into 10 g samples of either a 400 cp furfuryl alcohol resin or a coal tar pitch and blending for 15 minutes at 70°C.

Additionally, two furfuryl alcohol resin samples containing 17 and 23% by weight of an organometallic compound of V are listed in the following table. The resin base samples were cured to 200°C in air over a 72 hour cycle, baked to 900°C in argon over a 24 hour cycle, and graphitized to 2800°C in helium over a 10 hour cycle. With the exception of a cure, which they did not have, the pitch base samples were heat treated in a similar manner.

All samples were examined by both optical and electron microscopy. Metal carbide particle size values were determined by optical and electron microscopy. In the table, the heading Metal refers to the weight percent of metal in relation to the weight of the combined organometallic compound and the carbon precursor material.

Formation of Metal Carbide Particle Dispersion in Carbon Matrices Using Organometallic Additives

Carbon Precursors*	Organometallic Additives	Additive Conc (wt %) Compound	Metal	Approx Carbide Particle Size (μm)	Degree of Dispersion, Carbide Particles
PFA	Ti-oxyacetylacetonate	10	1.8	4 max, most <2	Fair
	Ti-oxyacetylacetonate	20	3.6	4 max, most <2	Fair
	Zr-acetylacetonate	10	1.9	4 max, most <2	Good
	Zr-acetylacetonate	20	3.7	4 max, most <2	Good
	Ta-ethoxide	10	2.8	<2	Very good
	Ta-ethoxide	20	5.6	<2	Very good
15V	Ta-ethoxide	20	5.6	Mostly <0.5	Good
	Zr-acetylacetonate	20	3.7	4 max, most <2	Fair
	Ti-oxyacetylacetonate	20	3.6	<6	Good
	V-acetylacetonate	20	2.9	<4	Good
PFA	V-acetylacetonate	17	2.5	Mostly <2	Good
	V-acetylacetonate	23	3.3	Mostly <2	Good

*PFA = furfuryl alcohol resin, 400 cp; 15V = coal-tar pitch.

As shown in the table, with each precursor material the tantalum organometallic yields finer and more uniform carbide dispersions than do the other organometallics. In the pitch-derived carbon, the TaC particles are usually under 0.25 μm in size, but they range up to 2 μm in the resin-derived carbon. Carbide particles of the other metals are fairly evenly distributed, but are usually significantly larger than those of tantalum.

Higher concentrations of dispersed carbides can be readily achieved using the methods of the process. In addition, carbide particle sizes can be influenced by controlling the maximum heat-treatment temperature. Thus, particle sizes can be kept small by limiting the maximum heat treatment temperatures in relation to the various carbide-carbon eutectic and peritectic temperatures or increased by exceeding the eutectic and peritectic temperatures.

Tungsten-Titanium Carbide Sols or Gels

According to *A.C. Fox and K.R. Hyde; U.S. Patent 3,795,522; March 5, 1974; assigned to United Kingdom Atomic Energy Authority, England* aqueous, stable, dilutable sols and aqueous, redispersible gels having a weight ratio of $TiO_2:WO_3$ of at least 0.15 containing tungsten and titanium and optionally other metals are made by mixing freshly precipitated tungstic acid or a tungstate with an aqueous solution of a titanium salt. The sols and gels may be calcined below 1800°C in the presence of carbonaceous material to give mixed tungsten-titanium carbides.

Boron Carbide Dispersed in Molybdenum Disilicide

R.K. Stringer, N.A. McKinnon and L.S. Williams; U.S. Patent 3,772,042; Nov. 13, 1973; assigned to Commonwealth Scientific and Industrial Research Organization, Australia describe an abrasion resistant composition comprising 40 to 60% by volume of boron carbide in particulate form dispersed in a matrix of molybdenum disilicide.

The preferred compositions contain 48 to 58% by volume of boron carbide. The boron carbide particles preferably range in size from 200 mesh to 400 mesh BSS.

The abrasion resistant compositions find particular application in nozzles for the passage of high velocity streams of, or containing, abrasive materials, for example gas streams containing abrasive particles such as are used in sand-blasting and like operations. They are also useful in grinding wheels, wheel dressing sticks, filament thread guides and the like.

The abrasion resistant composition may be produced by any suitable technique. For example a mixture of boron carbide and molybdenum disilicide, in particulate or powdered form may be hot-pressed, i.e., molded under heat and pressure, to the desired form. However, to ensure that the composite material has maximum mechanical strength and abrasion resistance the material is preferably produced by reaction pressing.

In this method particulate boron carbide is mixed with the finely divided elemental constituents of molybdenum disilicide and the mixture is placed in a mold of the desired shape. Pressure is applied to the material in the mold while the mixture is heated to the point where a rapid reaction between the molybdenum and silicon occurs. During the reaction a contraction in the volume of the mixture will occur and it is important to ensure that a uniform pressure is maintained on the mixture during this event.

Preferred reaction conditions are 1500° to 1650°C and at least 6,000 lb/in². The maximum pressure is limited only by the bursting strength of the graphite die material.

Example 1: An airpipe tip for use in a flotation cell for the treatment of lead/zinc ores was produced by mixing 58% by volume of 320 mesh boron carbide (B_4C) with Mo and Si in the stoichiometric proportions for $MoSi_2$. The mixture was placed in a suitably shaped die and heated to 1650°C for 45 minutes under a pressure of 8,000 psi, applied by a plunger. Airpipe tips of this type are subject to considerable wear due to the abrasive action of the ore particles in the surrounding liquid. The tip produced as above showed superior wear resistance to tips made of conventional materials.

Example 2: A grit or sand blast nozzle was prepared by mixing 58% by volume of B_4C of half 320 mesh, half 500 mesh with Mo and Si in the proportions for $MoSi_2$, pressing the mixture into a suitably shaped die at 8,000 psi and heating to 1650°C for 45 minutes while under pressure.

Example 3: A grit blast nozzle was prepared by the method of Example 2, using 54% by volume of 100 mesh B_4C and a temperature of 1500°C.

Example 4: A grit blast nozzle was prepared by the method of Example 2, using 58% by volume of 325 mesh B_4C of high boron content and a temperature of 1650°C. The nozzles of each of Examples 2 to 4 showed no appreciable wear after 400 hours of use. Conventional bonded tungsten carbide nozzles used under the same conditions were worn out after 40 hours use.

MgO-SiC Lossy Dielectric

L.E. Gates, Jr. and W.E. Lent; U.S. Patent 3,765,912; October 16, 1973; assigned to Hughes Aircraft Company describe a lossy dielectric for dissipating high power wave energy of 100 to 1,800 watts comprised of a magnesia matrix press-formed of a dry blended mix of calcined magnesium oxide and 1 to 80% by weight silicon carbide.

In general, the lossy dielectric composition consists of critically prepared dielectric matrix and silicon carbide mixtures to provide a lightweight lossy dielectric of uniformly reproducible strength and properties. The process consists of providing a matrix of selective materials whereby the dielectric properties are adjusted within certain limits to obtain improved operating characteristics for microwave attenuators or high power loads. The dielectric constant of the resulting material is 9 to 20 and the loss tangents 0.005 to 0.600, thus providing a range of property values capable of accommodating a wide spectrum of high power microwave energy absorption.

The proportions of matrix and conducting granules vary to provide the required dielectric properties. Attenuators require 1 to 35% by weight of conducting granules. Loads generally require conducting granules in greater amounts up to

80%. Thus, in the matrix the silicon carbide granules may vary from 1 to 80%. It is most important that the conducting granules be uniformly dispersed during the mixing operation so that all particles are separated and surrounded by dielectric matrix. Otherwise, the composite will act as a metallic conductor and will reflect rather than absorb energy. The initial preparation of preforming or molding and sintering operations is critical to attain the highest uniform density to maximize thermal conductivity, strength and hardness, and to minimize porosity.

Example 1: A magnesia-based ceramic material is fabricated as follows:

(1) Grind the following for 2 hours in a 1 gal porcelain mill with 6,000 grams of alumina grinding media: magnesium oxide (about 97% based on total dry weight), 800 grams; lithium fluoride, 22.8 grams; and distilled water, 1,100 cc.

(2) Remove from mill and dry for at least 20 hours at 250°F in an air convection oven.

(3) Micropulverize the dried material.

(4) Mix dry the following for 10 minutes in a paddle-type blender: micropulverized powder of step (3), 350 grams and silicon carbide, 280 mesh, 233 grams.

(5) Slowly add distilled water until a very thick but well blended paste is formed and continue blending for 30 minutes.

(6) Dry paste completely.

(7) Pulverize to pass a 14 mesh screen.

(8) Calcine in 600 gram batches to 1000°C for 1 hour.

(9) Hot press 425 grams of the calcined material in a 2-inch diametér induction heated graphite die for 10 minutes at 1550°C under three tons total pressure. Allow to cool slowly and remove.

The composition of Example 1 may be modified within the following limits:

Lossy Ceramic Compositions

Magnesium oxide	20 to 99%
Lithium fluoride	0 to 5%
Calcium pyrophosphate	0 to 20%
Manganese dioxide	0 to 10%
Boron phosphate	0 to 20%
Silicon carbide	1 to 80%

The ceramic material of Example 1 is suitable for use in microwave loads and as terminations in traveling wave tubes. With an eight-gram specimen, 1,800 watts of continuous wave power were dissipated without failure of the material. The best commercial material available, when fabricated to this geometry, failed catastrophically when subjected to about 15 watts of continuous wave power.

Similarly, the efficiencies of traveling wave tubes have been greatly increased by use of the composition described. The material is otherwise especially suited to applications where medium to low dielectric constants, high thermal conductivities and very high power dissipation are required.

Example 2: Modification of preparing the magnesia matrix can be made by altering the steps as follows:

    (1)   Grind for 2 to 16 hours a lossy ceramic forming composition consisting of the inorganic materials, magnesium oxide powder 20 to 97%, lithium fluoride 0 to 5%, calcium pyrophosphate 0 to 20%, manganese dioxide 0 to 10%, boron phosphate 0 to 20%, with distilled water therewith sufficient to form a paste of the composition combination;

    (2)   Dry and calcine the mixture of step (1) to 1000°C for one hour;

    (3)   Micropulverize the calcined mixture;

    (4)   Dry blend the mixture with 1 to 80% by weight particulates of silicon carbide of 100 to 300 mesh; and

    (5)   Hot press the dry blended mixture of step (4) under high pressure and heat the molded form for 10 minutes to one hour at 1400° to 1550°C.

Graphite-Alumina-Silicon Carbide

I. Komaru, K. Takeda and K. Yuki; U.S. Patent 3,753,744; August 21, 1973; assigned to Nippon Crucible Co., Ltd., Japan describe a graphite-alumina-silicon carbide base refractory having good erosion resistance, spalling resistance and durable for oxidizing attack at elevated temperature. This refractory contains more than 85 weight percent of main components of graphite (including combined carbon), alumina and silicon carbide (the weight ratio of graphite:alumina:silicon carbide being 10-38:60-80:2-18), and these three main components are combined with each other by a carbon bond which contains glassy material (2 to 6 weight percent) and forms the network structure.

The alumina, which may be obtained from the electrofusing or sintering process, preferably has a purity of at least 94 weight percent, and serves to increase the pyrometric cone equivalent, the load softening point, the mechanical strength, and the resistance to the erosive slag attacks for the resulting refractory. The thermal conductivity of the pure alumina is the highest among other components of the refractory, except for the graphite and the silicon carbide. The alumina particles are preferably composed of most parts of the coarse particles having a size of 4,760 to 297 microns and small parts of fine particles having a size of less than 105 microns. The favorable grain size of the graphite particle is that which passes through a sieve opening of 297 microns.

The silicon carbide disperses in the network structure of the carbon bond and improves the oxidation resistance and the mechanical strength of the resulting refractory.

The pitch or tar serves as the binder for the components in the forming or molding stage, but serves as the secondary binder together with the Si carbide and forms the thermal conductive material after the forming is burned in the reducing atmosphere and the volatile matters are expelled from the pitch or tar.

Dense Group IV-b or V-b Monocarbides

According to *J.M. Tobin and L.M. Adelsberg; U.S. Patent 3,758,662; Sept. 11, 1973; assigned to Westinghouse Electric Corporation* a fully dense, homogeneous, single phased refractory monocarbide is prepared by placing a quantity of Groups IV-b and V-b metal in a carbonaceous mold, heating the metal and mold in an inert atmosphere to a temperature above the melting point of the metal, but below the metal carbide-carbon eutectic temperature, until a full carbide is formed by diffusion of carbon from the carbonaceous material into the metal, cooling to room temperature, and removing the resulting metal carbide member from the mold.

The metal to be converted into the single phase metal carbide is charged into the carbon mold cavity preferably as a solid body conforming to the shape of the mold cavity. Excellent results are obtained if the solid metal body is heated to a temperature above its melting point and depending upon solid state diffusion of carbon to convert it to the metal carbide.

The metal may be introduced into the mold either as a solid body, or as an assembly of several loose particles or powders of such shape that the mold cavity is essentially completely filled, or as a molten mass, or a preformed compact (which may have been vacuum sintered) conforming to the shape of the mold cavity.

When metal powders are employed to fill the mold cavity, it is important to exclude fine particles. If any appreciable amount of the powder comprises particles less than 30 mils in diameter, it has been found that in the process of heating the mold and its contents to the melting point of the metal, particles below 30 mils carburize substantially to a metal carbide while in the solid state and consequently will not melt at the metal melting temperature.

Consequently, any metal particles that are uncarburized do melt and form a mushy mass with the appreciable amount of metal carbide powders, wherein contained gases are trapped and are unable to rise out of the heterogeneous mass. As a consequence the resulting carbide body will be full of voids and multiphase components.

The fully dense carbides are particularly useful for wire dies, lamp bulb filaments, phonograph record needle blanks, crucibles, and high temperature structural materials such as rocket nozzles. Moreover, the method of making fully dense carbides is conducive to the preparation of large single crystals of the metal carbides.

Example: The reaction between liquid zirconium and graphite at temperatures substantially above 1850°C but below 2890°C and preferably, between 2000° and 2800°C produced a fully dense reaction product, which formed between the graphite and the metal at such temperatures. In this process, a cylinder of zirconium metal (99.99% Zr) was placed in a ¼ inch inside diameter graphite crucible (99.9% C). The sample was heated by induction with a 5 kw–450 kc power source.

Temperatures inside the crucible were also compared with the Zr-ZrC eutectic (1850°C) and the Nb-NbC eutectic (2335°C) temperatures. The temperature was controlled by automatically adjusting the power input and variations of less than ±5°C were obtained during the carburization run.

After the system was purged with inert gas the sample was quickly heated to the melting point of zirconium (1857°C), held there for 1 minute, and then heated quickly (approximately 1 minute) to diffusion temperature of 2700°C and held at the latter temperature for 60 hours. Thereafter the cylinders were cooled at the rate of 1.4°C per minute by cooling the furnace in an inert atmosphere. However, any other suitable cooling rate can be employed.

Under the microscope the fully dense materials are characterized by a grain structure typical of single-phase metals with sharply delineated grain boundaries and completely lacking in voids, as compared with hot pressed carbides having relatively large areas of voids. The rate of formation of the full carbide, ZrC, from the surface at various temperatures is as follows:

Temperature, °C	Cm/min
2000	0.000075
2300	0.00015
2500	0.00055
2750	0.00094
2800	0.00096

Samples of the ZrC so produced were examined metallographically under the microscope. The products are homogeneous, fully dense materials characterized by a grain-structure typical of single-phase metals with sharply delineated grain boundaries and with no voids.

In a similar manner niobium carbide, hafnium carbide, and tantalum carbide members were produced. All were single phase bodies. All the other refractory metals and mixtures of two or more can be similarly converted to fully dense carbides.

Also an alloy of 90% Nb and 10% U (depleted) in the form of a cylinder ¾ inch in diameter with a central aperture of about ⅛ inch was also fully carburized in this way. In the same manner disks of niobium-zirconium (50% by weight of each) are converted to the carbides.

In a similar manner a compact of particles of the capacitory metal, of over 30 microns is prepared and placed in a graphite mold and carburized according to this example. A diamond polishing or grinding wheel may be used to provide a special surface finish or precise dimensions on the resulting metal carbide members.

Various advantages are obtained including the provision of any desired castable shape, preparation of the carbides at high temperatures thus reducing the carburization time, and the provision of carbides of refractory metals which have high density as well as extra high strengths and exceptionally satisfactory moduli of elasticity. The hardness of these carbides is between that of sapphire and diamond. The purity if often better than that of the starting metal charge. Oxygen and nitrogen contents of 20 to 40 ppm have been obtained.

The densities and lattice parameters of six high stoichiometric fully dense carbides prepared according to the process are listed on the following page.

Density and Lattice Parameters of Carbides Prepared as Fully-Dense Carbides

	Parameter A_0 (A)	Density (g/cm^2)
TiC	4.330	4.90
ZrC	4.700	6.64
HfC	4.641	12.7
VC	4.160	5.81
NbC	4.469	7.80
TaC	4.450	4.57

Cold Pressed Boron Carbide

D. Stibbs, C.G. Brown and R. Thompson; U.S. Patent 3,749,571; July 31, 1973; assigned to the United States Borax & Chemical Corporation describe a method of producing a hard shaped article which comprises sintering a cold-pressed compact of a powder of a compound of boron and carbon, particularly boron carbide which compact contains, as a densification aid, silicon, aluminum, magnesium, or a boride of a metal or mixtures of two or more of these materials.

For example, the aluminum or the metal boride can be combined with silicon. A combination of the metal boride and aluminum, with or without silicon, can also be used with good effect as a densification aid. The preferred metal borides are borides of metals lighter than zirconium, especially titanium diboride and chromium diboride.

The densification aid may be introduced into cold-pressed compacts of the powdered boron-carbon compound before these compacts are sintered. It can be combined with the powdered boron-carbon compound to form a mixture which is cold pressed and then sintered, or it may be impregnated into cold pressed compacts of the boron-carbon compound, which impregnated compacts are then sintered.

In the first method of introducing the densification aid, the rough mixture containing preferably 5 to 10% by weight of the densification aid is ball milled to effect intimate admixture and to further reduce the particle size and the mixture is then cold pressed. The cold-pressing will usually be carried out at ambient temperature and at a pressure of 20 British tons/in² (3,150 kg/cm²) to the desired shape, e.g., a disc or plate. The shaped compact is then sintered about 2000°C and preferably at 2100°C.

Alternatively the densification aid, when it has a melting point below 2200°C, can be introduced by impregnating a cold-pressed boron carbide compact. The compact may have been produced by cold-pressing at the abovementioned temperature and pressure. The densification aid can be introduced into the compact while in the molten state or vapor phase.

For example, it can be introduced in the molten state by immersing a cold pressed boron carbide compact in the molten material, or by dipping the cold pressed compact in molten material so that the densification aid rises by capillary action, or by pouring the molten densification aid over the cold pressed compact. For example, impregnation with molten silicon can take place at about 2000°C and preferably 2100°C. Alternatively the silicon may be introduced by vacuum deposition of vaporized silicon, but it is preferred to impregnate the compacts with liquid (molten) silicon. The compacts are preferably impregnated with sufficient silicon to provide 5 to 10% by weight of silicon. The silicon serves as a densification aid, so that on subsequently being heated to 2000°C or more the compacts sinter and densify (as is demonstrated by shrinkage) to achieve a density of 99% or more of the theoretical density.

The cold-pressing may take place at any of the usual temperatures and pressures for cold-pressing. The preferred conditions are about ambient temperature and a pressure of 20 to 25 tons/in² (3,150 to 3,925 kg/cm²).

Generally the sintering is carried out at 1900° to 2300°C, preferably 2150° to 2250°C. The quantity of densification aid in the mixture is 3 to 15% by weight, most preferably 5 to 10%.

Cold-pressed and sintered articles are not necessarily of the same strength and impact resistance as hot-pressed boron carbide articles, but they have a strength and impact resistance of the same order and are generally comparable. These articles may be used in place of hot-pressed articles, e.g., as shot-blast nozzle, ceramic armour or in wear-resistant applications such as bearing surfaces.

Example 1: Boron carbide powder having a particle size below 350 British standard mesh (below 45 microns) is mixed with aluminum powder and the mixture is then ball-milled in a vibrator to effect intimate mixing and to reduce the particle size to below 6 microns. A binder, e.g., camphor in ether or polyethylene glycol in acetone, is then stirred into the mixture to provide a temporary binder during cold pressing, and the mixture is then pressed at 20 British tons/in² (3,150 kg/cm²) into circular discs having a thickness of 0.5 inch (1.2 cm) and a diameter of 3 inches (7.5 cm).

The cold pressed shapes are then heated for 1 hour to a temperature below the melting point of aluminum and held at this temperature in an inert (argon) atmosphere and then the compacts are sintered at 2100°C which is sufficient to achieve rapid densification of the compacts without substantial loss of aluminum as vapor.

When a mixture comprising 90% boron carbide and 10% aluminum is cold pressed and sintered in this way, the dense compacts formed have a density of 99.2% of the theoretical maximum density, indicating that they are very hard.

Examples 2 and 3: The boron carbide powder of Example 1 was mixed with silicon powder and the mixture was ball-milled in a vibrator to effect intimate mixing and to reduce the particle size of the mixture further, e.g., to 6 microns. A temporary binder, as specified in Example 1, was added, and then the mixture was pressed into circular plates 0.25 inch (0.6 cm) thick and 3 inches (7.5 cm) in diameter at 20 British tons/in² (3,150 kg/cm²).

The cold-pressed plates were then heated for 1 hour to attain a temperature of 500°C in an inert atmosphere to evaporate all the binder and then were sintered at 2100°C for 13 minutes. A shorter sinter period, 5 minutes, was used for pellets of the boron carbide mixture, 0.5 inch in diameter. Two batches were prepared by this procedure, one containing 90% boron carbide and 10% silicon (Example 2), and the other containing 95% boron carbide and 5% silicon (Example 3). Plates from both mixtures exhibited shrinkage after sintering, demonstrating that densification had occurred. The density of the plates containing 10% silicon was 99.2% of the theoretical maximum density, while the density of the plates containing 5% silicon was calculated to be 100% of the theoretical maximum, indicating that plates of extremely high density had been made.

Carbide-Oxide Insulators in Hot Hydrogen Environments

R.E. Riley and J.M. Taub; U.S. Patent 3,740,340; June 19, 1973; assigned to the United States Atomic Energy Commission describe composite metal carbide-metal oxide materials of the composition MC-M'O_2 where M and M' may be Ti, Zr, Hf, V, Nb, Ta, Th, or U, and the metal carbide content may be 25 to 75 volume percent, have low thermal conductivities and adequate structural properties in a flowing hydrogen environment heated to 2000°C and higher.

The designation MC-M'O_2 is a general one and does not necessarily imply a stoichiometric monocarbide and/or dioxide. This is true also of any reference to a specific metal carbide or metal oxide as, e.g., ZrC and ZrO_2. These materials are thus highly suited for use as high temperature insulators in a hot hydrogen atmosphere.

Particularly, a nominal 25 volume percent ZrC-75 volume percent ZrO_2 composite has essentially the same thermal conductivity after 16 hours at 2300°C in flowing hydrogen as it possesses after its initial exposure to that temperature.

Silicon Carbide by Pyrolyzing Polyphenylene and Silicon

O.E. Accountius; U.S. Patent 3,719,452; March 6, 1973; assigned to North American Rockwell Corporation has found, that when the polyaromatic compounds which are normally in the form of a powder as formed, are compacted in a press under pressure, a coherent mass can be obtained. This compacted mass, is then subjected to high temperatures in a furnace whereby the material is pyrolyzed.

Particularly, the process pertains to polymers where the linkage is from ring to ring in the polymer chain. Upon compaction, it is believed that the benzene rings comprising the structure align themselves in an orderly pattern so as to present a crystalline-like substance. This cannot be attained with nonaromatic polymers, nor with aromatic polymers where compaction does not transpire. Upon the pyrolysis of the compacted polyaromatic compounds, the products are obtained which have chemical properties similar to graphite.

However, the material is harder than graphite and possesses other distinguishing characteristics. The resultant pyrolyzed products are layered structures which, as indicated, are obtained through the initial compaction of the benzene rings.

The material can be used wherever graphite is normally used, such as electrodes, rocket engine throat inserts, rocket engine nozzles, brush stock, and projector carbon, just to mention a few. Of equal or greater significance is the fact that through the process one can incorporate filler material using the pyrolyzed polymers, a binder therefor.

Thus, oxidation-resistant carbon-base bodies can be formed by compacting the polymer prior to pyrolyzation with additives such as metallic borides, carbides, and nitrides which oxidize at elevated temperatures to give an oxidizing resistant coating. The pyrolyzed material can be strengthened by pressing metallic, ceramic, and various polymeric fibers and the like in the matrix prior to pyrolysis.

Due to the initial pressing operation, products of various sizes and shapes can be molded. For example, carbon base chemical resistant structural materials such as pipe, bricks, and the like can be formed by molding, pressing, or extruding mixtures of powder, carbon, or graphite, with these polymers with subsequent pyrolysis.

Other possibilities lie in the field of formation of silicon carbide bodies formed by compacting powdered silicon with the polymers in the proper proportions so that at carburization temperatures the body is converted to silicon carbide.

Example 1: 1.68 g of para-polyphenylene polymer was placed in a two-punch steel die. The material was molded in a press at a pressure of 33,500 psi. The resultant specimen was a bar 2.35 x 0.31 x 0.11 in. The compacted bar was then placed in a ceramic tube furnace which was heated by electrical resistance. The temperature of the furnace was 1000°C. The residence time of the bar in argon atmosphere within the furnace was one hour. Upon removal from the furnace the bar was inspected and determined to have been completely pyrolyzed.

Example 2: The above procedure was repeated utilizing 0.48 g of para-poly-phenylene. The pressing pressure was 122,000 psi. Instead of the bar being formed, a cylinder having the dimensions of 0.5 in. dia. x 0.12 in. in length was made in accordance with the mold used in the pressing apparatus. The firing temperature was 1000°C with a residence time of one hour in the furnace; this sample was also maintained in an argon atmosphere. The pyrolyzed para-poly-phenylene carbonaceous product of the process was obtained.

Example 3: The process of Example 2 was repeated forming the same size cylinder using the same amount of polymer. However, in this example a pressing pressure of 10,000 psi was utilized. Once again a carbonaceous product of the process was obtained.

Example 4: To illustrate the utilization of the products of the process, the bar made from Example 1 was used in a dry cell battery. A commercial D-size flashlight dry cell battery, having a 1½ volt rating, was utilized. The carbon electrode in the dry cell was removed and replaced with the bar from Example 1. The battery was then tested using a voltmeter, and the 1½ volt output was maintained.

Example 5: 3.4 grams of silicon of –325 mesh and 2 grams of para-polyphenylene were thoroughly mixed. The mixture was subjected to a pressing pressure of 120,000 psi to form a cylinder 0.5 in. in diameter by 0.16 in. in length.

The compacted cylinder weighed 0.87 gram. The cylinder was then subjected to a heating temperature of 1400°C. The resultant product was beta-silicon carbide as determined by x-ray diffraction techniques.

NITRIDES

Silicon Nitride-Metal Oxide Ceramics

According to *H. Masaki; U.S. Patent 3,950,464; April 13, 1976; assigned to Toyota Chuo Kenkysusho KK, Japan* sintered articles based on silicon nitride and having the properties of low porosity, high strength and high corrosion resistance are obtained by sintering the silicon nitride in combination with metal oxides which are such that when heated alone to a suitable sintering temperature spinels are formed. Spinels are crystals of the cubic system and have the general formula $RO \cdot R'_2O_3$ where R and R' are metal elements and the pure spinel consists of one mol of divalent oxide and one mol of trivalent oxide.

Further, it is known that spinels can be formed between a divalent metal oxide and a tetravalent metal oxide such as that of titanium or tin or the like. Where the spinel is made from a divalent and a tetravalent metal oxide, the usual ratio is 2 mols of a divalent oxide to 1 mol of the tetravalent metal oxide. Tests have shown that sintered articles of low porosity and high strength are not always obtained when silicon nitride powder is combined with oxides in such a ratio as to form a spinel. However, the desired properties can be obtained with a restricted group of oxides where the ratio of the oxides to each other and the ratio of the oxides to the silicon nitride lies between specific limits.

A powder mixture containing 60 to 92 mol % of silicon nitride with the remainder being metal oxide is prepared. The metal oxide consists of at least one member selected from a first group consisting of MgO, ZnO and NiO, and at least one member of a second group consisting of Al_2O_3, Cr_2O_3, Y_2O_3, TiO_2 and SnO_2. The mol ratio of member or members of the first group to member or members of the second group is 1:1 to 9:1. In one form of the process the mixed metals are heated to 1600° and 1800°C for 2 to 3 hours.

Alternately, the members of the two groups of metal oxides are first heated high enough and long enough to form a spinel, then cooled and reground, after which the metal oxides in the form of a spinel are mixed with the silicon nitride and sintered as above. Sintering is carried out in an inert atmosphere such as nitrogen or argon.

The product formed when the members of the two groups of oxides are not in the true spinel ratio is not exactly a spinel but rather a solid solution having a spinel-like structure. Such spinel-like materials in combination with silicon nitride give a sintered product which has the improved properties desired to an extent which is quite comparable with what is achieved by using the materials in the ratio necessary for giving a pure spinel.

While the process can be carried out by mixing the metal oxides together with the silicon nitride, or compounds being in the form of fine powders, and then sintering, a preferred method is to combine one or more members of the first group of metal oxides with one or more members of the second group of metal oxides in finely powdered form and to sinter the metal oxides alone. The sintered metal oxides which have formed a spinel-like structure are then finely ground and mixed with the silicon nitride and formed into a compact to be treated as above.

Using this method, the product obtained after sintering has lower porosity and higher strength than the product obtained by the first method. It is believed that the reason is that the metal oxides react more effectively to form the desired spinel-like structure so that the conversion is more complete and in the sintering operation, the dissolution process by which the spinels are dissolved in the silicon nitride to aid the sintering operation progresses more effectively.

Example: 60 to 92 mol % of silicon nitride powder having a maximum size of less than 300 mesh and having a purity of 98% is mixed with 4 to 36 mol % of magnesium oxide of the same mesh and high purity and 4 to 36 mol % of alumina of similar quality. The mixture is compacted at 500 kg/cm² and sintered for 2 to 3 hours at 1600° to 1800°C in an inert atmosphere, preferably of argon or nitrogen. By this means, samples of 40 x 20 x 6 mm are produced. Specific sample compositions with their properties are shown below. Also, for comparison, the properties of compositions with silicon nitride obtained with aluminum oxide alone and with magnesium oxide alone are given for comparison in the same table (Samples 101 and 102).

Sample number	Quantity (mol %)			Sintering Temperature(°C)	Specific gravity (g/cm³)	Porosity	Bending strength (kg/mm²)	Oxidation resistance (mg/cm²)	Thermal-expansion coefficient (20–400°C) (×10⁻⁶/°C)
	Si_3N_4	Al_2O_3	MgO						
1	92	4	4	1750	3.01	7	31.0	2.8	2.5
2	70	4	26	1750	2.70	15	24.0	5.2	2.4
3	60	4	36	1750	2.75	14	24.8	6.3	2.4
4	90	5	5	1750	3.05	4	40.0	1.5	2.4
5	79	6	15	1780	3.03	5	38.5	1.8	2.4
6	79	6	15	1600	2.70	15	23.0	10.6	2.4
7	65	15	20	1750	3.00	6	32.5	2.0	2.4
8	76	20	4	1750	2.86	10	29.2	3.6	2.4
9	60	36	4	1750	2.90	9	30.6	2.8	2.4
101	85	15	0	1780	2.53	20	16.0	10.0	2.4
102	75	0	25	1780	2.40	25	15.5	15.0	2.4

Noting the values given in the table, by sintering mixtures of powder containing 60 to 92 mol % of silicon nitride together with mixtures of metal oxides where the MgO content ranges from 36 (sample 3) to 4 mol % (sample 9), it is apparent that the porosity is 4 to 15%, the bending strength is 23 to 40 kg/mm² and the oxidation resistance is 1.5 to 10.6 mg/cm².

These values are definitely superior to those of the comparison samples which showed a porosity of 20 to 25%, bending strength of 15 to 16 kg/mm² and an oxidation resistance of 10 of 15 mg/cm². The superiority of the product is particularly evident in the silicon nitride ceramic of sample 4 where the porosity is 4%, the bending strength is 40.0 kg/mm² and the oxidation resistance is 1.5 mg/cm², these values being nearly equal to those of sintered articles of silicon nitride formed by the hot-press method.

Silicon Nitride-Aluminum Compound Ceramics

O. Kamigaito and Y. Oyama; U.S. Patent 3,903,230; September 2, 1975; assigned to Toyota Chuo Kenkyusho KK, Japan describe a process in which silicon nitride, aluminum nitride and aluminum oxide in the form of fine powders when thoroughly and uniformly mixed in suitable proportions are heated to 1500° to 2000°C to produce ceramics which are high in abrasion resistance and corrosion resistance and can be used at temperatures at or in excess of 1400°C. The compositions comprise 10 to 80 mol % of silicon nitride, 10 to 90 mol % of alumina and 0 to 70 mol % aluminum nitride.

Preferably, the aluminum nitride content is at least 2 mol % of the composition. During the heating most of the alumina and aluminum nitride are occluded in silicon nitride and the silicon nitride base ceramics products are formed.

Aluminum Nitride-Boron Nitride Materials

K. Reinmuth; U.S. Patent 3,854,967; December 17, 1974; assigned to Elektroschmelzwerk Kempten GmbH, Germany describes refractory, corrosion-resistant materials comprising 10 to 60% aluminum nitride and 90 to 40% boron nitride containing 5 to 25% boron oxide. These materials are prepared by compressing the mixture of ingredients at 1500° to 2200°C and 30 to 200 kg/cm².

Under the conditions of the above described reaction the binding agent is an intermediate reaction product of the above-stated ingredients and during the reaction between the aluminum nitride and the boron oxide phase presumably nitrogen and oxygen at least partially become exchanged so that, depending upon the sintering conditions, a mixture of the following character results: boron nitride, Al_2O_3, B_2O_3, aluminum oxynitride having the formula $Al_{(8/3+x/3)}O_{(4-x)}N_x$ (where x has a value between 0 and 4), aluminum oxide and perhaps some other oxide/oxynitride mixed phases. This mixture of materials hereafter is simply referred to as binder.

As the reaction product obtained under the above conditions is corrosion-resistant and also resists high temperatures and temperature changes it is particularly suitable for use in the production, processing and testing of metals in the molten state.

Examples 1 through 4: In the hot pressing of the mixtures of aluminum nitride and boron nitride use was made of boron nitride powder containing boron oxide finely distributed therewith (10 m²/g BET-surface). Preferably, use is made of boron nitride powder which retains boron oxide from its method of preparation. Less favorable surface values are obtained when boron oxide is added to boron nitride which does not contain the specified amounts of boron oxide.

However, such mixtures of boron nitride containing boron oxide may be used for the preparation of shaped articles even though the strength values will generally be somewhat lower than when the preferred mixture of boron nitride and boron oxide described above is used. The following table shows the properties of the articles made with the above general procedure.

On the basis of x-ray diffraction patterns it will be noted that the products of Examples 1 and 2 consist of aluminum oxide and boron nitride. However, in spite of having the same chemical compositions the product of Example 1 shows better mechanical properties. Furthermore, this material may become still more densified.

A significant difference between the products of Examples 1 and 2 consists in the striking tendency of the product from Example 2 to orient the layer planes of boron nitride vertical to the direction of pressing. This orientation, which occurs at the boron nitride sintering, makes itself felt in the values of some properties which show a directional dependency, like the transverse rupture strength, the thermal expansion and the thermal conductivity.

In producing the product of Example 3 the aluminum nitride assay was decreased and simultaneously the molar ratio AlN/B_2O_3 became shifted in favor of the boron oxide by the addition of boron nitride poor in boron oxide. The resulting product, which still has an impressive transverse rupture strength, shows a more pronounced orientation due to the excess of B_2O_3. If the aluminum nitride assay is lowered still more, as in Example 4, the strength drops still more. This indicates that if it is desired for the boron nitride to become practically solidified without any orientation by the reaction products from aluminum nitride and boron oxide, it is necessary to use at least 10%, by weight, of aluminum nitride in the reaction mixture and at least 2 mols of aluminum nitride per mol of boron oxide.

	Examples			
	1	2	3	4
Raw material mixture, % by wt				
BN (15% B_2O_3)	80	–	55	20
BN (0.2% B_2O_3)	–	80	37	75
Al_2O_3 (10 micron smallest)	–	20	–	–
AlN (100 micron smallest)	20	–	8	5
Mol ratio AlN/B_2O_3 in mixture	2.8	–	1.6	2.8
Sintering temperature, °C	2100	2100	2100	2100
Sintering pressure, kp/cm^2	100	100	100	100
Density, %	83	82*	75*	75*
Transverse rupture strength, kp/mm^2	6.1**	3.2**	3.2**	0.42**
Hardness, kp/mm^2	10.4	7.8	7.5	2.4
Ratio bending strength, vertical/parallel	1.9/1	7/1	4/1	2/1
Ratio thermal expansion, vertical/parallel	1/1.2	1/55	–	–
Ratio thermal conductivity, vertical/parallel	1/1	5/1	–	–

*The maximum density which can be produced (maximum compacting)
**Bending strength measured with samples which were cut vertical to the direction of pressing

Boron Nitride Bonded with Hard Material

M. Wakatsuki, K. Ichinose, R. Mori and T. Aoki; U.S. Patent 3,852,078; Dec. 3, 1974 describe a hard mass comprising a matrix and a dispersed phase substantially comprising more than 10% by weight of polycrystalline cubic system boron nitride as the matrix and granular diamond homogeneously dispersed within and bonded by the polycrystalline cubic system boron nitride matrix.

The hard mass contains at least one granular hard material selected from the borides, carbides or nitrides of titanium, zirconium, hafnium, vanadium, niobium, tantalum, chromium, molybdenum and tungsten, boron carbide, silicon carbide, aluminum nitride, silicon nitride, alumina, silica, iron oxide, nickel oxide, chromic oxide, garnet and agate dispersed in the matrix in addition to the dispersed diamond.

These hard masses are prepared by maintaining a mixture of hexagonal system boron nitride powder the main part of which is less than 3 microns in diameter in terms of the primary particles and the powder of a hard material above 1100°C at 60 to 100 kbars. The combination of the temperature and pressure is within the thermodynamically stable area of cubic system boron nitride, to convert the hexagonal system boron nitride into a polycrystalline cubic system boron nitride matrix with the powder hard material dispersed uniformly in the matrix.

Preparation of Silicon Nitride

W.J. Arrol; U.S. Patent 3,839,540; October 1, 1974; assigned to Joseph Lucas (Industries) Limited, England describes a method of manufacturing a silicon nitride product whereby a compact containing silicon and a fluxing agent is heated in a nitriding atmosphere to produce a silicon nitride body of dimensions greater than the required finished product, and of density less than the required finished product. The body is then hot pressed to produce a silicon nitride product of the required dimensions and the required density.

Preferably, the compact containing silicon and the fluxing agent has a density of 1.1 to 1.7 g/cc. Preferably, the silicon nitride body as a density of 1.7 to 2.7 g/cc. Preferably, the final, hot pressed product has a density of 2.7 to 3.2 g/cc. Preferably, the silicon and the fluxing agent are each present in the compact in discrete, powder form. Preferably, the fluxing agent is magnesium oxide or a magnesium compound thermally reducible to the oxide; the fluxing agent which is contained by the compact is a magnesium silicate. Preferably, the magnesium silicate is $MgO \cdot SiO_2$. Alternatively, the magnesium silicate is $2MgO \cdot SiO_2$, or a mixture of $MgO \cdot SiO_2$ and $2MgO \cdot SiO_2$.

Alternatively, the compact is formed from a mixture of silicon in powder form and a fluxing agent in liquid form, the liquid fluxing agent being converted on heating to the solid flux required for the hot pressing operation. Preferably, a removable binder is present in the compact of silicon and the fluxing agent, and is removed from the compact by heating the compact prior to the nitriding step. Preferably, the binder is introduced as a powder, a solution, or as a dispersion.

Example: 1 kg of fine silicon powder was mixed using a high speed rotating disc with 3 liters of a 2% mixture in water of a magnesium oxide powder fluxing agent. The mixture was then filtered and transferred to a paddle mixer together with a binder in the form of 416 cc of a dispersion of an acrylic material in water, preferably in the form of a latex. The material was mixed for 6 hours and heat was supplied during the mixing operation to drive off the dispersant and produce a powder.

The powder was then rolled to form a compact of dimensions greater than the required finished product, and of density less than the required finished product. In particular, the density of the compact was arranged to be between 1.1 and 1.7 g/cc. The compact was then loaded directly into a nitriding furnace and heated to a temperature not exceeding 1300°C, typically 1280°C, in an atmosphere consisting of 90% nitrogen and 10% hydrogen by volume to remove the acrylic binder and nitride the silicon to form a porous, substantially all α-phase silicon nitride body. The porous, reaction bonded body thus produced was also arranged to be of dimensions greater than the required product and of density less than the required product.

In particular, the density of the reaction bonded, silicon nitride body was 1.7 to 2.7 g/cc, and in the example was 2.1 g/cc. The external surface of the body was then spray-coated with boron nitride powder and the coated body was subsequently introduced between graphite dies of the required finished shape, and was hot pressed to finished dimensions and density at 4,000 psi and 1750°C. In this particular example the density of the final product was 3.2 g/cc. It is to be appreciated that the increase in density of the silicon nitride body during the hot pressing process also results in the strength of the final product being increased.

Hexagonal Silicon Aluminum Oxynitride

G.Q. Weaver; U.S. Patent 3,837,871; September 24, 1974; assigned to Norton Company describes a quaternary compound silicon aluminum oxynitride having a hexagonal phenacite crystal structure. The quaternary compound is formed by reacting silicon oxynitride with an appreciable percentage of aluminum. The percent of the product which is converted to the hexagonal quaternary compound is believed to be 6 times the percentage of aluminum in the product, up to 15 parts Al/100 parts Si_2ON_2 when the silicon oxynitride in the product is largely converted to the hexagonal quaternary compound $Si_{2-x}ON_2$.

As more aluminum is added, the relative proportions of aluminum and silicon in the compound seem to change until the formula $SiAlON_2$ is reached at which point no additional aluminum can be accommodated in the crystal lattice. When the mixture is hot processed, the product has high strength and less than 1% porosity.

Example: Silicon oxynitride powder having the following properties was used as a starting material: 95% Si_2ON_2, remainder SiO_2 and β-Si_3N_4, 2 μ average particle size, about 0.1% CaO. 20 grams of this powder was mixed with 10 grams of aluminum having the following properties: atomized Al powder 99+% pure, 4.5 μ average particle size, low oxygen content, of the type normally used as rocket fuel. The mixture was added to a 1 quart tungsten carbide lined ball mill containing tungsten carbide balls. 450 cc of isopropanol was added and the mixture was milled for 20 hours.

The slurry was removed, the alcohol was driven off by heating to 90°C and the dried powder was screened through a 40 mesh screen. 10 grams of this powder was then loaded into a graphite mold of 1" i.d. and 3" long. Graphite plungers were placed in the mold and the powder was subjected to simultaneous heat and pressure in accordance with the following schedule: room temperature to 1200°C, pressure increased from 100 to 3,000 psi; pressure held constant at 3,000 psi while temperature raised to 1780°C. When the furnace had cooled, the piece was removed from the mold and was tested for strength, density, chemical composition and crystal structure. Test results are set forth as follows: cross bending strength (average of 40 tests), 67,000 psi; density, 3.10 g/cc; and phase composition by x-ray diffraction, Si_2ON_2 + hexagonal (phenacite) material.

Hot Pressed Silicon Nitride

For many applications of silicon nitride it is extremely important that the product have high strength at room temperature and high strength (or rupture modulus) at elevated temperature. This is particularly true where it is to be used as a high temperature structural material (e.g., turbine blade), bearing or other wearing surface, for such applications as gas turbines, Wankel engines and the like.

Accordingly, *D.W. Richerson and M.E. Washburn; U.S. Patent 3,836,374; Sept 17, 1974; assigned to Norton Company* describe a hot pressed silicon nitride product having high strength at room temperature as well as high strength at elevated temperature. The product has a flexural strength in excess of 100,000 psi at 20°C, in excess of 90,000 psi at 1200°C, in excess of 35,000 psi at 1375°C, and preferably above 45,000 psi at 1375°C. A preferred form of the product includes between 0.25 and 2.0 mg in the form of complex silicate, which silicate also contains a limited amount of iron, aluminum and calcium. The complex silicate has a liquidus above 1400°C. The product has a density of 3.1 to 3.3 g/cc, the total oxygen content of the product being less than 5%.

Aluminum Nitride and/or Silicon Nitride Composites

According to *K. Komeya, A. Tsuge, H. Inque and H. Murata; U.S. Patent 3,833,389; September 3, 1974; assigned to Tokyo Shibaura Electric Co., Ltd., Japan* heat resistant and strengthened composite materials are obtained by mixing, and sintering in a nonoxidizing atmosphere, the combination of powders of aluminum nitride and/or silicon nitride, with powders of an oxide of lanthanum, cerium, scandium, yttrium, and/or yttrium aluminum garnet, and with powders or whiskers of silicon carbide, boron nitride and/or carbon. The composite material so produced is characterized by high shock resistance and excellent mechanical strength.

Ordinary sintering is carried out at 1600° to 2000°C if the first component is AlN, and 1300° to 1900°C if the first component is Si_3N_4. The upper limit of the sintering temperature is determined by the sublimation and decomposition temperatures of the first component nitride and the SiC, BN or C. The lower sintering temperature limit is the progressive reaction temperature.

Ordinary sintering is preferred for complex shapes because of its lower cost, whereas hot press sintering is desirable where a completely closed object is intended to be prepared. Hot press sintering may be at 1300° to 2400°C and 20 to 1,500 kg/cm².

When ordinary sintering is effected, it is preferred to heat in an inert, nonoxidizing atmosphere, such as nitrogen or argon. Sintering can be effected in a single heating step or in multiple heating steps. The powder components may have particle sizes of 0.4 to 40 μ, and the whisker components may have diameters of 0.5 to 40 μ with lengths of 40 to 100 μ.

Example: 90% by weight of aluminum nitride powder, having an average particle size of 1.2 microns, 10% by weight of yttrium oxide powder, having an average particle size of 1.6 microns, and 0, 5, 10, 20, 30, 50% by volume of silicon carbide powder, having an average particle size of 7 microns, or silicon carbide whiskers having a diameter of 10 microns, a length of 40 to 100 microns, and 5% by weight of stearic acid, as a binder, were mixed.

Each mixture was shaped into a large number of columnar pieces, 13 mm in diameter and 20 mm long, at a molding pressure of 5,000 kg/cm². One of the test pieces thus formed was hot pressed in a hot pressing apparatus. Then the test piece was placed into a crucible made of aluminum nitride, and was packed with aluminum nitride powder. The test pieces were heated by the following rates of temperature rise in nitrogen gas atmosphere which was prepared by passing the gas through the crucible at a rate of 800 l/hr: room temperature to 300°C for 2 hours, 300° to 400°C for 4 hours, and 400° to 1800°C for 1 hour. After sintering for 2 hours at 1800°C, the sintered test pieces were permitted to self-cool. During sintering, the binder was decomposed and evaporated off. The physical properties of the test pieces sintered are shown in the table below.

Proportion of raw materials						
AlN	Y₂O₃	SiC powder	SiC whisker	density (%)	Bending strength (kg/mm²)	heat shock test***
weight %		volume %				
90*	10*	0**	0**	98.4	37.5	less than 10 times
90*	10*	5**	0**	97.3	34.0	20–40 times
90*	10*	10**	0**	95.2	30.1	50–100 times
90*	10*	20**	0**	92.7	25.5	more than 100 times
90*	10*	30**	0**	88.2	17.5	more than 100 times
90*	10*	50**	0**	85.1	15.2	more than 100 times
90*'	10*	0**	5**	95.8	35.2	50–100 times
90*	10*	0	10**	93.2	28.4	more than 100 times
90*	10*	0**	20**	90.5	21.8	more than 100 times
90*	10*	0**	30**	85.2	13.7	more than 100 times
90*	10*	0**	50**	80.3	10.5	more than 100 times

*Represents the amount charged as AlN-Y₂O₃ system
**Represents the amount of SiC added to AlN-Y₂O₃ system
***Represents the number of heat shocks before cracks appeared

As shown in the above table, if the amount if SiC added increases, the density decreases. Therefore, it is preferable that the amount of SiC used be less than 30% by volume if it is in the form of a powder, and less than 20% by volume if it is in the form of fibers or whiskers.

These samples were cut and shaped into 10 mm diameter by 5 mm (low cylinder shape), and were finished by mirror polishing. They were then subjected to a heat shock test by being placed in an electric furnace heated to 1000°C (in an atmosphere of air). After 5 minutes they were removed from the furnace, and cooled rapidly by being dipped into water.

A definite increase in heat shock resistance was noted when the composition contained SiN. The dimensions of the heating chamber, which was of the elevator type electric furnace, with a silicon carbide heater, was 300 x 300 x 400 millimeters.

Silicon Nitride Sintered with Yttrium Compound

G.E. Gazza; U.S. Patent 3,830,652; August 20, 1974; assigned to the U.S. Secretary of the Army describes the preparation of high strength, high density silicon nitride by the addition of 1.0 to 3.5% by weight of yttrium in the form of a compound to silicon nitride powder and pressing the material at 1750° to 1800°C and at uniaxial pressures of 6,000 to 7,000 psi. The yttrium compound may be yttrium oxide, yttrium chloride or yttrium nitrate. When the yttrium compound is yttrium oxide, the amount of yttrium oxide is 1.25 to 4.45 by weight of the silicon nitride.

Example: 30 grams of −100 mesh silicon nitride and 0.6 gram (2% by weight) of yttrium oxide powder (−325 mesh) were mixed together with a small amount of ethyl alcohol to form a slurry. The slurry was placed in a mill jar together with several tungsten carbide balls and mixed for 24 hours. The slurry was then removed and dried under an infrared lamp. 18 grams of the dried powder was placed in a graphite die whose walls and plunger face were earlier coated with a boron nitride slurry to prevent a reaction between the die and the silicon nitride powder. The loaded die was heated, under 1 atm of nitrogen gas to 1500°C.

A uniaxial pressure of 6,500 psi was applied and the temperature was raised to 1750°C and held for 2 hours. Pressure was released after the 2 hour hold time and the die was slowly cooled. The hot pressed specimen was removed from the die and cleaned on its surfaces. The bulk density of the specimen was determined to be 3.25 g/cc based upon the liquid immersion technique. Bend specimens were machined from the sample and tested at room temperature in four point bending. Based upon the testing, an average modulus of rupture value of 102,000 pounds per square inch was calculated. A piece of the specimen was polished and etched and magnified 11,600 times through a Phillips electron microscope.

High Density Silicon Nitride

G.K. Layden; U.S. Patent 3,821,005; June 28, 1974; assigned to United Aircraft Corporation has shown that the addition of a certain combination of sintering aids to silicon nitride powders will not only permit complete densification of the powders at reasonable conditions of temperature and pressure, but will also provide articles which exhibit superior strengths at elevated temperatures. These sintering aids are certain mixtures of aluminum and/or gallium together with phosphorus and/or arsenic where the metal ion ratio (aluminum + gallium)/(phosphorus + arsenic) is unity or near unity.

Accordingly, there is added to the silicon nitride powders, such additives as $AlPO_4$, $GaPO_4$, $AlAsO_4$, $GaAsO_4$, AlP, GaP, $AlAs$, and/or $GaAs$. In addition many other substances will serve the desired purpose when introduced in the appropiate mixtures, such as $(3AlN + P_3N_5)$ and $(3GaN + P_3N_5)$.

Example 1: 5 wt % $GaPO_4$ was added to α-Si_3Ni_4 powder and the mixture was ball milled for 72 hours using Al_2O_3 grinding balls and a methanol grinding medium. The mixture was then dried and a 12 gram sample was hot pressed in a graphite mold in a nitrogen atmosphere at 4,000 psi and 1650°C. The sample was then ground smooth and cut into several test specimens approximately 0.1" by 0.2" by 1.5". The modulus of rupture (three point bend strength) of the specimens was determined at both room temperature and at 1300°C. The values were: room temperature, 100,000 psi; and at 1300°C, 42,000 psi.

Example 2: 5 wt % $AlPO_4$ was added to Si_3N_4 and samples were prepared and tested as described in Example 1. The results were: room temperature, 80,000 psi; and 1300°C, 50,000 psi.

Silicon Nitride Bonded with Butyl Methacrylate and Trichloroethylene

According to *E.R.W. May; U.S. Patent 3,819,786; June 25, 1974; assigned to National Research Development Corporation, England* silicon nitride components are formed by hot milling and then shaping a dough-like material comprising silicon powder and a suitable organic chemical binder consisting of butyl methacrylate and trichloroethylene and subsequently subjecting the form to a nitriding treatment. The process is carried out by the steps listed below.

Production of Flexible Sheet: The materials required are silicon powder able to pass 400 mesh sieve BS with purity as required. The binder is 35 to 40% butyl methacrylate and 60 to 65% trichloroethylene. The silicon powder and binder are mixed together in the ratio of 65% powder to 35% binder to the consistency of a stiff dough, spread on trays to a depth of approximately ¼", and allowed to stand at room temperature until 20% (by weight) of the trichloroethylene present in the binder has evaporated.

Semidried cake is then repeatedly hot milled in a roller type mill, roll temperature 45° to 50°C, until a consolidated sheet of the required thickness is obtained. A sheet so produced is very flexible when warmed to 50°C, while it is fairly stiff when cold.

Formation of Components: The flexible sheet is cut into strips or other suitable shapes and formed into components using form tools heated to 45° to 50°C. One example is the use of toothed rolls to produce corrugated strips, and the use of dies or formers to produce shapes which may be domed, or box-like.

Spirals, and similar shapes are easily handmade by wrapping strips of the flexible sheet around warmed formers. Tubes may be made from rectangular sheets, the seam joint being effected by rolling until the bond line is eliminated.

Prenitriding Treatment: The components are then thoroughly freed from remaining solvent. One suitable method is to place formed components in a suitable vessel and heat under reduced pressure. The time required is 2 hours at 130°C for components with a thickness of 0.02" to 40 hours at 130°C for components with a thickness of 0.2".

Large, delicate articles may require support to prevent collapse during the early stages of the process. Too rapid a rate of temperature rise in the apparatus encourages the formation of blisters on the surface of components. A satisfactory schedule for the prenitriding process is: temperature rise 15°C/hr from room temperature to 130°C, and pressure in the vessel maintained at 20" Hg. After the prenitriding treatment components are hard, dry, easily handled without damage, and will not distort during subsequent nitriding.

Nitriding Process: Stage 1, Removal of Binder Prior to Nitridation — Furnace temperature is raised to 1000°C at 50°C/hr. A continuous flow of nitrogen is maintained throughout this period to ensure that decomposition products are removed from the furnace.

Stage 2, Nitridation — A static pressure of nitrogen of 1 psi is maintained in the furnace during the nitriding process. Nitriding times are governed by the cross-sectional area of components to be nitrided. A typical schedule is as follows.

Thickness of component, inch	0.2
Time at 1350°C, hr	20
Time at 1450°C, hr	10

Some properties of this silicon nitride are: specimen dimensions, 0.2" by 0.2" by 2"; bulk density (green), 1.25 g/cc; bulk density (nitrided), 2.24 g/cc; and cross breaking strength at room temperature, 17,500 psi (3 point loading).

Machinable Refractories Containing Boron Nitride

According to *A. Lipp; U.S. Patent 3,813,252; May 28, 1974; assigned to Electroschmelzwerk Kempten GmbH, West Germany* machinable refractory articles are produced by incorporating 10 to 20% by weight, of boron nitride (based on the weight of the article) into a refractory comprising aluminum nitride, or silicon nitride, or preferably aluminum boride, rendered electrically conductive by the presence of an electrically conductive material selected from graphite, boron carbide, titanium carbide, zirconium carbide, chromium carbide, silicon carbide, titanium boride, zirconium boride, chromium boride, silicon boride, beryllium boride, magnesium boride and calcium boride, the ratio of refractory material to electrically conductive material being 20 to 80% by weight of the former to 80 to 20% by weight of the latter.

In preparing the machinable refractory articles the operation is carried out most effectively when the particle size of the refractory material is 40 to 5 microns, and preferably 15 microns for aluminum nitride and silicon nitride; the electrically conductive material is 90 to 15 microns and finer, and preferably 50 microns and finer; and the boron nitride is 20 to 5 microns or smaller.

A convenient method of producing the machinable refractory comprises thoroughly homogenizing a mixture of the above specified refractory material, electrically conductive material and boron nitride in the proportions and particle sizes above specified and then forming the mixture into the desired shape by hot-pressing at 1500° to 1900°C and 50 to 400 kg/cm².

Example 1: A thoroughly homogenized mixture of 27% by weight of aluminum nitride, 63% by weight of titanium boride and 10% by weight of boron nitride, the titanium boride being of a particle size of 50 microns and finer and the particle size of the aluminum nitride being 15 microns and smaller and that of the boron nitride being 20 microns and smaller, was hot-pressed at 1800°C and 200 kg/cm². The machinability was found to be materially improved by the incorporation of the boron nitride.

The hot-pressed article, in the form of a crucible, was found to be eminently satisfactory for the vaporization of aluminum and other metals, and to be free from the disadvantages of the prior art crucibles used for this purpose. They were found to give satisfactory operating results for, on an average, 50% longer operating time than the best commercially known crucibles used for such purposes and at the same time to require less energy consumption than previous types of evaporation crucibles.

Example 2: A mixture of 60% by weight of aluminum boride, 25% by weight of titanium carbide and 15% by weight of boron nitride, the aluminum boride being of a particle size of 40 microns or finer, the titanium carbide being of a particle size of 50 microns or finer, and the boron nitride being of a particle size of 10 microns or finer, was hot-pressed in a boron nitride form under 100 kg/cm^2, within the above specified temperature range to produce sintered geometrical shapes of desired form. The machinability of the shapes was appreciably improved by the incorporation of the boron nitride. The products thus produced were less expensive and more easily produced than those produced by prior art methods.

Example 3: A mixture of 55% by weight of zinc boride having a particle size of 40 microns or smaller, 25% by weight of aluminum nitride having a particle size of 10 microns or smaller and 20% by weight of boron nitride having a particle size of 10 microns and finer, was subjected to hot-pressing in a graphite mold under 100 kg/cm^2, at a temperature within the range above specified to produce machinable geometric forms of desired size and shape. The products thus formed were found to be less expensive, more practical to use and readily machinable.

Silicon Nitride Protected with Silica and Boric Acid

According to *J.W. Henney and J.W.S. Jones; U.S. Patent 3,811,928; May 21, 1974; assigned to United Kingdom Atomic Energy Authority, England* shaped ceramic objects of silicon nitride are protected from oxidation by a coating comprising silica and boric oxide. The coating may be formed by heating a shaped mixture of silicon nitride and boron nitride in air to convert part to silica and boric oxide and causing a silica/boric oxide glass to flux and protect the object from further oxidation. The boron may be incorporated during oxidation but is preferably added to the silicon before nitriding.

In particular, Fe, Mg, Al, Al_2O_3, MgO, CaO, BaO, ZnO, or ZrO_2, may be added to the silicon nitride, or more preferably to the silicon prior to nitriding, in an amount of up to 10% by weight on the weight of the final product. It is believed that CaO and BaO have the effect of suppressing the volatilization of B_2O_3. It is believed that the listed materials can have the effect of crystallizing what would otherwise normally be a SiO_2/B_2O_3 glass. Provided that the crystalline material does not undergo a phase and volume change on cooling (this is the disadvantage of pure SiO_2), this modification can have the effect of improving the high temperature properties of the ceramic.

However, it is important that the coefficient of thermal expansion of the second phase must be a reasonable match for the Si_3N_4 ($2.5 \times 10^{-6}/°C$) if the body is to retain a high strength. The thermal expansion of SiO_2 itself is very low (apart from the sudden volume change at 230°C) but is substantially increased by small addition of most other glass-making additives. B_2O_3 is peculiar in that additions do not increase, and may even decrease, the thermal expansion of SiO_2. The SiO_2/B_2O_3 coating should preferably not have a coefficient of thermal expansion above $3 \times 10^{-6}/°C$.

Thermal Shock Resistant Boron Nitride

According to *W.M. Fassell; U.S. Patent 3,811,900; May 21, 1974; assigned to Philco-Ford Corporation* highly ordered boron nitride flakes are mixed homogeneously with a refractory material and formed into a refractory article. The boron flakes intersect and arrest microcracks formed in the article by thermal shock or other stress producing conditions.

Refractory materials particularly useful include magnesia, alumina, beryllia, thoria, silica and zirconia. Other refractory materials such as mullite, corundum, spodumene, cordierite, etc. also can be used with good results. A good combination of economy, thermal shock resistance, and strength is achieved by mixing homogeneously 1 to 20 wt % of the boron nitride flakes into the refractory materials. High strength and limited thermal shock resistance can be acquired with 2 to 5 wt % of the boron nitride flakes.

Increased thermal shock resistance with slightly decreased strength is achieved by increasing the weight proportion of the boron nitride flakes. Mixtures containing above 20 wt % of the boron nitride flakes can be used in applications requiring extremely high thermal shock resistance and relatively low strength. The mixtures preferably are hot-pressed at a temperature of at least two-thirds of the mixture melting point to densities exceeding 80% of theoretical.

Pyrolytic boron nitride flakes having high purity, high density and a highly ordered crystal structure can be obtained by heating a gaseous mixture of boron trichloride and ammonia to at least 2000°C in the presence of a graphite substrate. The ammonia and the boron trichloride gases usually are in approximately equal quantities. Pyrolytic boron nitride deposits on the graphite substrate. The boron nitride is ground to the desired flake size and mixed homogeneously with an appropriate amount of the refractory material. Hot-pressing techniques can be used to form the resulting mixture.

Microscopic investigations reveal that subjecting the refractory articles to thermal shock conditions produces microcracks. The boron nitride flakes intersect the microcracks and effectively halt propagation, thereby preventing massive thermal shock failure of the article. Complete intersection of a microcrack by a boron nitride flake appears to be essential to achieving the high thermal shock properties. This criteria seems to establish the lower limit of 0.002" for the greater surface dimension of the boron nitride flakes, since flakes below this value have a relatively low probability of completely intersecting microcracks. The best combination of thermal shock resistance and strength is provided by flakes having a greater surface dimension of 0.003" (76 microns) to 0.030" (760 microns).

Example 1: 10 weight parts pyrolytic boron nitride are ground to a flake having a greater surface dimension of 0.010" and mixed with 90 weight parts alumina (99.18% pure). The mixed powder is placed in a disc shaped graphite die and induction heated to 1650°C. A compacting force of 3,000 lb/in² is applied to the heated powder for 20 minutes. The die and pressed disc are cooled to room temperature before the disc is removed from the die.

The resulting disc is ¼" thick and has a diameter of 1½". Its density is 98% of theoretical. One edge of the disc is heated with an oxygen-natural gas torch until rounding of the edge is visible (2000°C). The disc then is dropped in cold water. No significant cracks can be detected until after 10 repetitions of this test.

Example 2: 10 weight parts pyrolytic boron nitride are ground to a greater surface dimension of 0.016" (400 microns) and mixed with 90 weight parts fine grained beryllia having an average grain diameter of 15 microns. Hot-pressing the mixture as described in Example 1 produces a disc having a nominal density of 2.75 g/cm^3.

Various tests show that the material has remarkable thermal shock resistance. Propagation of thermally induced shock waves is arrested by the boron nitride flakes, thereby enabling the material to withstand severe heating without failure. Specimens of this material have withstood simulated reentry environmental heating such as 1,300 Btu/ft^2-sec at a total enthalpy of 2,600 Btu/lb and a model stagnation pressure of 6 atm. Increasing the boron nitride proportion to 20 wt % produces even better themal shock properties.

Low Pressure Sintering of Boron Nitride

According to *A. Muta, Y. Hayakawa and M. Manaka; U.S. Patent 3,720,740; March 13, 1973; assigned to Hitachi Seisakusto KK and Hitachi Kasei Kogyi KK, Japan* boron nitride (BN) powder is subjected to cold-pressure molding to produce a shaped molded article, the thus molded article is then confined in a sintering mold where the BN article is subjected to heat-sintering under conditions such that the free expansion of the article which occurs during heating is restricted by the sintering mold. A boron nitride article of high density and strength is produced where the only pressure supplied during the sintering operation is that of the substantially nonmoving walls of the sintering mold.

The internal dimensions of the sintering mold should be as close as possible to those of the cold-pressed BN article, so that the clearance between the article and the walls of the mold is as near to zero as practical. Since the coefficient of the deformation due to the expansion of a cold-pressed BN article is 5% at 2100°C, the clearance between the mold used for sinter-molding and the article to be sintered is 5% or less of each dimension of the molded article in any direction. The mold used for sintering the BN article is made of a material such as graphite, having a low coefficient of expansion, and in particular a coefficient of expansion which is lower than that of boron nitride.

Various means may be employed for cold-press molding the BN article from BN powder prior to sintering Generally, the cold-pressure molding is carried out under a pressure of 1 to 4 tons/cm^2 and more preferably 2 to 3 tons/cm^2. Prior to cold-press molding the BN powder to form the cold-pressed BN article, a binder may be combined with the powder if desired, particularly organic binders such as polyethylene glycol and the like.

The crystal structure of the boron nitride article formed by cold-press molding is highly strained and distorted due to the application of high pressure. Accordingly, the distortion and strains on the crystal structure of the BN are relieved by heating the article at high temperature while the article is confined so that its free expansion which has been found to take place when the article is heated, is restricted. It has been found that by carrying out the heat-sintering while the article is confined so that it can expand no more than about 5% in any direction, and preferably less, an article of high density and high strength is produced.

Example: Cold-pressure molded articles were prepared by molding 23 grams of BN powder by means of a hydrostatic press under pressures of 2 and 3 tons/cm^2, respectively. Each molded sample was then machined into a cylinder of 24 mm diameter and 40 mm length. Two machined samples respectively molded under the pressures of 2 and 3 tons/cm² were selected and inserted into respective graphite molds, each of which was of cylindrical cup shape with a cylindrical wall thickness of 6 mm and a bottom thickness of 20 mm, and then each mold was closed by a screw plug with a flange of 15 mm thickness and a threaded plug of 10 mm length which was screwed into corresponding screw threads tapped in the mold at its open end.

The two molds containing the machined samples were then heated in a Tanmann furnace to 1500°C. In each case, the clearance between the BN molded sample prior to heating and the graphite mold was 0.2 mm or less in all directions. On the other hand, the remaining BN press-molded samples were sintered at 1500°C according to the conventional method without being placed in molds.

The measured values of specific gravities and transverse strengths relating to the BN sinter-molded articles produced by the above described two methods are shown in the following table in which the molded samples designated by numbers 3R and 4R are those produced by the conventional method.

Molded Sample No.	Molding Pressure ton/cm²	Bulk Specific Gravity of Pressed Article	Bulk Specific Gravity of Sintered Article	Transverse Strength of Sintered Article, kg/cm²
1	2	1.546	1.417	181.8
2	3	1.618	1.491	231.8
3R	2	1.543	1.346	141.2
4R	3	1.596	1.370	180.0

Removal of Carbon from Metal Nitrides

According to *C.A. Morgan, S.G. Arber and O.W.J. Young; U.S. Patent 3,718,490; February 27, 1973; assigned to United States Borax & Chemical Corporation* a refractory metal nitride containing up to 30% by weight of carbon is reacted with boric oxide in an atmosphere of nitrogen at 1200° to 2050°C. Removal of carbon takes place according the the equation

$$B_2O_3 + 3C + N_2 \rightarrow 2BN + 3CO$$

Thus, the quantity of boric oxide should be sufficient to react with all the carbon present. The boric oxide may be present in the form of B_2O_3 added as such, or may be derived from a boron compound which yields B_2O_3 during the reaction, such as boric acid. Aluminum nitride is the preferred refractory metal nitride.

The final product will therefore contain boron nitride as well as the refractory metal nitride. This is advantageous for many industrial processes, particularly those for the manufacture of vacuum evaporator bars, which require the joint presence of boron nitride and aluminum nitride during the process.

Thus, this process provides a single reaction which both removes carbon and adds boron nitride to the product. Another advantage of the process is that, in the case of aluminum nitride, the resultant product contains very little residual Al_2O_3. During the initial reaction stage alumina is carbothermically reduced according to the equation

$$Al_2O_3 + 3C + N_2 \rightarrow 3CO + 2AlN$$

A content of carbon of about 4% in the aluminum nitride is usually adequate to reduce unnitrided alumina. Extra carbon may be added to increase the amount of boron nitride in the product. A larger amount of carbon may be added when it is desired to remove all unreacted alumina. This high excess of carbon is then removed by reaction with boric oxide.

Impure refractory metal nitride such as aluminum nitride (containing up to 30% carbon) and boric oxide (or a compound which will yield boric oxide) are ground and mixed together. A small amount of paraffin preferably is used as a binder. This mixture is then compacted and fired in an atmosphere of nitrogen at 1600° to 1900°C. The process can be used to remove carbon from other metal nitrides, such as the nitrides of zirconium, vanadium, tantalum, hafnium, and niobium, especially those obtained by a carbothermic process.

Example 1: Boric nitride (15.2 lb) containing a total of 6.1% free and combined carbon was mixed with boric oxide (7.3 lb) and compacted. The compacted material (18.6 lb) was heated in an atmosphere of nitrogen for 3 hours at 1650° to 2000°C. The resultant product gave an analysis of: N, 55.7%; B, 43.4%, C, 0.12%. Theoretical for BN—B, 44%; N, 56%.

Example 2: Titanium nitride (100 grams) containing 3.5% free and combined carbon was mixed with boric oxide (15 grams) and compacted. The compacted material was heated in a nitrogen atmosphere for 3 hours at 1600° to 2000°C. The resultant product analyzed as follows: Ti, 74.6%; N, 19.1%; B, 1.7%; C, 0.2%.

Liquid Nitride Coating

R. Watanabe and M. Watanabe; U.S. Patent 3,709,723; January 9, 1973; assigned to Tokyo Shibaura Electric Co., Ltd., Japan describe liquid inorganic refractories and a process for imparting fire resistance on various base materials by applying the liquid compositions to the surfaces of the various base materials and by subsequently hardening the applied liquid compositions. The inorganic refractory liquid compositions contain as main constituents (1) metal nitride selected from aluminum nitride, boron nitride, titanium nitride, iron nitride, thallium nitride, and zirconium nitride, and (2) phosphoric acid or ammonium phosphates. The liquid compositions are applied to the surface of a base material, thereafter hardened at room temperature to 150°C so as to obtain an inorganic refractory coating.

If a foaming agent such as aluminum powder is mixed in the composition, the better foamed body will be obtained. Further, ferric oxide, volatile fluid such as water, alcohol, etc., and heat resistant inorganic substances such as silica, alumina, etc. may be added to the composition.

The refractory liquid compositions are used as paints, adhesives, castable refractory cement, fire-proof construction material, fire-proof layered material, or fire-proof electrically insulating material. Further, other ordinary inorganic refractory materials, e.g., mica scrap, asbestos, silicon carbide, metallic powders, etc. may be used as additives, in addition to silica and alumina mentioned above.

Example 1: 19 wt % aluminum nitride was added to 43 wt % phosphoric acid (85% concentration) and 38 wt % silica, and then the mixture was sufficiently stirred. The mixture is a viscous fluid and has best extensibility and adhesiveness for use as paints and adhesives. This mixture was coated on plates of metal, glass, asbestos and then dried at 50°C for 10 minutes to form a thin rigid coating with aluminum phosphate as a main constitutent. This coating was heated above 1000°C with an oxyhydrogen burner, but it neither burned nor cracked. The adhesive strength between the base material and the coated film was so great that even a sharp-edged tool could not scrape off the coating.

Example 2: 12 wt % ethyl alcohol was added to the mixture of 22 wt % aluminum nitride, 33 wt % phosphoric acid, and 33 wt % silica, and a resultant viscous fluid was coated on the same base material as in Example 1. After drying at 100°C for 10 minutes, there was obtained a thin rigid coating containing aluminum phosphate as a main constituent and having excellent heat resistance and adhesiveness as was the case with Example 1.

In case alcohol was not added, the mixture hardened when left alone at room temperature for 24 hours. In case that alcohol was added and then left alone for many hours, the mixture did not harden and maintained an unreacted condition until the alcohol had completely vaporized. This characteristic of the mixture is of great advantage in the case of preservation. Needless to say, alcohol completely vaporizes upon hardening and does not affect incombustibility adversely.

BORIDES

TiB_2-BN-TiH_2 Electrically Conductive Boat

Electrically conductive boats are commonly used in the vacuum deposition of metals, e.g., aluminum, onto suitable substrates, such as paper or plastic film. Electric current is passed through the boat in order to heat it to a temperature at which metal will evaporate. The useful life of such boats is generally quite short because of, among other things, the high temperature of operation. The combination of high temperature, corrosiveness of the metal being evaporated and thermal cycling that may occur during the life of the boat can cause cracks to occur in the boat.

Accordingly, *E.M. Passmore; U.S. Patent 3,928,244; December 23, 1975; assigned to GTE Sylvania Incorporated* describes an electrically conductive refractory boat, for the vacuum deposition of metals consisting of 5 to 65 vol % of BN, 20 to 50 vol % of TiB_2, 1 to 5 vol % of TiH_2, and 0 to 30 vol % of AlN.

Example 1: 254 grams (41 vol %) of TiB_2, 178 grams (57 vol %) of BN and 14 grams (2 vol %) of TiH_2 were thoroughly mixed together and then hot-pressed in a vacuum chamber inside a graphite die mold at 2050°C and 4,000 psi for 4 hours to yield a disc $4^3/_{16}$" in diameter by $^5/_8$" thick. The disc had a density of 3.06 grams per cubic centimeter, which is 97.7% of the absolute theoretical density of 3.235 g/cc.

A similar disc was prepared of a prior art composition using the same hot-pressing process, the composition consisting of 267 grams TiB_2, 169 grams BN and 9 grams of H_3BO_3. Two evaporation boats having overall dimensions of 0.375" by 0.750" by 3.00" were machined from each composition. Cavities were machined in the boats to hold the metal evaporant. The boats were then tested under identical conditions by self-resistance heating to the evaporation temperature range for Al (1400° to 1700°C) in a vacuum chamber, and 0.060" diameter Al wire was fed thereinto.

5 grams of Al were so evaporated at a rate of 1.0 g/min, after which the rate was increased successively to 2.0, 3.0, 4.0, 5.0, and 6.0 g/min with the evaporation of 50 grams at each rate. Thus, the test consisted of the evaporation of a total of 300 grams of Al at successively increasing rates from 1 to 6 g/min. After the test, the boats were examined for evidence of cracking. In the two boats made in accordance with this process, one had no cracks and the other had only very slight cracks. However, the two prior art boats were both cracked to a substantial extent.

It is believed that the increased resistance to cracking may be due to the reduced porosity, as shown by the higher percentage of theoretical density, that results from the use of TiH_2. The composition of this process had a density that was 97.7% of theoretical absolute density, while the prior art boat's density was only 94.5% of theoretical absolute density.

Example 2: In order to determine whether TiH_2 also increased the density of composites containing AlN, a second pair of composites was made in a similar manner to that described above. A composite was made by mixing together 165 grams (31.5 vol %) TiB_2, 113.5 grams (43.2 vol %) BN, 88.5 grams (23.3 vol %) AlN and 12 grams (2 vol %) TiH_2 and hot-pressing for 4 hours at 2050°C under 4,000 psi pressure.

Simultaneously, a prior art composite consisting of 176 grams TiB_2, 113.5 grams BN, and 88.5 grams of AlN was also hot-pressed under the same conditions. Again, the composite of this process had a significantly higher percentage of absolute theoretical density than the prior art composite, 97.0 versus 94.5%.

These tests show that the enhanced density conferred by the addition of TiH_2 is attained whether or not AlN is included as a constituent of the composite. In fact, the benefits conferred by the addition of TiH_2 are also not limited to composites containing TiB_2, but are specifically applicable to any composite containing BN or to the latter alone.

Example 3: The use of TiH_2 in the fabrication of BN is provided by the following example. 636 grams of a BN powder which had been previously shown to be incapable of consolidation into a sound, dense refractory body by hot-pressing (or any other means) was mixed thoroughly with 328 grams of TiH_2 powder. This composition equals 80 vol % BN, 20 vol % TiH_2. The mixed powders were hot-pressed as described above at 2050°C and 2,500 psi for 4 hours into a sound refractory body 5.03" diameter by 1.09" thick. The density was determined to be 2.53 g/cc, corresponding to 93% of the absolute theoretical density of 2.72 g/cc.

Attempts to make sound BN bodies using lesser TiH_2 additions were not successful. One of these comprised the equivalent of 10 vol % TiH_2 (half that of the above composite) and exhibited extensive cracks normal to the hot-pressing direction after simultaneous consolidation under the same conditions used above. Crucibles were machined from both composites, and the latter showed much thermal shock cracking in tests involving external heating and evaporation of Al.

In contrast, the former (20 vol % TiH$_2$) composite showed no cracking. A third composite comprising 5 vol % TiH$_2$ was also made at the same time under the same conditions, and completely fell apart (disintegrated) after removal from the graphite die. Thus, it is evident that the TiH$_2$ addition is the bonding agent, and that it must be present in sufficient amount (more than 10 vol % TiH$_2$ for this grade of BN) in order to attain a sound body.

TiB$_2$-B$_4$C-SiC-Si Composites

According to *W.O. Bailey, C.H. McMurtry and B.R. Miccioli; U.S. Patents 3,859,399; January 7, 1975; and 3,808,012; April 30, 1974; both assigned to The Carborundum Company* hard, dense composite ceramic bodies of titanium diboride, boron carbide, silicon carbide and silicon, are produced by preparing a substantially homogeneous initial mixture of granular titanium diboride and granular boron carbide in a proportion of 5:95 to 95:5 and a temporary binder; forming the initial mixture into a desired shape by pressing, extruding, investment or slip casting or any other suitable method; wetting the temporary binder, if necessary, to impart sufficient coherence to the shaped green body to permit further processing; and siliconizing the coherent green body by heating it in contact with silicon to a siliconizing temperature above the melting point of silicon.

Thereupon, the silicon in the molten state, infiltrates the body and undergoes a rather complex reaction with some of the boron carbide, producing some silicon carbide in situ. The titanium diboride apparently does not undergo any reaction. To the extent that interstices exist in the body between the titanium diboride, the remaining boron carbide and the newly-formed silicon carbide, the interstitial space is permeated by free silicon. The silicon carbide and free silicon bond the other materials and an extremely hard and dense body is formed.

The temporary binder employed may be such as to be completely dissipated during the siliconizing heating cycle; or it may be a carbonizable material which will leave a carbon residue in the body upon heating, in which case the silicon will also react with substantially all of the residual carbon to produce silicon carbide, and accordingly the resulting body will generally have a somewhat higher silicon carbide content and a somewhat lower free silicon content. The same result may be obtained by incorporating in the initial mixture a small amount, i.e., up to 10% of the combined weight of the titanium diboride and boron carbide, of finely divided carbon of any suitable variety such as powdered graphite. If desired, both finely divided carbon and a carbonizable binder may be included in the initial mixture.

The hard dense composite ceramic bodies consist of 2 to 80% titanium diboride, 2 to 70% boron carbide, 5 to 30% silicon carbide, and 3 to 20% free silicon, the precise composition of a given body depending primarily upon the composition of the initial mixture. Such composite bodies have a 2.6 to 4.1 SG, increasing with increasing titanium diboride content.

In addition to being hard and dense, the composite ceramic bodies possess many other desirable properties, being refractory, tough, wear-resistant, abrasion-resistant, and resistant to most acids and alkalis. The oxidation resistance of the bodies tends to increase with increasing titanium diboride content, bodies of high titanium diboride content having particularly outstanding oxidation resistance. These desirable properties render the bodies useful in a wide variety of wear-resistant and other applications, including, extrusion dies, sandblast nozzles, cutting tool tips, abrasives, suction box covers for paper-making machines and the like.

The bodies are characterized by a high Young's modulus of elasticity ranging from 3×10^6 kg/cm^2 to 4×10^6 kg/cm^2 which, together with their other desirable properties, renders the bodies, in suitable shapes, highly useful as personnel, vehicular and aircraft armor. It has also been found that the bodies, especially those having a relatively high titanium diboride content, are quite electrically conductive and extremely resistant to corrosion by molten aluminum and aluminum alloys, thus they find utility as current conducting elements for use in contact with molten aluminum and alloys thereof, such as electrodes for refining aluminum. They also find utility as various parts of pumps used for pumping molten aluminum and alloys thereof, such as pistons, cylinders, impellers, bearings, and the like.

When the bodies are to be used in armor or aluminum applications, it is preferred that the proportion of titanium diboride to boron carbide in the initial mixture be at least 65:35; and it is also preferred, especially for armor, that this proportion not exceed 85:15, so that sufficient boron carbide is present to result in the production of enough silicon carbide to contribute to a particularly strong bonding phase.

The preferred composite bodies are produced from initial mixtures containing titanium diboride and boron carbide in this preferred range of proportions, and consist of 45 to 75% titanium diboride, 10 to 30% boron carbide, 8 to 30% silicon carbide, and 3 to 20% free silicon. Such composite bodies have a specific gravity of 3.1 to 3.8, being sufficiently light to be useful as armor even in many applications where light weight is an important factor.

Example: A mixture is prepared consisting of 8,500 grams of granular titanium diboride having a particle size of 45 microns and finer and 1,500 grams of granular boron carbide comprising 45% with a median particle size of 100 microns and ranging from 70 to 140 microns, 23% with a median particle size of 65 microns ranging from 40 to 100 microns, and 32% with a particle size of 10 microns and finer. Thereto is added 1,100 grams of a temporary carbonizable binder consisting of 600 grams of a liquid thermosetting phenol-formaldehyde resin and 500 grams of furfural, the latter also serving to add plasticity to the mix.

The mixture is blended until it is substantially homogeneous and the resulting mix, or initial mixture, is passed through a coarse screen to break up any agglomerates. 1,800 grams of the mix is placed in a steel mold 9" square (22.8 cm square) and compressed at 6,000 psi (420 kg/cm^2) to form a plate 9" square by 0.82" (2.08 cm) thick. The piece is placed in an oven and heated at 60°C for 16 hours, at 100°C for 24 hours, and finally at 180°C for 24 hours, to set the temporary binder, the resulting coherent green body having a specific gravity of 2.85.

The green body is placed on a graphite supporting plate in the chamber of an induction heated furnace, and 1,530 grams of granular silicon is distributed evenly over the surface of the green body. The furnace chamber is evacuated to a pressure of 50 microns of mercury, and the power source to the induction coils of the furnace is turned on. The reduced pressure is maintained throughout the heating cycle.

An optical pyrometer sighted on the piece is used to ascertain the temperature of the piece as the temperature rises. The temperature of the piece is brought to 1470°C; the silicon melts, and the molten silicon infiltrates the piece quite abruptly and reacts with some of the boron carbide and with substantially all of the residual carbon from the binder to produce silicon carbide. The power is immediately turned off and the furnace and its contents are permitted to cool to room temperature.

The ceramic plate thus produced has a specific gravity of 3.5, an electrical re-sistivity of 5×10^{-4} ohm-cm, a flexural strength of 1,400 kg/cm^2, a Young's modulus of elasticity of 3.4×10^6 kg/cm^2, and a shear modulus of 1.7×10^6 kg/cm^2. Elemental analysis for total titanium, total boron, and free silicon indicates that the piece consists of 66% titanium diboride, 12% boron carbide, 11% silicon car-bide, and 11% free silicon. X-ray diffraction analysis using monochromatic copper K-alpha radiation indicates that boron carbide present is of two types, the first type having a diffraction pattern corresponding to normal B_4C, and the second type having a diffraction pattern of boron carbide with an expanded lattice, this second type apparently being a boron carbide type solid solution.

The x-ray diffraction analysis also indicated the presence of titanium diboride, silicon carbide, and free silicon. The material exhibits excellent oxidation resis-tance, a portion of the plate showing substantially no weight change upon ex-posure to air at 1000°C for 125 hours. No effect is observed upon immersion of a portion of the plate in molten aluminum at 700°C for 100 hours, the same result being obtained when the aluminum is replaced with an aluminum alloy containing 3.5% copper, 7.5 to 9.5% silicon and 0.1 to 0.5% magnesium or with an aluminum alloy containing 1.6% copper, 2.5% magnesium, 5.6% zinc and 0.3% chromium. The effectiveness of the piece as armor is evidenced by the fact that it successfully resists a 0.50 caliber steel core armor piercing projectile fired at point blank range.

Diborides of Ti, Zr, Hf, Nb and Ta

According to *E.V. Clougherty, L. Kaufman and D. Kalish; U.S. Patent 3,775,137; November 27, 1973; assigned to Man-Labs, Incorporated; and E.V. Clougherty and D. Kalish; U.S. Patent 3,775,138; November 27, 1973* diboride materials containing TiB_2, ZrB_2, HfB_2, NbB_2, TaB_2 or mixtures thereof are produced by adding 10 to 35% by volume of SiC, and preferably 20% by volume SiC. Im-proved properties include lower fabrication temperature, grain size control, oxi-dation resistance, mechanical integrity, strength to density ratio, thermal stress resistance, and the like.

Example: A number of hot-pressed articles in the shape of cylinders were made with mixtures of commercially available zirconium diboride powder and commer-cially available silicon carbide powder. Zirconium diboride powder, 325 mesh or finer, and the silicon carbide powder, 5 micron average particle size, were mixed and dry milled in the proportions of 88% by weight zirconium diboride and 12% by weight silicon carbide, this being equivalent to 80% by volume zir-conium diboride and 20% by volume silicon carbide.

The mixed powders were transferred to a hot-press, and were hot-pressed at 4,000 psi for 2 hours. Various pressings were made at 2150°C, 2100°C, at 2050°C and at 2000°C. In all cases, densities higher than 95% of theoretical density were achieved, and in most cases virtually 100% density was achieved. In comparative procedures, zirconium diboride powder alone was pressed under the same conditions, and in each case, the zirconium diboride hot-pressed article was less dense than the ZrB_2-SiC hot-pressed article.

Hot-pressed ZrB_2-SiC articles as thus produced were tested for oxidation resistance by placing them in a furnace heated to a controlled temperature. At the desired temperature, air was introduced into the furnace, passed through a tube heated to the furnace temperature to bring the air to approximately the controlled furnace temperature. The air was then passed across the surface of the test cylinder. The test procedure was checked to avoid an oxidation test which would be gas supply limited. The test cylinder was held at the temperature for one hour.

At the end of the desired test time, air was replaced by a controlled inert atmosphere. Finally, the furnace and sample were cooled. A measure of oxidation resistance is the number of thousandths of an inch of oxide on the surface of the article after such exposure to oxidizing conditions. Under these test conditions, oxide thickness of at least 0.100" resulted in the control article of zirconium diboride alone, oxidized at 1900°C or higher, whereas oxide thickness of 0.002 to 0.010 result at test temperatures of 1800° to 2000°C for the ZrB_2-SiC hot-pressed products.

Examples 2 through 4: The procedure of Example 1 was repeated for the production of other hot-pressed mixtures of zirconium diboride and silicon carbide as follows: Example 2, 90 vol % ZrB_2, 10 vol % SiC; Example 3, 85 vol % ZrB_2, 15 vol % SiC; Example 4, 65 vol % ZrB_2, 35 vol % SiC. The products had good mechanical integrity, being capable of being removed from the mold without cracking. All the products were of fine grain size. All the products had good oxidation resistance according to the oxidation test used in connection with Example 1.

The results obtained with zirconium diboride-silicon carbide hot-pressed articles are essentially unchanged if up to one-fourth of the zirconium diboride is replaced with one of the other diborides including TiB_2, HfB_2, NbB_2 or TaB_2. Specifically, if it is economically feasible to replace up to one-quarter of the less expensive zirconium diboride with more expensive hafnium diboride, the results will be excellent.

METAL CASTING REFRACTORIES

SLURRY DEWATERING WITH CLAY AND FLOCCULANT

In casting molten metals to form ingots it is customary thermally to insulate the ingot head metal while allowing the body of the ingot to cool and solidify. Molten metal from the head then feeds to the body of the ingot to compensate for the shrinkage on solidification and minimize the formation of pipe. This thermal insulation is generally achieved by lining the inside of the ingot mold at its head (or the head box, in the case of a mold with a head box) with a plurality of preformed slabs or a layer of refractory heat insulating material. For convenience these materials are commonly referred to as hot tops or hot top linings.

According to *V.E. Mellows; U.S. Patent 3,933,513; January 20, 1976; assigned to Foseco Trading AG, Switzerland,* a refractory heat insulating slab is formed comprising by weight 84 to 35% of a heavy refractory filler, 6 to 35% of a refractory fibrous material other than asbestos, 0.5 to 10% of a high swelling clay, 1 to 10% of a binding agent, 0.01 to 2% of a flocculating agent for the clay and 0 to 10% of a lightweight refractory filler, the slab being free or substantially free of cellulosic fiber.

Slabs of refractory heat insulating material are prepared by a process of dewatering against a perforate surface a quantity of an aqueous slurry of the ingredients just defined to form a damp solid layer, and removing and drying the layer. This process can be carried out by conventional methods without difficulty. It is found that if flocculating agent for the clay is omitted, then dewatering the slurry is difficult and time-consuming, and in extreme cases where fibers other than asbestos are employed it is virtually impossible to produce a homogeneous insulating slab of uniform dimensions.

Preferably the ingredients are adjusted so as to give a final dry slab of density 0.4 to 1.1 g/cc. The heavy filler, that is, one having a bulk density of over 1.2 g/cc, is preferably not silica (in order to avoid any silicosis hazard) but rather is a refractory silicate such as zircon or olivine, or an oxide such as zirconia, alumina or magnesia. The refractory fibrous material may be a pure alumino-silicate or calcium silicate material, or it may be glass fiber, rock wool, mineral wool or slag wool. The high-swelling clay is preferably a bentonite type, e.g., sodium bentonite.

The binding agent may be synthetic resin (urea or phenol formaldehyde resins are preferred) or starch. The flocculating agent/surfactant for the clay may be chosen from a wide variety of materials. For bentonite-type clays, quaternary ammonium salts are preferred flocculating agents, e.g., dialkyl quaternary ammonium chlorides where the alkyl group contains from 6 to 20 carbon atoms.

The lightweight refractory filler, i.e., one having a bulk density of less than 0.3 gram per cc, if used may be, for example, kieselguhr, expanded perlite or vermiculite, fly ash floaters, hollow alumina microspheres, calcined diatomite, calcined pozzolana or calcined rice husks.

Example 1: An aqueous slurry of 73.0% silica flour, 4.0% lightweight filler, 14.5% inorganic fiber, 6.0% organic binder, 2.0% sodium bentonite and 0.5% dialkyl quaternary ammonium chloride was formed in a hydropulper. The slurry was pumped into a mold having one surface made of pervious mesh until sufficient slurry to form a dry slab 17 mm thick had been charged. Superatmospheric (40 lb/in^2) pressure was then applied to the surface of the slurry which expressed all the surplus water from the slab. The dewatering time was 30 seconds. The green slab formed was transferred to a drying oven at 150°C.

The final dried cured slab produced possessed a density of 0.80 g/cc and a transverse strength of 200 lb/in^2. The overall dimensions and surface finish of the cured slab were fairly smooth indicating that the slurry mixture was homogeneous during the forming process.

Example 2: The procedure outlined in Example 1 was repeated with corresponding acceptable results using 80.0 parts by weight fine olivine sand, 2.5 parts by weight sodium bentonite, 7.5 parts by weight slag wool, 5.0 parts by weight organic binders and 1.0 part by weight dialkyl quaternary ammonium chloride. The density of the final slab was 1.15 g/cm^3. The dewatering time needed was 75 seconds for a slab of thickness 27 mm.

CARBON-SILICON-HYDRAULIC CEMENT BLOCKS

J.R. Parsons and H.L. Rechter; U.S. Patent 3,923,531; December 2, 1975; assigned to Chicago Fire Brick Company describe castable shapes with carbon or graphite which can be used in the most severe iron and steelmaking applications, such as a skimmer in a blast furnace trough or cupola spout, requiring only a dryout, and which perform comparably to conventionally formed carbon and graphite refractories in the ability to withstand moving streams of molten iron and slags but with superior resistance to erosion and oxidation.

The castable formulations consist of 20 to 80% by weight of carbon or graphite, preferably three-fourths inch and finer although larger size carbon can be used; 5 to 25% by weight silicon metal, preferably of high purity, in a mesh size suitably of –20 mesh or finer, preferably –200 mesh; 5 to 30% by weight of a hydraulic cement, for example, calcium aluminate and calcium silicate cements, and 0 to 65%, preferably 20 to 50% by weight, of other refractory additives, such as refractory oxides and silicates, which are beneficial to bond strength and/or oxidation resistance, including aluminum oxide, silica, zircon, pulverized firebrick and clays. Up to 20% by weight of pitch may optionally be added to provide better bonding than obtained with the hydraulic cements at temperatures below the reaction temperature of carbon and silicon.

Refractory shaped articles such as tile and brick are prepared by blending the constituents with water, for example 12 to 20%, suitably in a cement mixer, a small bowl type for small batches or a large paddle mixer for larger quantities. The mix is then cast into molds with, for example, a 9 x 2½ inch cavity 4½ inches deep. The shaped articles are air set, for example, overnight and oven dried at 200°F for 2 or more days to dry out sufficiently for shipment and use.

Firing of the shapes takes place on use in furnaces at high temperatures. Typical firing carried out on tests of brick specimens was at 2550°F, a 5-hour hold. Bulk density, modulus of rupture, compressive strength and linear change was determined at room temperature. The final weight loss was used to determine carbon losses, correcting for normal firing weight losses with castables and calcining of clays. This exposure of five sides in an oxidizing kiln provides the most severe test conditions for these compositions.

IMPELLER FOR MOLTEN PIG IRON

K. Kanbara, S. Nagai and H. Yanagi; U.S. Patent 3,854,966; December 17, 1974; assigned to Nippon Steel Corporation, Japan describe castable refractories for use as an impeller to stir molten pig iron to which an alkaline additive is added comprising chamotte, aluminous cement and optionally chromium oxide. The chamotte is obtained by burning fire clay consisting chiefly of kaolinite, and the amounts of the aluminous cement and chromium oxide are, respectively, 12 to 25% and 3 to 15% by weight per total amount of the refractories. The castable refractories contain 5 to 25% by weight of chamotte having a particle size of 0.8 to 5.0 cm^3.

The chamotte obtained by burning fire clay consisting chiefly of kaolinite comprises 49 to 54% of SiO_2 and 42 to 48% of Al_2O_3. The chamotte desirably possesses a relatively low porosity. The aluminous cement used comprises less than 4% of SiO_2, 50 to 80% of Al_2O_3 and 17 to 35% of CaO. The chromium oxide used possesses a purity higher than 97% and a particle size smaller than 0.05 mm.

In using such castable refractories to obtain a structural product, it is desirous to add a minimum amount of water necessary for the hydration of the aluminous cement and to give a proper moisture content.

The amount of water necessary in working the process castable refractories containing relatively coarse chamotte particles as above mentioned lies suitably in the range of 0.4 to 0.7 times the amount of aluminous cement contained in the refractories. The mixture is, after kneading sufficiently, worked up to a structural product with a desired shape.

CORROSION-RESISTANT METAL POURING TUBES

N.W. Roudabush; U.S. Patent 3,846,145; November 5, 1974; assigned to The Chas. Taylor's Sons Company describes a composition for preparing refractory articles, such as metal pouring tubes and the like, having increased resistance to metal and slag corrosion, comprising 45 to 75% of a fusion product, 10 to 30% of kyanite, 3 to 15% zircon, 3 to 13% calcined alumina and 0.5 to 10% chromium oxide. The fusion product is a fused aggregate of 1 part of zircon to 1 part of alumina, which is first prepared by admixing 0.5 to 1.5 parts of alumina with each part of zircon, melting the mixture in an electric arc furnace, solidifying the fused product and grinding the fused product until most of it is -8+40 mesh. The other dry ingredients preferably should be as follows: -35 mesh kyanite, -40 mesh zircon, -325 mesh calcined alumina and -325 mesh chromium oxide.

The entire mixture is suspended in a colloidal silica solution, thus forming a slurry having a mud-like consistency. The colloidal silica solution should be added in an amount from 5 to 15% by weight of the total ingredients employed and the solution should contain 20 to 40% silica. It is desirable to have present in the mixture 1 to 7% silica as the binder for the mixture. The slurry is then trickled into a vacuum chamber to remove the air bubbles. The slurry is then pumped into a mold having the configuration of the pouring tube for example, and is heated to cause the colloidal silica solution to gel and thereby set the slurry into the configuration of the mold. After setting, the molded article is dried at a low temperature and then fired to form the refractory article.

The refractory product obtained comprises 45 to 72% of a fusion product, 10 to 29% kyanite, 3 to 14% zircon, 3 to 11% calcined alumina, 0.5 to 10% chromium oxide and 1 to 7% silica, the fusion product containing one part of zircon for each part of alumina.

Example: A fusion product containing 1 part of alumina for each part of zircon was used. The ingredients were mixed in the following amounts in the size ranges indicated:

	Mesh	Percent
Fusion product	-8+10	26.6
	-10+38	22.0
	-40	10.4
Kyanite	-35+325	10.2
	-325	10.2
Zircon	-40	10.6
Calcined alumina	-325	8.0
Chromium oxide	-325	2.0

This mixture was then added to 10% of a colloidal silica solution containing 30% of silica. The slurry formed a mud-like consistency. This thick slurry was then placed into a vacuum tank at a vacuum of 28 inches of mercury to remove the air bubbles. The slurry was then added to a pouring tube mold, heated to 150°F to gel the colloidal silica into the configuration of the pouring tube. The green pouring tube was then dried at 150°F for 12 hours and then fired at 2650°F for 36 hours.

For comparison purposes a pouring tube was prepared as a control having the same composition as the tube prepared above except that the control tube contained no chromic oxide.

These pouring tubes were then tested for corrosion resistance to both metals and slags and it was found that the pouring tube containing the chromic oxide was superior to the pouring tube containing no chromic oxide. In addition the pouring tube containing the chromic oxide was used for more than twice as many heats as the tube which did not contain the chromic oxide.

SILICA-ALUMINA PREFORM CORE

Preform cores comprise the interior molds employed in the manufacture of hollow metal castings such as, for example, jet engine blades and vanes. *R.B. Forker, Jr., M.C. Carson and R.C. Washington; U.S. Patent 3,839,054; October 1, 1974; assigned to Corning Glass Works* describe a method of manufacturing a preform core by compounding a batch consisting of 2 to 20% of finely divided aluminum metal and 80 to 98% of finely divided silica or an equivalent thereof, shaping the batch into a preform core body of the desired configuration, firing at 500° to 1200°C for 1 to 12 hours to form aluminum oxide and silicon, cooling to 275°C to form α-cristobalite and finally flash-firing the body at least once at 1000° to 1650°C for ¼ to 12 hours to promote the oxidation of the silicon remaining in the body to silicon dioxide and to encourage the development of mullite as well as α-cristobalite and α-alumina crystal phases therein.

The preform core body consists, in weight percent as calculated from the batch, of 50 to 96% SiO_2, 4 to 37% Al_2O_3, and not more than about 15% silicon metal, and comprises crystalline phases of α-cristobalite, α-alumina and mullite. The body is refractory, inert, leachable and strong enough to withstand handling, wax impregnation, casting and other mechanical stresses to which preform cores are commonly subjected.

Example: The following components were used in the quantities indicated.

	Density (g/cc)	Weight (g)	Volume (cc)	Weight %	Volume %
Fused silica	2.2	1026	467.2	68.7	59.9
Powdered aluminum	2.7	115	42.6	7.7	5.5
Vehicle	1.46	228	156.2	15.2	20.0
Deflocculant	0.92	34.0	37.0	2.3	4.7
Binder	1.20	91.5	76.2	6.1	9.8

Green cores prepared from the above composition are heated in an electric furnace at 350°C per hour to 800°C, held at 800°C for 2 hours, cooled at the furnace rate (25°C per hour) to 250°C, and then removed from the furnace and cooled to room temperature. The fired cores are then flash-fired by plunging into an electric furnace operating at 1300°C for 2 hours, removed from the furnace and cooled to room temperature. Usually the cores are covered to prevent over-rapid cooling. Finally, the cores are again flash-fired by placement in a furnace operating at 1650°C for one-half hour and then removed and cooled to room temperature while covered.

X-ray examination of the fired core material shows a major crystal phase of α-cristobalite and minor phases of α-alumina, mullite and silicon. Typical modulus of rupture values determined on these cores are 1,760 psi, with shrinkage on firing of 1.2% by volume being observed.

The refractoriness and inertness of the cores is tested by contact with molten cobalt-based superalloy during casting at 1500°C. This alloy is quite reactive, being characterized by a coefficient of thermal expansion of about $181.6 \times 10^{-7}/°C$ and having the approximate composition, in weight percent:

Cr	21.3
Ni	10.0
W	6.8
Ta	3.5
C	0.55
Zr	0.48
Ti	0.19
Co	57.5

No evidence of temperature failure after pouring at 1500°C is observed, nor does any significant reaction between the core and the molten alloy occur to deleteriously affect the quality of the casting.

The cores produced show excellent leaching characteristics, with 72% volume dissolution of the core material being accomplished by a 30-minute immersion in a boiling KOH solution containing 45% KOH by weight.

ALKYL SULFATE FOAMING AGENT FOR FOUNDRY MOLDING COMPOSITIONS

K.E.L. Nicholas; U.S. Patent 3,826,658; July 30, 1974; assigned to British Cast Iron Research Associates, England describes pourable foundry molding compositions comprising refractory filler material, water, a bonding agent and an anionic alkyl sulfate foaming agent that produces a foam structure in the composition of an inherently limited life, ensuring dispersion of the foam shortly after pouring. The composition comprises 100 parts of refractory filler material, 1 to 6 parts of water, 6 parts of sodium silicate solution as bonding agent and 0.1 part of a 2-ethylhexyl sulfate foaming agent.

The composition can be made self-hardening by the addition of a hardening agent such as calcium silicate which reacts exothermically with the sodium silicate solution and evolves a gas. The free escape of the gas from the molded composition is ensured by arranging that the early dispersion of the foam occurs, and thereby renders the composition highly permeable, before the hardening reaction commences and produces the gas.

In the mixture which comprises the 2-ethylhexyl sulfate foaming agent and sodium silicate solution as the binder, the exact time for which the foam structure lasts can be controlled as required by adjusting the ratio of the sodium silicate to water, and by adding controlled amounts of a second more stable foaming agent. By these means it is possible to arrange that the foam structure has substantially dispersed before the hardening reaction commences between the sodium silicate solution and a hardener included in the mixture. Clearly, the hardener can be a material such as calcium silicide which reacts exothermically with the sodium silicate and evolves a gas, as the gas can then escape from the highly permeable mixture.

Further, early collapse of the foam structure in both mixtures allows them to be hardened by the passage of carbon dioxide gas through them soon after being poured into the core and molding boxes. Gassing takes just a few minutes and then the cores and molds can be stripped from their boxes and used shortly thereafter.

Even if the mixtures are used without employing an exothermic or gas hardening reaction, the early collapse of the foam structure still helps the molded mixture to harden and dry out more quickly and allows the escape of steam and gas from the mold once molten metal is poured into it, all of which features allow the molds to be used sooner after the mixture is poured into the molding boxes.

Example: To 100 parts of clean, washed silica sand there were added 3 parts of water, 6 parts of sodium silicate solution of specific gravity 1.375 and molecular ratio 3:1, and 0.1 part of the sodium salt of 2-ethylhexyl sulfate. Air was entrained in this mixture by a suitable mixing or stirring operation to produce a pourable fluid slurry. To this slurry there were then added 3 parts of a finely ground (200 mesh) hardening agent, the preferred material being calcium silicide. The mixture was immediately cast into a flask or core box and after 2 or 3 minutes the foam collapsed, and after this time interval an exothermic reaction commenced between the sodium silicate and calcium silicide with the evolution of large volumes of hydrogen gas.

As the foam collapsed the slurry lost fluidity and simultaneously the sand mass became freely permeable, permitting the free escape of the hydrogen gas to atmosphere. At normal temperatures the reaction was found to be substantially complete within 30 minutes and the sand mass was bonded throughout to form a hard body.

TRIHYDROXYBIPHENYL IN FOUNDRY GREEN MOLDING SANDS

According to *R.E. Melcher; U.S. Patent 3,816,145; June 11, 1974; assigned to Whitehead Brothers Company* trihydroxybiphenyl is useful as an additive for improving the properties of foundry green molding sands or clays. The trihydroxybiphenyl is incorporated into the sand in the form of an aqueous solution. Alternatively, the solution may be applied to the surface of a green sand mold for use as a facing agent.

The use of aqueous trihydroxybiphenyl affords the following advantages:

(1) The molding sand has good flowability. Thus the sand is readily formed and compacted around mold patterns of complicated design. Moreover, the sand can be readily employed in automatic molding machines. In addition, the good flowability permits achievement of a desired sand hardness and apparent bulk density with the expenditure of less compacting energy than other green molding sands. Further, the danger of overramming a

portion of the mold due to variations in compacting energy is reduced due to more uniform compaction of the sand. Thus the rammed sand is more homogeneous with respect to density and therefore strength.

(2) The compacted sand possesses desirable green strength characteristics at lower moisture contents than are achieved with conventional green molding sands.

(3) The additive acts as a facing agent, and prevents burn-on or the fusing of quartz sand grains to the surface of the casting, and promotes excellent finish and peel.

(4) The additive reduces shifting of the sand during the casting process, whether it be mold wall movement or enlargement of the mold cavity, or whether it be a localized shifting of the sand resulting in such casting defects as rat-tails, scabs and buckles.

(5) The additive permits casting to be effected at lower pouring temperatures and promotes increased fluidity of the metal during casting.

(6) The additive yields adequate dry compression strength and yet excellent shakeout is obtained even with green molding sands employing Western bentonite as the clay binder.

(7) Finally, the additive is employed at relatively low levels, which in turn minimizes the formation of gas during casting.

PRODUCING FOUNDRY MOLDS AND CORES

A.M. Lyass, P.A. Borsuk, Z.G.O. Usubov, V.G. Kuznetsov, N.Y. Kagan, J.A. Razumeev, V.M. Bortnik and I.V. Korenbljum; U.S. Patent 3,804,641; April 16, 1974 describe a method of producing foundry molds and cores by (1) preparing a mixture containing 100 parts by weight of molding sand, 1 to 5 parts by weight of lignosulfonates of an alkali, alkaline earth metal, ammonium or mixtures thereof as a binder, and 3 to 10 parts by weight of a material containing alkali metal aluminate; (2) making molds and cores from the mixture; and (3) holding it for air hardening.

An intermediate product obtained in the production of aluminum oxide from bauxite or nepheline ores is preferable for use as the material containing an alkali metal aluminate. The above product is produced by sintering bauxite or nepheline ores mixed with sodium carbonate and limestone at 1200°C. Next the resulting product, hereinafter referred to as a bauxite sinter or a nepheline sinter, is cooled and ground to grain size of 0.3 to 1 mm.

The most widely-known binder containing sodium, calcium and/or ammonium lignosulfonate is the waste product resulting from the production of cellulose from wood pulp by the sulfite process. Hereinafter the product will be referred to as a sulfite alcohol waste liquor.

The lignosulfonate content in the liquid sulfite alcohol waste liquor may be characterized by its specific gravity. Accordingly, the molding mixture contains commonly an aqueous solution of the sulfite alcohol waste liquor of a specific gravity of 1.10 to 1.27 g/cm^3, the lignosulfonate content of the solution ranging from 25 to 55% by weight which corresponds to 1 to 5% of the weight of the molding sand.

A method of hardening the binder—lignosulfonate of an alkali, alkali-earth metal, ammonium or their mixture—is of particular importance for fluid molding sands which include a foaming agent introduced in the mixture in quantities large enough to pass the mixture to a fluid state. The amount of the foaming agent ranges within 0.4 to 1% of the weight of the molding sand.

The foaming agents may be alkylaryl sulfonates, alkyl sulfonates, primary and secondary alkyl sulfates, the products of oxyethylation of alcohols, phenols, amines or quaternary ammonium compounds of long-chain fatty amines. The most suitable foaming agent is an anion-type surface-active compound, such as sodium alkylaryl sulfonate ensuring the production of a fluid molding mixture featuring high flowability and the requisite foam stability in the mixture, i.e., the property to retain mobility during the length of time needed to pour the mixture into core and molding boxes. An advantage of the fluid molding sand is its high flowability by which virtue the conventional sand compacting technique can be dropped to be replaced by pouring the sand into the core boxes and on patterns during the production of the molds and cores.

The overall strength of the cores can be increased by carbamide introduced into the molding mixture. The best results are obtained when adding from 0.8 to 2.0% of carbamide (of the weight of the molding sand). The carbamide made from a fluid self-hardening sand containing the foaming agent can reduce appreciably sand humidity; the amount of water added to the sand can be also decreased by 1.0 to 2.0% of the weight of the molding sand.

Example 1: 100 parts by weight of quartz sand are mixed with 4.0 parts by weight of a sulfite alcohol waste liquor of a specific gravity of 1.24 to 1.26 g/cm^3 containing accordingly 48 to 52% by weight of lignosulfonate, the liquor being preliminarily tempered with 1.0 part by weight of water in which 0.8 part by weight of powdered carbamide were dissolved.

Within 1.5 to 2 minutes 2.5 parts by weight of a bauxite sinter containing 40% by weight of sodium aluminate are introduced into the molding mixture. The molds and cores made from the mixture featured the following compression strength (kg/cm^2): after air hardening for 1 hour, 3.5; after air hardening for 3 hours, 7.0; and after air hardening for 24 hours, 13.5.

Example 2: To prepare a fluid molding sand added to the composition disclosed in Example 1 is 0.6 part by weight of a foaming agent, sodium alkylaryl sulfonate. After the molding mixture has passed to the fluid state 2.5 parts of a hardening agent, a bauxite sinter, containing 40% by weight of sodium aluminate, are introduced into the sand. The molding mixture ensures the following compression strength (kg/cm^2) of the molds and cores produced thereof: after air hardening for 1 hour, 3; after air hardening for 3 hours, 6.5; and after air hardening for 24 hours, 12.0.

GRAPHITE-ALUMINUM PHOSPHATE COMPLEX MOLDING COMPOSITIONS

J.D. Birchall and J.E. Cassidy; U.S. Patents 3,804,648; April 16, 1974 and

3,950,177; April 13, 1976; both assigned to Imperial Chemical Industries Limited, England describe a composition which comprises (a) graphite, (b) a binder consisting of either a complex phosphate of aluminum containing at least one chemically bound molecule of a hydroxy compound, ROH, where R is an organic group of hydrogen and an anionic group of a strong inorganic acid, other than an oxyphosphorus acid, or of a carboxylic acid, or an aluminum phosphate in which the ratio of the number of gram atoms of aluminum to the number of gram atoms of phosphorus is at least 1:1; and (c) a dispersant for the binder, the binder being present in an amount of 0.5 to 25% by weight of the composition.

The compositions may be used for molding to produce shaped articles but also for purposes in which the hardening of the composition, its ready adherence to its surroundings and the heat-resistance of the product can be utilized. It may be used, for example, as a mortar, cement or filler for binding ceramics or for high temperature applications in furnace walls and linings.

Suitable percentages by weight of the compositions for the components may be as follows: 10 to 95% graphite, 0.5 to 25% binder and 1 to 50% dispersant. Preferably, the binder is present in an amount of 2 to 10% by weight of the composition. Generally, the weight ratio of graphite to binder may be from 20:1 to 1:1

Any grade of graphite may be used, for example, baked carbon, electrographite, impervious carbon and impervious graphite, electrode graphite, unpurified, thermally-purified and chemically-purified nuclear graphite, pyrographite and porous graphite.

Examples of complex phosphates include: (a) that containing chlorine and ethyl alcohol and having the empirical formula $AlPClH_{25}C_8O_8$. It is designated aluminum chlorophosphate ethanolate (ACPE); (b) that containing chemically-bound water and chlorine and having the empirical formula $AlPClH_{11}O_9$. It is designated aluminum chlorophosphate hydrate (ACPH); and (c) that containing bromine and ethyl alcohol having an empirical formula $AlPBrH_{25}C_8O_8$. It is designated aluminum bromophosphate ethanolate (ABPE).

The dispersant, generally a liquid dispersant, is preferably a solvent for the aluminum phosphate or complex although the binder may be dispersed in the dispersant, for example, as a suspension, sol or gel.

Suitable solvents for the complex phosphates are polar solvents, e.g., methanol, ethanol, isopropanol, butanol, ethylene glycol monoethyl ether, water or a mixture of two or more such solvents.

Example 1: 15 grams of a synthetic graphite mixture were mixed with 2.5 grams ACPE and 1 cc of water. The mixture was placed in a stainless steel mold and a pressure of 24.5 kg/cm² applied. The block was then removed from the mold and dried for 1 hour at 140°C after which it was fired for 1 hour at 300°C. The final block has a crushing strength of 42 kg/cm² and a specific gravity of 1.40.

Example 2: To 15 grams of the mixed synthetic graphite (as described in Example 1) was added 2 grams ACPH and 1 cc of water. After thorough mixing, the mixture was placed in a stainless steel mold and a pressure of 24.5 kg/cm² applied.

The block was removed from the mold and dried for 1 hour at 140°C after which it was fired at 300°C for 1 hour. The final block had a crushing strength of 40 kg/cm² and a specific gravity of 1.44.

Example 3: 14 grams of molochite mix (45 parts of particle size 3.2 to 6.4 mm, 10 parts of particle size 0.25 to 0.5 mm of particle size less than 0.075 mm were mixed with 6 grams of the mixed synthetic graphite (as described in Example 1). To the mixture was added 2 grams ACPH and 1 gram water. After thorough mixing the mixture was placed in a stainless steel mold and a pressure of 49 kg/cm² applied. The block was removed from the mold and dried for 1 hour at 140°C after which it was fired at 500°C for 1 hour. The crushing strength of the block was 98 kg/cm² and the specific gravity 1.85.

Example 4: To a mixture of 14 grams of the mixed molochite of Example 3 and 6 grams of the mixed synthetic graphite (as described in Example 1) was added 2 grams ACPE and 1.5 grams water. The mixture was placed in a stainless steel mold and a pressure of 49 kg/cm² applied. The block was removed from the mold and dried at 140°C for 1 hour and then fired for 1 hour at 500°C. The final block had a crushing strength greater than 44 kg/cm² and a specific gravity of 1.79.

EXOTHERMIC ANTIPIPING COMPOSITION

B.C. Rumbold and J.E. Cartwright; U.S. Patent 3,804,642; April 16, 1974; assigned to Foseco International Limited, England describe an exothermic antipiping composition which comprises a refractory, heat insulating material, an exothermic component and acid treated graphite. The exothermic component is preferably a mixture of an oxidizable metal (e.g., aluminum, magnesium, silicon, ferrosilicon, calcium silicide) and an oxidizing agent therefor (e.g., alkali metal nitrates or chlorates, iron oxide, ammonium perchlorate), though in some cases the exothermic component may consist merely of a finely divided, easily oxidizable material such as aluminum.

In this latter case, the oxidizable material preferably constitutes 10 to 50% by weight of the antipiping composition. Preferably the proportion of acid treated graphite in the antipiping composition is 1 to 50% by weight, most preferably 3 to 20% by weight.

The refractory heat insulating material may be, for example, a refractory filler such as grog, silica flour, alumina, bauxite, magnesia, clays such as ball clay or china clay or any suitable refractory silicate; light weight fillers such as vermiculite, perlite and pumice may also be used. Carbonaceous materials may also be used. The grain size of such material is preferably such that it does not segregate from the mixture and such that the mixture can be pelletized, granulated or tabletted as required. The refractory heat insulating material may constitute 10 to 70% by weight of the composition.

Where the composition contains finely divided aluminum as the exothermic component, or part of it, the composition may also advantageously contain a fluoride, known per se for its use in controlling the energy of aluminothermic reactions.

The antipiping compositions are preferably in powder form, but may be granulated or formed into tablets; a small proportion of a clay or other binder may be included in the composition to aid pelletization. It is found that the antipiping compositions of the process are of particular value in the casting of molten metals, particularly ferrous metals.

The antipiping compositions are clean and produce little dust in use, they are relatively economical to manufacture and use, and can easily be formed into easily handleable tablets, pellets or granules. Carbon pickup by the molten metal is negligible, and there is only a low contact area of the antipiping composition with the molten metal.

It is possible to obtain satisfactory results even where the molten metal is subject to turbulence (e.g., jolting of an ingot mold in a train of such molds) by incorporating in the antipiping composition a material which will tend to sinter or melt at the molten metal temperature, and so bind the layer to a more coherent form. Suitable materials are expanded perlite, expanded vermiculite, fly ash and pumice.

Example: 400 grams of a mixture comprising 30% aluminum (–100 mesh BSS), 12% acid treated graphite, 6% wood flour, 4% expanded perlite, 2% sodium fluoride and 46% alumina were placed on a 23-cm square hot plate maintained at 1420°C. The mixture ignited after 30 seconds thus minimizing chill, expanded and formed a layer 50 mm thick on the plate. The rate of heat loss after 14 minutes was found to be two-thirds of that obtained using 1,000 grams of a conventional proprietary exothermic antipiping composition in an identical fashion. After 40 minutes the total heat lost was half that lost when using the conventional proprietary exothermic antipiping composition.

Trials made at four steelworks have shown that there is an average increase in yield of 1% of sound metal in ingots of 1½ to 20 tons, using an exothermic antipiping compound as defined in the foregoing example, at a rate of two-thirds the quantity by weight of a conventional commercially available antipiping compound.

CHROMITE SAND-CHROMITE FLOUR MOLD OR CORE

P.F. Stephens; U.S. Patent 3,788,864; January 29, 1974; assigned to Bethlehem Steel Corporation describes a refractory chromite sand-chromite flour mold or core having low density and good thermal stability which imparts a chilling effect to the molten metal cast therein to minimize penetration of the sand and to allow the passage of gas to decrease susceptibility to gas porosity and to produce large ferrous and nonferrous castings having surfaces which require a minimum amount of cleaning.

The refractory mold and/or core suitable for casting molten ferrous metal and molten nonferrous metal of the high lead and/or high tin copper alloys characterized by having a permeability of not more than 10 when tested by standard AFS permeability test comprises (a) 50 to 90% chromite sand having a particle size of AFS 55/65 to AFS 85/90, (b) 10 to 50% chromite flour having a particle size of AFS –325, and (c) 4.5 to 7.0% of an inorganic binder consisting of a mixture of bentonite and fireclay in which the weight ratio of bentonite to fireclay is 2:1 to 1:2.

Example 1: A refractory chromite sand-chromite flour mix was made by mixing together 160 pounds of chromite sand having a particle size of AFS 85/90 and 40 pounds of chromite flour having a particle size of AFS -325. After mulling for 3 minutes an inorganic material consisting of 5 pounds of bentonite and 6 pounds of fire clay was added to the chromite sand-chromite flour mix. Sufficient water to make up 3% of the final mix was then added, and the materials were mixed for 3 minutes. The resultant wet mix was then used to make a spade core which was used to form the coupling on a 45-inch slab mill steel roll. The roll was made to the following specification:

| | - - - - - - - - - - - - - - - - - - -Percent of- | | | | | | | |
	C	Mn	P	S	Si	Ni	Cr	Mo
Minimum	0.60	0.75	(*)	(*)	0.30	1.35	0.30	0.30
Maximum	0.70	0.95	0.05	0.05	0.45	1.55	0.45	0.40

*Maximum.

The cast steel roll weighed 67,000 pounds as shipped. Coupling was smooth as removed from the mold and required practically no cleaning.

Example 2: The refractory mix was used to manufacture center cores used in casting tin bronze nuts weighing 4,400 pounds as shipped. The dry refractory mix was made by mixing 240 pounds of refractory chromite sand having a particle size of AFS 85/90 and 60 pounds of chromite flour having a particle size of AFS -325. The inorganic material, containing 9 pounds of a 52% aqueous solution of sodium silicate having a silica (SiO_2) to sodium (Na_2O) ratio of 2:1, and 6.9 pounds of 75% ferrosilicon, was added to the refractory mix. The materials were mixed for 5 minutes. A mold made from the resultant mix was used to form the casting. No problems were encountered in shaking-out the casting. The casting was found to have excellent surfaces which were free of burned-in sand. No burned-on sand from the mold was seen on the casting surface, thus indicating that there was no metal penetration of the mold.

SILICA-ALUMINA-ZIRCONIA POURING TUBE

According to *T.M. Smith; U.S. Patent 3,782,980; January 1, 1974; assigned to The Babcock & Wilcox Company*, fired refractory articles, such as pressure pouring tubes, are composed of a fused refractory aggregate and a refractory binder material for bonding the aggregate together. The aggregate consists of 42 to 48% alumina (Al_2O_3) and 17 to 23% silica (SiO_2) generally in the form of mullite crystals ($3Al_2O_3 \cdot 2SiO_2$), and 33 to 39% unstabilized zirconia (ZrO_2) dispersed throughout the mullite structure.

The aggregate forms 50 to 80% while the binder forms 50 to 20% of the refractory article. The binder originally contains 10 to 90% of 100 mesh zircon sand, 10 to 90% 325 mesh alumina and 0 to 50% raw clay. In addition, suitable plasticizers and organic materials are added for forming purpose and are eliminated during heat treatment of the article.

FUSION BONDED VERMICULITE MOLDING MATERIAL

According to *P.M. Brown, H.C. Duecker and D.C. de Vore; U.S. Patent 3,778,281;*

December 11, 1973; assigned to W.R. Grace & Co., fusion bonded vermiculite molding material prepared from vermiculite, sodium borate, sodium metaphosphate and clay is used to prepare a variety of materials and structures including, among others, a strong, lightweight decorative fire resistant board, a superior foundry mold and a lightweight synthetic aggregate.

A free-flowing lightweight vermiculite molding material can be prepared by proper coating and gentle blending of the vermiculite, clay and borate-phosphate flux. This is accomplished by either of two methods. In the first method, the vermiculite and clay are gently blended, a flux solution is prepared and this solution is gently blended into the dry components.

In the second method, the flux solution is prepared and the clay is thoroughly dispersed in this solution. The resultant slurry is then gently blended with the vermiculite. The method of blending the components is critically important. It is essential that each vermiculite particle be homogeneously coated with the other components. Furthermore, it is essential that in the blending the lightweight friable vermiculite particles are not compacted. These blending and coating conditions are essential to prepare a molding powder, which when formed, produces a strong lightweight body.

It is necessary to adhere to these conditions of gentle homogeneous blending of these materials to obtain the fine microstructure created upon firing. This fine microstructure leads to a continuous matrix and products of exceptional strength. By selectively choosing the particular type of vermiculite to be used (coarse or fine) the density of the finally obtained product and the strength of the end product can be varied. These characteristics can also be affected by the relative percentages of each of the ingredients described above. For example, by selectively using the above-described ingredients, a fused vermiculite board can be prepared at temperatures as low as 1300°F and the resulting product has a vermiculite matrix bonded together by a continuous second phase. The actual obtained products have a strength of 30 psi to 2,000 psi and densities of 12 lb/ft^3 to 80 lb/ft^3.

The use of the flux, i.e., borax and sodium metaphosphate, allows fusions at less than 1300°F. Other low-melting phosphates can also be used. An example is sodium pyrophosphate. Hence, sodium metaphosphate also includes sodium pyrophosphate. A fused lightweight, strong vermiculite board having varying densities and varying strength dependent upon the composition of the product and the firing times can be prepared from these molding materials.

WOOL FIBERS WITH CRUSHED COKE OR FLY ASH FILLER FOR INSULATION

J. Wilton; U.S. Patent 3,770,466; November 6, 1973; assigned to Foseco International Limited, England describes refractory fiber compositions which can be used as slabs or sleeves for the lining of hot tops or the heads of ingot molds in the casting of metals at high temperatures. The compositions consist of 30 to 55% by weight of crushed coke or fly ash, 40 to 60% by weight refractory fiber and 4 to 8% of weight binder. The compositions, in a shaped form, can be effectively used in casting copper based alloys or iron without noticeable shrinking or other undesirable effects. In contrast, shaped compositions prepared with other refractory fillers shrink at comparable filler concentrations and have poor insulating characteristics when more filler is employed.

The synthetic fibrous refractory materials are preferably silicates. Particular application is found with respect to calcium silicate mineral wool fibers such as slag wool and rock wool since these fibers are not ordinarily useful by themselves in casting applications above 800°C. However, by adding the above noted quantities of fly ash or crushed coke to such fibrous refractory materials, the useful operating temperature of shaped compositions prepared therefrom can be significantly elevated.

Any of the known binders such as thermosetting resins, (e.g., phenol-formaldehyde) and binders such as clay (bentonite and ball), starch, wheat flour, phosphate-modified starches, and colloidal silica can be used. Water-soluble binders such as urea-formaldehyde can also be used, either alone or in combination with insoluble binders, though the insoluble binders alone are generally not preferred when the compositions are prepared from an aqueous slurry.

Sleeves, slabs and the like are prepared preferably by felting onto a mesh former from a slurry. By this means slabs and sleeves may be produced easily and quickly and of low density, for example of density 0.15 to 0.6 g/cc, and particularly 0.35 g/cc.

Example 1: The fibrous refractory material was a calcium silicate slag wool with a maximum use temperature of 815°C and analyzing: CaO, 38%; Al_2O_3, 14%; SiO_2, 36%; MgO, 8% with 4% miscellaneous impurities (TiO_2, MnO, Fe_2O_3 and the like).

A slurry of 3% by weight solids content was made, the solid material dispersed in the slurry having the following composition by weight: 35% crushed coke, 57% fibrous refractory (as above), 8% resin binder (1:2 mixture of urea-formaldehyde and phenol-formaldehyde resins).

A former consisting of a cylindrical chamber having mesh walls on its curved surface was connected by a hose to a vacuum pump. The former was immersed in the slurry and a vacuum pressure of 3 inches Hg applied via the pump. This was continued for 15 seconds, water being sucked away at 12 to 15 gallons per minute. There was thus formed on the former a sleeve of height 6 inches, outside diameter 5 inches and inside diameter 4 inches.

The sleeves thus made were stripped from the former by reversing the air flow and then dried at 175°C for 1½ to 2 hours. The product has a density of 0.3 to 0.35 g/cc and a permeability of above 57 AFS units.

A sleeve prepared as just described was used to line the feeder head in casting cast iron, which was poured at 1350°C, and the cooling curve of the casting was determined. It was found that the sleeve gave a 20-minute delay between the pouring of the metal and its solidification, this being more than double that which was found when using a conventional sleeve of bonded sand.

Example 2: Various sleeves were made up by the process described in Example 1 having compositions in the ranges of: 30 to 55% coke dust (–100 mesh BSS) or fly ash, 40 to 60% calcium silicate fiber and 4 to 8% urea-formaldehyde/phenol-formaldehyde resin mixture (1:2). These sleeves were found to be of particular value in casting iron and copper base alloys.

Example 3: Sleeves having 37% calcium silicate fiber, 55% coke dust (–100 mesh BSS) and 8% urea-formaldehyde/phenol-formaldehyde mixture (1:2), produced by the method described in Example 1, performed well as riser sleeves for cast iron castings.

Example 4: Slabs having 34% aluminosilicate fiber, 60% coke dust (–100 mesh BSS) and 6% urea-formaldehyde/phenol-formaldehyde resin mixture (1:2) were found to be suitable for use in lining the feeder heads of both copper base and cast iron castings.

INJECTION MOLDABLE REFRACTORY WITH ETHYLCELLULOSE

R.A. Horton; U.S. Patent 3,769,044; October 30, 1973; assigned to Precision Metalsmiths, Inc. describes an injection moldable refractory mixture containing comminuted refractory material as the major component, a sublimable binder, and a nonsublimable binder which is at least in part ethylcellulose. These compositions may be used for molded refractory articles, such as cores employed in metal casting processes.

When ethylcellulose is added to a refractory mixture which includes a high vapor pressure, sublimable binder, it is especially effective in preventing bleeding during injection, i.e., separation of the sublimable binder, and it also results in improvements in the physical properties of the molded refractory articles.

Preferred compositions include comminuted refractory material as the major component, a sublimable binder capable of being sublimed out of articles molded from the mixture to produce a porous structure, and at least 0.8% by weight of a nonsublimable binder capable of forming a preliminary bond to hold the molded articles intact prior to subsequent hardening operations. The nonsublimable binder is comprised of ethylcellulose in an amount of from 4.5 to 100% by weight with the ethylcellulose being present in a minimum amount of 0.5% by weight of the total weight of the mixture.

The pore structure of refractory cores or other articles molded from the foregoing composition may be impregnated with a strengthening and hardening agent. The preferred material used for impregnation is a refractory binder in a carrier liquid. In an alternate procedure, the molded refractory articles may be fired and sintered subsequent to sublimation.

In addition to preventing bleeding or separation of the sublimable binder during injection, the ethylcellulose has been found to be very effective in increasing the strength and hardness of the molded articles prior to their being impregnated or sintered. This increased strength frequently permits injection molded articles to be ejected and handled sooner, thereby shortening the mold cycle times.

The ethylcellulose is compatible with the various subliming binders used in preparation of the refractory mixtures, and it goes into solution rapidly at low temperatures. This makes it easier to prepare the refractory mixtures and reduces losses of the volatile, subliming binder, during preparation, thereby producing more consistent results.

Any of the refractory materials which are conventionally used for making ceramic cores and the like can be employed in preparing the mixture. Typical refractories include fused and crystalline silica, zircon, zirconia, alumina, calcium zirconate, various aluminum silicates, tricalcium phosphate, nepheline syenite and the like.

Typical sublimable binders include para-dichlorobenzene, which has a conveniently low melting point, good molding properties, low cost and the capability of being readily sublimed, urethane (ethyl carbamate), acetamide, naphthalene, benzoic acid, phthalic anhydride, camphor, anthracene and para-chlorobenzaldehyde.

Example: A core batch was prepared of the following composition: 75% refractory blend, 21% p-dichlorobenzene and 4% ethylcellulose, 20 cp. The refractory blend consisted of the following ingredients: 59% zircon flour (45% plus 325 mesh), 39% fused silica (100% minus 200 mesh, 75% minus 325 mesh) and 2% nepheline syenite.

The core batch was prepared by first blending together the refractory powders and then warming the refractory blend to approximately 200°F. The p-dichlorobenzene was melted and the ethylcellulose dissolved in it. All of the ingredients were combined using a mixer having a whip type agitator. The core batch was granulated by cooling to solidification while continuing agitation. Alternatively, the blended mixture can be poured into slabs which are cooled and then broken up in a separate operation.

A variety of cores for commercial parts were molded from the core batch using a standard plastic injection molding machine. The machine was of the plunger type with horizontal injection and vertical mold opening and closing. The injection temperature was 155°F and the injection pressures 500 to 700 psi depending upon core configuration, although higher pressures could have been employed satisfactorily.

p-Dichlorobenzene was removed by placing the injection molded cores in a vacuum chamber operating at room temperature under a reduced pressure of 27 inches of mercury below atmospheric pressure for a length of time suitable to remove substantially all of the subliming binder. The time for removing the subliming binder was 4 to 10 hours depending upon the sizes and shapes of the cores.

After the subliming operation, some of the cores were impregnated under a vacuum with an aqueous colloidal silica sol. The sol was a fine particle size sol containing 30% by weight colloidal silica particles having a particle dimension of 7 to 8 mμ. The time for impregnation varied from 45 to 60 minutes. Subsequent tests have shown that much shorter impregnation times are practical and often desirable. As little as 2 minutes is often sufficient and even a shorter time may be feasible for very small cores.

Following impregnation, the cores were rinsed quickly in water and placed in the freezing compartment of a standard home refrigerator until frozen hard. The cores were then removed and allowed to thaw and dry in a gentle current of air. The cores were ready for use immediately after drying, but also could be stored for indefinite periods after use if desired. Other cores were sintered instead of being impregnated. The sintering operation was carried out by firing the cores at 2000° to 2300°F.

The cores were incorporated into disposable patterns in two ways. Simple cores were inserted directly into openings in the patterns. In instances where this was not possible or desired, the cores were inserted into a pattern injection mold and wax was injected around the cores. The patterns containing the cores were assembled into setups of the type conventionally used in making ceramic shell molds, and the setups were processed according to normal ceramic shell practice.

The ceramic shell molds were fired at 1600° to 2000°F and various molten steel alloys were cast into the molds against the cores to produce castings. The cores were then removed from the castings in a molten caustic bath.

RAPID FUSE CASTING IN LIQUID COOLED METAL MOLDS

According to *J. Duchenoy, G. Gasparini and Y. Marty; U.S. Patent 3,763,302; October 2, 1973; assigned to L'Electro-Refractaire, France*, a charge of molten refractory oxides, such as mixtures of magnesia and chromium oxide, containing a gas liberating ingredient are poured into a liquid cooled mold via a swirling casting jet in a funnel where small fragments of fused cast scrap of like composition are added and homogeneously mixed with the melt. The solid fragments, the gas liberating ingredient and the liquid cooled metal mold cooperate to cause a solidified shell to form next to the mold surface of such a thickness that the part can be removed from the mold within a few minutes after casting and thereafter cooled outside the mold at a conventional rate without cracking or distortion of shape.

SILICON CARBIDE-ALUMINA ARTICLES FOR USE WITH MOLTEN METALS

Refractory articles for use with molten metals, that is articles such as protective sheaths or protection blocks for pyrometry equipment, degassing units, nozzles and stoppers, cut-off gates of melting furnaces, conveyors and holders such as launders, runners and tundishes, molds and ingot bases, and lances and other components for dipping in molten metal, should have particularly a good resistance to thermal shock and resistance to attack by the molten metal being handled or by its slag. For certain applications a high strength and hardness are also desirable.

Accordingly *W.J. Steen; U.S. Patent 3,759,725; September 18, 1973; assigned to Morgan Refractories, Limited, England* describes the preparation of articles, which in use are in contact with molten ferrous metals, of silicon carbide particles and more finely divided alumina, with or without silica additions provided by, for example, adding clay, the shaped article being fired so that the alumina coats and bonds the silicon carbide particles.

By firing a mixture of silicon carbide with another refractory material which does not react with molten ferrous metal and which is present in a sufficient proportion and has such a particle size as to coat protectively and bond the silicon carbide particles, the articles produced have good thermal shock properties and also good resistance to corrosion by molten ferrous metals.

Accordingly the process includes firing a green form produced from a particulate mixture having as its principal constituents alumina and silicon carbide, the alumina being present in such proportion and having such a particle size that in the fired article it coats and bonds the silicon carbide particles.

The mixture contains as principal constituents 80 to 50% by weight silicon carbide and 20 to 50% by weight alumina, the silicon carbide having a larger particle size than the alumina, at least a substantial proportion of which should pass a 200 mesh sieve but may be as fine as 10 μ. Measurable amounts of silicon carbide passing a 200 mesh sieve should be avoided. The presence of larger alumina particles may be accepted provided there are sufficient fine particles to coat the silicon carbide. The green form of the refractory article may be made by any standard technique such as casting, extrusion or dry pressing, and the green form is fired to sinter the alumina.

In addition to the above constituents, the mixture may be given an increased silica content of up to 10% by weight of the silicon carbide/alumina mix. The added silica may react with alumina to form mullite. A suitable vehicle for the mix for casting or extrusion is water. Other constituents, which may be present in trace quantities, for example as impurities, without detracting from the stability of the article in use, are silicon, titania and sodium, potassium and calcium oxides.

Apart from being coarser than the alumina, the silicon carbide should preferably not be too fine as the thermal shock resistance decreases with decrease in particle size and the resulting increase in surface area of the silicon carbide increases the risk of oxidation or other reactions and also increases the required proportion of alumina. Preferably the silicon carbide should pass a 4 mesh sieve, but not more than 5% should pass a 120 mesh sieve.

Too close packing of the particles in the article, which may be due to the use of too fine particles or due to excessive pressures in producing the green form prior to firing, should preferably be avoided, but the packing should also not be such as to give an apparent porosity of the finished article of more than about 25%, an apparent porosity of about 21 to 22% being preferred. The apparent porosity is that attributable to open pores at the surface of the finished article.

Examples 1 through 3: A mixing for casting in a plaster mold is prepared with water from a mix having the following composition: silicon carbide, A, percent by weight, substantially all of which passes an 8 BSS sieve and is retained on a 120 BSS sieve, alumina, B, percent by weight, all of which has a particle size less than 10 μ, and clay, C, percent by weight, the clay being formed of equal parts of ball clay and china clay and having a composition consisting of 5 parts by weight alumina, 6.5 parts silica and 1 part impurities, and substantially all particles passing a 300 BSS sieve.

After casting in the plaster mold, the green form so produced is dried and then fired at 1200° to 1500°C, preferably 1350°C. The resulting article has the silicon carbide bonded together and coated by alumina.

The casting mix for Example 1 contains 75% A, 15% B and 10% C. The casting mix for Example 2 contains 65% A, 25% B and 10% C. The casting mix for Example 3 contains 55% A, 45% B and 0% C. In each case a deflocculant was added, the amount being 0.1% by weight of the total weight of the mix.

In a test, blocks formed in accordance with these examples were immersed in superheated oxidized steel at 1620°C for 6 hours. At the end of this period, there had been no measurable pickup of carbon or silica by the steel.

PROTECTIVE COATING FOR MATERIALS EXPOSED TO MOLTEN ALUMINUM

E.S. Gamble; U.S. Patent 3,754,949; August 28, 1973; assigned to Olin Corporation describes a composition which may be used as a coating on articles to be placed in molten aluminum, which comprises 30 to 70% by weight of a refractory material consisting of tricalcium phosphate, 10 to 30% by weight aluminum oxide, 2 to 8% calcined clay, 3 to 10% raw air floated kaolin clay and 12 to 45% magnesium orthoborate. From 7 to 25% boric acid plus 5 to 20% light magnesium oxide may be employed in place of the magnesium orthoborate and which converts to magnesium orthoborate upon the addition of water.

This coating protects a substrate from virtually any attack by molten aluminum and aluminum alloys for long periods of time. The coating adheres tenaciously, is thermally shockproof, nonwetting and easily cleaned after use. When applied to oxide and other refractory crucibles and ceramic bodies, it also exhibits the property of inhibiting penetration of molten metal and oxides, thus extending life and minimizing contamination of the melt.

An additional advantage is the abllity of the coating composition to reduce to a substantial extent the rate of oxygen uptake by the underlying metal to which it is applied despite the exposure of the coated metal to high temperature oxidizing atmospheres. A further advantage is that the coating material does not hydrate when in use and consequently the danger of a molten aluminum steam explosion is eliminated.

Graphite in an amount of from 4 to 30% by weight may also be included, if desired, in order to increase the applicability of the coating composition as, for example, by painting. The graphite may also be premixed with the other nonaqueous ingredients if desired.

Example: The following dry coating ingredients were prepared by mixing the following finely divided powders in a suitable blender until uniformly mixed:

	Percent by Weight
Tricalcium phosphate, commercial bone ash, −325 mesh particle size, 97% $Ca_3(PO_4)_2$ min	32.0
Aluminum oxide powder, −200 mesh	14.4
Calcined kaolin clay, −325 mesh	3.2
Kaolin clay, air floated, water washed, −325 mesh	3.2
Boric acid powder, fine	17.6
Magnesium oxide, light	9.6
Graphite	20

The above powders were intimately mixed and then tempered with clean water to prepare an aqueous slurry containing 60% by weight water. This slurry was brushed on degreased nodular cast iron specimens and air dried. After air drying for 5 minutes, the coated samples were oven dried at 300°F to remove all remaining moisture.

The coated metal part was then immersed in molten 5A grade commercially pure aluminum at 1380°F, removed and reinserted over a 4-hour period at frequent intervals. After each removal the coated specimens were easily cleaned of any adhering aluminum film and the substrate found unattacked in each case and at the conclusion of the test.

CLAY-GRAPHITE POURING SPOUT

F.M. Mitchell III, U.S. Patent 3,752,372; August 14, 1973; assigned to South-wire Company describes a pouring spout for pouring molten metal in a continuous casting system which includes an elongated tubular spout of oxidation-resistant material, including an enlarged, tapered support shank at the upper end of the pouring spout, which comprises less than half the length of the pouring spout, and combining with a complimentary tapered portion on the pouring pot to give a secure fit between the pouring spout and the pouring pot, a lower portion of the pouring spout being exposed to the atmosphere and comprising over half the total length of the pouring spout.

The oxidation-resistant materials comprise 28 to 70% carbon, 0 to 36% silica, 0 to 22% alumina, 0 to 52% silicon carbide, 0 to 20% silicon, 0 to 20% glaze and 0 to 5% other refractory substances. The glaze generally has the following composition: 0 to 3% feldspar, 0 to 4% borax, 0 to 5% cryolite, 0 to 3% calcium silicate and 0 to 10% silicon.

Successful pour spouts have been constructed having a tapered bore with a diameter of 1¼ inches at the large end and approximately $^{13}/_{16}$ inches at the small end, i.e., the included angle of the taper being 1° or more, and having a straight bore with a $^{13}/_{16}$-inch diameter. The radius of the rounded inlet on both the tapered bore and straight bore pour spouts must be equal to or greater than the radius of the uppermost portion of the bore.

Before molten metal is allowed to flow through the pouring spout, the pouring spout should be preheated to a temperature approximately equal to the temperature of the molten metal to prevent the initial metal flow through the pouring spout from solidifying before it reaches the casting wheel. After the molten metal is allowed to flow through the pouring spout, the preheating function can be terminated since the molten metal will maintain the temperature of the pouring spout. The average life span is 12 to 15 hours.

Example 1: A pouring spout of the following composition was utilized in a continuous casting system: 70% carbon (graphite), 20% silica (SiO_2), 8% glaze and traces of refractory substances. The dimensions of this pouring spout, having a tapered bore, were as follows: large end overall diameter—3 inches, large end of bore diameter—1¼ inches, small end overall diameter—1¼ inches, rounded inlet radius—$^{11}/_{16}$ inch, small end of bore diameter—$^{13}/_{16}$ inch, overall length 17¾ inches. The pouring spout was preheated to a temperature equal to the temperature of the molten copper to be poured. Then the molten copper was poured through it into the casting wheel. The pouring spout remained in continuous service for 14 hours and 45 minutes before the casting operation was terminated.

Example 2: A pouring spout of the following composition was utilized in a continuous casting system: 50% carbon (graphite), 20% silica (SiO_2), 20% silicon,

8% glaze and traces of refractory substances. The dimensions of this pouring spout were the same as the dimensions of the pouring spout of Example 1. The pouring spout was preheated to a temperature equal to the temperature of the molten copper to be poured. Then the molten copper was poured through it into the casting wheel. The pouring spout remained in continuous service for 12 hours before the casting operation was terminated.

GLASS POLISHING RESIDUE FOR INSULATING COMPOSITIONS

N.F. Tisdale, Jr. and J.F. McCarthy; U.S. Patent 3,732,177; May 8, 1973; assigned to The Union Commerce Bank describe an exothermic insulating composition and articles made therefrom, particularly an article for use as or in hot tops placed on the top of ingot molds during ferrous metal ingot casting operations. The composition has the following parts by weight: 50 to 82% of finely divided refractory material, 4 to 12% asbestos fiber, 0 to 12% of a cellulosic material, 0 to 12% of a powdered reactive metal, 0 to 14% of metal oxide, and 3 to 12% of a binder resin.

The finely divided refractory material comprises, by weight of total composition, at least 10% finely divided glass polishing residue, 0 to 72% finely divided sand, 0 to 20% silica flour and 0 to 20% diatomaceous earth. Preferably the finely divided refractory material comprises 60 to 79% by weight of the total composition; the glass polishing residue comprises at least 15%, and the finely divided sand comprises 0 to 64%. It is preferred that the finely divided refractory material have a particle size below 60 mesh, preferably below 70 mesh, i.e., less than 0.25 mm.

The glass polishing residue imparts equal or better insulating properties at particle sizes substantially larger, e.g., –60 or –70, than the very fine particle sizes, e.g., –200 mesh, generally required to obtain optimum insulation with materials such as finely divided sand, silica flour and the like. As a general rule greater porosity and, therefore, better insulation is achieved with smaller particle size refractory materials. The glass polishing residue with a larger particle size range imparts as good or better insulation properties without the disadvantages of greatly reduced filtration rates normally caused by smaller particle size refractory additives. Thus processing of the exothermic insulating compositions is considerably easier, faster and more economical.

Binder requirement for a given weight percent of refractory mix is also reduced by smaller overall surface area of the larger particle sizes with resulting economies. In addition to excellent insulation capacity, other physical properties of the finished product are superior including resistance to cracking and spalling, resilience, resistance to breakage and damage during handling and storage, and disintegration after use.

The asbestos fiber may be either solid or tubular and any of the shorter-length, more economical fibers may be used as is desired. The cellulosic material may be wood flour, ground-wood, pulp, shredded paper and like materials, and preferably wood flour.

The powdered reactive metal may be aluminum, boron, aluminum-silicon, boron-silicon, calcium-silicon, ferro-silicon, ferro-silicon-aluminum, and of these, aluminum,

aluminum-silicon and calcium-silicon are most preferred. It is preferred that the particle size approximate that of the finely divided refractory material.

The metal oxide may be any metal oxide that is compatible with the other components of a composition, exothermically reactive with the reactive metal at the teeming temperatures of the molten metal being cast, and substantially nonreactive at mixing, storage and handling conditions of the exothermic insulating compositions. Those metal oxides which are preferred are manganese dioxide and/or iron oxides.

The binder resin may be the silicates, polymers such as phenol-formaldehyde, urea-formaldehyde, polyvinylacetate and/or furfuryl alcohol. Adjustment of the pH of solutions of the compositions by the use of additives may be necessary or desirable to promote or effect the cure of the binder resin to adhere the composition in a desired shape or form.

Example: A composition made up of 65 parts of finely divided glass polishing residue from a glass polishing operation, 5 parts wood flour, 8 parts of asbestos fiber, 8 parts of powdered aluminum, 9 parts of manganese dioxide and 7 parts of phenol-formaldehyde was mixed thoroughly with sufficient water to form a slurry having 15% by weight solids. Sufficient of the mixed slurry composition was added to a mold to form a rectangular hot top 12 inches square and 10 inches high and a major portion of the water was removed from the slurry by pressure. The wet molded composition was removed from the mold and cured at 380°F for 2 hours 15 minutes. The prepared hot top was used under ordinary teeming operating conditions and performed effectively both for insulation and resistance to breakage.

A portion of the prepared composition was molded also into a test panel and tested for its insulating properties. It was found to have an insulating K value of under 1.7 at 2500°F.

INSULATING CASTABLE OF ACID AND/OR ALUMINA, CLAY AND CEMENT

J.A. Crookston and G.W. Charles; U.S. Patent 3,718,489; February 27, 1973; assigned to A.P. Green Refractories Co. describe an insulating refractory castable useable above 2800°F, specifically at 3000°F, which has a bulk density of 82 to 88 lb/ft³ in place. The product comprises 60 to 70% of a size graded acid and/or alumina refractory aggregate, 5 to 15% expanded clay aggregate of a density of 9 to 11 lb/ft³ and 15 to 35% calcium aluminate cement.

The nonbasic size graded aggregate may be calcined alumina, calcined South American bauxite, calcined diaspore, burley diaspore, kyanite and other fire-clay, and silica-alumina refractory materials. When the aggregate contains superduty fire clay, the clay must have a Pyrometric Cone Equivalent of at least Cone 33. This aggregate has a typical size gradation as follows:

	Preferred	Range
Held on 10 mesh	11.0%	11.0 ± 3.0%
Pass 10 on 28 mesh	56.0%	56.0 ± 3.0%
Pass 28 on 65 mesh	23.0%	23.0 ± 3.0%
Pass 65 mesh	10.0%	10.0 ± 3.0%

The clay aggregate has a typical chemical analysis as follows:

	Preferred	Range
Silica (SiO_2)	56%	52 to 58%
Alumina (Al_2O_3)	40%	38 to 42%
Iron oxide (Fe_2O_3)	2%	1.7 to 2.2%
Lime (CaO)	1%	0.6 to 1.1%
Alkalies ($Na_2O + K_2O$)	1%	0.8 to 1.2%

It has a size gradation as follows:

	Preferred	Range
Held on 4 mesh	Trace	0 to trace
Pass 4 on 10 mesh	12	12 ± 3%
Pass 10 on 28 mesh	88	88 ± 3%
Pass 28 on 65 mesh	Trace	0 to trace
Pass 65 mesh	Trace	0 to trace

The expanded clay aggregate is made from Missouri plastic clay preferably, but it can be made from flint clay, kaolin and ball clay.

The calcium aluminate cement has a typical chemical analysis as follows (by weight on the basis of an oxide analysis):

Silica (SiO_2)	0.1%
Alumina (Al_2O_3)	79.0%
Iron oxide (Fe_2O_3)	0.3%
Lime (CaO)	18.0%
Magnesia (MgO)	0.4%
Alkalies (Na_2O)	0.5%
LOI ($1100°C$)	1.5%

A typical castable usable at 3000°F will contain in the aggregate portion raw kyanite, calcined fire clay and calcined alumina. A 3000°F castable has a screen analysis of 0.5 to 5% +10 mesh, 38 to 48.5% –10+65 mesh and 51 to 57% –65 mesh.

Examples 1 and 2: In the following examples, Example 1 is an insulating refractory castable useful at 3000°F and Example 2 is an insulating refractory castable useful at 2800°F. In formulating both of these compositions, the products are dry blended and bagged. The bagged product then is mixed with water and poured into place.

The 631-K is a Missouri superduty flint clay which has been calcined above 2600°F. The terms 100/F, 325/F, 35/F and 9/F mean that the materials have been ground to 100 mesh, 325 mesh, 35 mesh and 9 mesh respectively, and the materials include all fines below that mesh. BM-48 mesh is 631-K material ball milled to minus 48 mesh size. Refcon cement is a calcium aluminate cement having a typical analysis as follows:

Silica (SiO_2), %	5.3
Alumina + titania ($Al_2O_3 + TiO_2$), %	56.0
Iron oxide (Fe_2O_3), %	1.59

(continued)

Lime (CaO), %	36.30
Magnesia (MgO), %	0.16
Alkalies (Na_2O + K_2O), %	0.03
LOI, %	0.62

Composition Properties

	Example 1	Example 2
Expanded aggregate (9-11 pcf), %	10.0	8.0
Raw kyanite 100/F, %	25.0	25.0
CA-25 cement, %	25.0	15.0
Calcined Missouri superduty flint clay (631-K) 9/F, %	27.5	–
Calcined alumina, 325/F, %	2.5	2.5
Raw kyanite, 35/F, %	10.0	10.0
Calcined Missouri superduty flint clay (631-K) BM-48 mesh, %	–	32.0
Refcon cement, %	–	7.5
Moisture, % (added)	18.5	22.0
Bulk density, pcf		
After drying at 220°F	90	85.9
After reheating at 1500°F	85	82.2
Modulus of rupture, psi		
After drying at 220°F	300	230
After reheating at 1500°F	210	178
After reheating at 2000°F	230	183
After reheating at 2500°F	470	478
Cold crushing strength, psi		
After drying at 220°F	1250	1066
After reheating at 1500°F	920	912
After reheating at 2000°F	810	825
Linear change, % (from mold size)		
After reheating at 1500°F	-0.11	-0.18
After reheating at 2000°F	-0.20	-0.26
After reheating at 2500°F	+2.20	+1.50
After reheating at 2700°F	–	-0.40
After reheating at 2900°F	+1.50	–
Screen analysis, %		
Retained on 10 mesh	3.0	1.5
-10+65 mesh	40.0	15.1
Pass 65 mesh	57.0	83.5

FURNACE LININGS

RELINING LADLES, SOAKING PITS AND FURNACES

C.B. Murton; U.S. Patent 3,897,256; July 29, 1975; U.S. Patent 3,856,538; Dec. 24, 1974; and U.S. Patent 3,737,489; June 5, 1973 describes a method for relining metallurgical ladles, soaking pits, and furnaces at temperatures of at least 250°F. The method consists of applying a refractory mixture to be deposited as a lining upon the walls of the furnace or ladle to a desired thickness, whereby maximum adherence is obtained through melting of the organic binder present in the refractory mixture and moisture removal is indicated by a change of color of the lining.

The mixture has a particle size of 8 mesh and consists of, by weight, ½% to 4% sodium silicate, ½% to 4% organic binders, 20% to 80% of clay, 12% to 72% quartzite, and (in most cases) 4% to 6% of water. The advantage of the composition is the reduction of the amount of material needed to provide an additional heat in steel ladle lining life from 0.9 ton to 0.7 ton, a reduction of 22%.

After a heat of liquid metal is poured from a ladle, the ladle is normally inverted to pour out any remaining slag. Thereafter, the stopper rod and nozzle are removed and replaced. The ladle is then ready for reuse, after allowing sufficient time for a reset nozzle to dry out. When the ladle has been used a sufficient number of times, such as 15 to 20 heats, the lining is normally worn thin enough to require replacement. After 70% of the normal lining life has been used, this process is employed to apply a replacement lining on the inner surface of the ladle.

The process comprises: holding a ladle to be relined at a temperature of at least 250°F; applying a mixture of refractory material upon the inner surface of the ladle to the desired thickness; and holding the ladle for a period of time sufficient to expel any included moisture, to allow the organic binder in the mixture to melt and carbonize.

The organic binders may include such materials as pitch, tar, rosins, polyvinyl chlorides, and polyethyl tetrachloride. Such binders have melting points of 250° to 400°F.

ALUMINA-SILICON CARBIDE-CARBON MONOLITH

K. Takeda; U.S. Patent 3,892,584; July 1, 1975; assigned to Nippon Crucible Co., Ltd., Japan discovered that the erosion of relatively large-sized particles in compositions of refractory materials due to molten metals and slag is relatively slow since large-sized particles have small specific surface areas, whereas the erosion of small-sized particles and the refractory matrix, which bonds the coarse particles to each other, is relatively rapid, whereupon the coarse particles, though not eroded, are gradually washed out from the surface of the relatively rapidly eroding matrix.

The monolithic refractory materials consist of 55 to 75% by weight of alumina, 10 to 30% by weight of silicon carbide and 5 to 20% by weight of carbon. A chemical binder in an amount of 5 to 12% of the combined weight of the mixtures of alumina, silicon carbide and carbon is present.

These monolithic refractory materials show excellent erosion resistance against molten iron and slag which can be used, for example, as linings of runners or troughs in blast furnaces for transferring molten metal, slag and the like.

The alumina is preferably fused alumina, a sintered alumina or a mixture thereof having a 94% by weight or greater alumina content. Alumina of high purity has various favorable merits e.g., high refractory properties, a high softening temperature under load, a high mechanical strength, a high erosion resistance against molten metals and various kinds of slag. An alumina where 75% or more of the alumina is coarse particles having the particle size of 1.5 mm or greater is used.

The silicon carbide has various merits such as high erosion resistance against various kinds of slag, high thermal conductivity and a low thermal expansion. The silicon carbide serves to fill the voids of the alumina particles to form a refractory matrix, and therefore the particle size diameter of the silicon carbide is preferably 3.4 mm or less. In particular, it is necessary that 60% or more of the silicon carbide be fine particles having a particle size of 0.15 mm or less.

As the raw material containing carbon, natural graphite and amorphous carbon which is solid at ordinary temperatures and contains volatile matter, such as pitch, are employed together, or the natural graphite and amorphous carbon can be used singly. The raw material containing carbon is advantageous since it is eroded and wetted by molten metals and slag only with difficulty.

The natural graphite used may be any one which is generally used as a component for forming refractory materials and may have any shape, e.g., as flakes, veins, etc., so long as the ash content thereof does not exceed 25%. However, too-large-sized particles of carbon are not suitable since they have poor dispersibility. In short, pulverized carbon particles having a particle size of 0.3 mm or less are preferred.

The amorphous carbon containing volatile matter, such as pitch, becomes liquid after being heated, and penetrates into the voids or pores of the particles of the other components to adhere thereto, and with a further elevation of the heating temperature the carbon releases the volatile matter to solidify and form a carbon bond. The amorphous carbon is used in the form of pulverized particles having a particle size of 0.5 mm or less so as to improve the dispersion upon mixing the raw materials. As the chemical binder, a sol is used having a solids content of more than 5% by weight such as a silica sol, alumina sol, mullite sol, etc.

The chemical binder does not contract at drying and thus the refractory products have high erosion resistance. In addition, the use of such a chemical binder is extremely advantageous for the formation of a refractory layer of large size. The chemical binder in the form of a sol does not lose its liquidity in the stamping processing of the composition containing the binder, and, therefore, the stamping pressure is evenly transmitted even into the deep portion of the refractory composition. Further, the sol chemical binder is thixotropic in nature, and when an external vibrating force is applied to the mixture containing the chemical binder, the mixture is softened and shows fluidity. Consequently, so-called vibration casting or solid casting can easily be conducted.

PERICLASE AND/OR CHROME ORE RAMMING MIXTURE

N. Cassens, Jr.; U.S. Patent 3,879,208; April 22, 1975; assigned to Kaiser Aluminum & Chemical Corporation describes a refractory composition suitable for ramming, gunning or casting which has good intermediate temperature strength and can be rammed to high densities.

The refractory composition having good strength after heating to intermediate temperatures and which can be rammed to densities of 180 lbs per cubic foot (pcf) or higher consists of sized refractory aggregate including at least 10%, based on the total weight of the composition, MgO-containing grain having at least 50% MgO and passing a 100 mesh screen, from 0.1 to 1.5%, on the dehydrated basis, aluminum sulfate, from 0.2 to 1.5% of an organic acid or salt thereof, and up to 1%, expressed as B_2O_3, of a boron compound.

The refractory aggregate used can be any such material, but preferably will be a nonacid aggregate such as periclase or periclase and chrome ore. Particularly the refractory aggregate is all periclase containing at least 85%, and preferably about 95% or more, MgO. The aggregate is sized to achieve maximum density ranging in size from a 4 mesh screen down to a 325 mesh screen. From 20 to 40% of the aggregate will pass a 100 mesh screen.

At least 10% of the composition, and usually from 20 to 40%, will be aggregate passing a 100 mesh screen and containing at least 50% MgO. This material can be a prereacted magnesia-chrome grain containing the requisite amount of MgO, but preferably is periclase containing at least 85%, and preferably about 95% or more, MgO. In a particularly preferred embodiment, all the refractory aggregate is periclase containing a 95% MgO, about 30% of which passes a 100 mesh screen.

The aluminum sulfate $[Al_2(SO_4)_3]$ is used in the form of a powder, of which substantially all passes a 100 mesh screen. The material can be used in one of its hydrated forms, for example that which contains 14 waters of hydration $[Al_2(SO_4)_3 \cdot 14H_2O]$.

The organic acid used is a material such as citric acid, succinic acid, maleic acid, and the like. As already indicated, the corresponding salts, for example, the sodium salts of these acids, can also be used. The boron compound used is also in granular form, for example, substantially all –28 mesh, and preferably is a soluble boron compound, for example, one of the sodium borates, or, particularly boric acid (H_3BO_3).

The composition is mixed dry and packaged for shipment to the user, who adds the requisite amount of water or other tempering liquid at the time of use. For example, if the material is to be rammed, from 2 to 5% water will be used, whereas in casting 4 to 8% water will be used, depending on the specific sizing of the compositions and its precise composition. In gunning, for example through a gun where water is added to the refractory at the nozzle, from 10 to 20% water will be used. The composition can also be used as a slurry, with up to 30% or more water added, although it is not specifically designed for this application. In addition, the composition can be used to form refractory shapes, for example by pressing, or as a mortar, if desired.

Three compositions were made from seawater periclase sized so that substantially all passed a 6 mesh screen and 30% passed a 100 mesh screen, and having the following typical chemical composition: 2.1% SiO_2, 1.1% CaO, 0.4% Fe_2O_3, 0.3% Al_2O_3, 0.3% Cr_2O_3, and (by difference) 95.8% MgO. The first composition (A) was bonded with 0.8% hydrated aluminum sulfate [$Al_2(SO_4)_3 \cdot 14H_2O$] and 0.6% citric acid. The second composition (B) was bonded with 0.8% hydrated aluminum sulfate, 0.6% citric acid and 0.7% boric acid. The third composition (C) was the same as (B) except that 1.04% borax was used instead of the boric acid. The aluminum sulfate contains $14H_2O$, is 100% –30 mesh and 97% –100 mesh.

Pellets formed at 10,000 psi from these three compositions with 3.75% water added had densities, as formed, of 186.4, 186.2, and 186.1 pcf, respectively, and cold crushing strengths after drying at 150°C of 7,382, 9,939, and 8,603 psi, respectively. After heating to 850°C, composition A pellets had a cold crushing strength of 1,088 psi, B of 2,174 psi, and C of 2,631 psi. The hot load failure temperature (under a 25 psi load) of A was 1700°C, of B 1600°C, and of C 1523°C. Rammed blocks of composition B (with 2.75% water) had a formed density of 184.0 pcf, and cast blocks of the same composition (5.5% water), 179.6 pcf after drying.

Compositions A, B and C were subjected to an explosion test wherein they were each mixed with 5.75% water, on the dry basis, cast into one-gallon cans, and after 4 hours of drying at room temperature and 6 hours' drying at 150°C, placed in a furnace which was heated to 800°C in 1 hour and to 1100°C in the second hour. This test has been found to be a very sensitive measure of the tendency of refractory ramming mixes to explode in use during heatup. None of the three compositions exploded during this test.

The preceding examples can be compared with a composition made from the same periclase but bonded with 0.9% chromic acid and 0.7% boric acid. Pellets of this later composition had a formed density of 183.4 pcf, a cold crushing strength of 8,318 psi after drying at 150°C, and of 2,254 psi after heating to 850°C. Its hot load failure temperature was 1648°C. Thus, it can be seen that these compositions have properties comparable to those of the prior art composition using the potentially hazardous chromic acid bond. This comparison composition also passed the explosion test described above.

Compositions A, B, and C can also be compared with compositions made from the same periclase but using only a single bonding agent, 0.8% aluminum sulfate [$Al_2(SO_4)_3 \cdot 14H_2O$] in Composition X, 0.67% boric acid in Y, and 0.6% citric acid in Z. These compositions had formed densities of 180.4, 180.9, and 181.6 pcf, respectively, but cold crushing strengths of only 1,883, 698, and 2,813 psi after 150°C drying, and only 814, 810 and 312 psi after heating to 850°C. Thus, it is all the more surprising that a combination of these three materials leads to such high intermediate temperature strengths.

Compositions A, B, and C can also be compared with a composition made from the same periclase and bonded with 1% aluminum sulfate and 0.67% boric acid. This composition had a formed density of 180.4 pcf and a cold crushing strength after drying at 150°C of 5,133 psi. After heating to 850°C, the cold crushing strength was 1,045 psi.

GUNNABLE REFRACTORY FOR ARC FURNACE WALLS

J.R. Parsons and H.L. Rechter; U.S. Patent 3,846,144; Nov. 5, 1974; assigned to Chicago Fire Brick Company describe a gunnable composition suitable for the gunning maintenance of the walls of arc melting furnaces.

The refractory is high in graphite or calcined coal content, with excellent oxidation resistance. It develops rapid strength on hot or cold gunning, so that it is not knocked off the wall during charging of scrap. Metallic silicon and bauxite, preferably calcined refractory grade bauxite, are present for oxidation protection of the graphite. Bonding is provided by a combination of calcium aluminate and silicate cements which achieve an accelerated set on hot or cold surfaces. The silicon provides good bonding on firing.

When the bauxite is not present, oxidation is apparently excessive. The addition of clays can result in loss of material by steaming off the hot walls. The cement preferably should be 2 parts calcium aluminate to 1 part calcium silicate, preferably a highly early type for most rapid strength development. Graphite should be at a minimum of 40% to dominate the refractory properties; that is, impart infusible characteristics, with a 6 mesh or 1/8 inch size best for gunning. Grain sizings over 1/8 inch usually cause much more rebound, and finer graphite sizings result in dusting. The composition can be formulated with the following percentages by weight:

	Percent
Graphite, or calcined coal, ¼-inch diameter maximum to 100 mesh minimum sizing	40–60
Bauxite, ¼ inch maximum	20–40
Silicon, 0 to 5% Fe, 60 mesh maximum	8–20
Calcium aluminate	6–16
Calcium silicate	3– 8

The bauxite is preferably calcined to remove the water and is of the high alumina refractory type containing less than 15% of the oxides of iron, titanium and silicon. The calcium aluminate may be in the form present in an aluminous cement containing at least 30% by weight of alumina.

The calcium silicate may be present as an ingredient of Portland cement. There is no minimum grain size for the bauxite and silicon, 100 mesh and finer being suitable. The graphite and calcined coal are preferably present in 30 mesh or larger size.

Example: The following formulation successfully provides protection for arc furnace walls eliminating the need of daily maintenance:

	Percent
Mexican graphite, –6, +30 mesh	45
Calcined refractory bauxite, 1/8-inch diameter maximum	30
Silicon, –100 mesh	10
Calcium aluminate cement (aluminous cement)	10
High early cement (Portland cement)	5

This material was applied pneumatically to the hot walls of a cooling arc furnace in thin layers as well as in several inches of thickness in eroded areas. The relatively coarse structure of the deposit was permeable by steam, preventing the loss of material by being pushed off by steam, a common occurrence in hot gunning. The quick firming action of the accelerated cement reaction prevented slumping. Upon completion of gunning, the furnace could be immediately charged and in operation, and a subsequent maintenance program would allow at least two days between furnace shutdowns, but generally longer since this material can be used for emergency repair of the refractories, not feasible with clay-based materials.

ALUMINA FOR PUSHER-TYPE REHEATING FURNACES

Pusher-type reheating furnaces are used in metallurgical operations for bringing metal ingots of great weight as for example a ton or more, or metal billets of considerable length, for example, of 6 meters or longer to the rolling temperature. The ingots or billets are moved by means of a pushing machine over the floor of the furnace, which is provided with a refractory lining, and are there heated to over 1,200°C. The refractory lining on the floor of the furnace is thus, of course, subjected to exceedingly severe stress, not only by the fact that the workpieces mechanically attack the lining, but also by the action of the scale that falls from and is knocked off the workpieces and into contact with the lining materials.

Another source of trouble is the casting crusts which flake off from the workpieces as they are passed through the furnace and which contain the compounds which melt at comparatively low temperatures and chemically attack the furnace lining. These low-melting compounds originate from the additives which are introduced into the ingot molds when the ingots are cast, so as to keep the runner gate fluid. Further, the bur on the top end of the ingot produced when the top of the ingot is sheared off to eliminate the pipe mechanically attacks the floor and the skidways. Lastly, it is conventional in many furnaces to move the ingot over a tripper so that the ingot is reversed, i.e., turned over and the dark spot formed on the ingot by the water-cooled skid is compensated for.

In addition to this mechanical wear-and-tear on the furnace lining there must be added the thermal stress on the lining, which also adversely affects the metal, as both when the furnace is charged with the ingots and billets and when they are removed, a door has to be opened through which the cold air flows into the furnace which suddenly chills the hot furnace lining.

Accordingly *M. Blanke and K. Hass; U.S. Patent 3,844,803; Oct. 29, 1974; assigned to Dynamit Nobel Aktiengesellschaft, Germany* describe cast refractories which can be employed as linings for the floors and skidways of reheating furnaces consisting of 97 to 99 wt-% Al_2O_3, 0.3 to 0.8 wt-% SiO_2, 0.2 to 0.4 wt-% TiO_2, 0.05 to 0.11 wt-% Fe_2O_3, 0.33 to 0.94 wt-% CaO or MgO or mixtures with 0.22 to 0.24 wt-% Na_2O and/or K_2O.

The special advantage associated with the use of this material consists in the fact that the eutectic point between Al_2O_3 and FeO is then situated above the operating temperature utilized in reheating furnaces, namely at 1340°C. Further bricks having this composition cannot be adversely affected by the formation of cristobalite, as they contain only an extremely small percentage of SiO_2. They are therefore more resistant to temperature variations than fusion-cast bricks having contents of 20% SiO_2.

The products, are also more wear-resistant when employed as linings for pusher-type reheating furnaces. As no infiltration of FeO occurs, the scale cannot get burnt into the bricks, even under the application of great pressure by the ingots. Even a possible fall of an ingot onto the draw floor does not harm the lining since fusion-cast products have a cold compressive strength greater than 2,000 kg/cm^2 as compared with a specific loading of 1.5 kg/cm^2 on the floor of the pusher-type reheating furnace.

When this material is used, therefore, it is no longer necessary to reinforce the floor as hitherto required by means of heat-resistant sheet bars and plate bars which entails considerable cost and in the end is not very satisfactory.

Bricks and shaped bodies consisting of 98% by weight of Al_2O_3, 0.3% by weight of SiO_2 and the balance of TiO_2, Fe_2O_3 and the above alkaline earth oxides and alkali oxides, have proven valuable as materials for use in the lining of pusher-type reheating furnaces.

Twelve tons of calcined alumina, 52 kg of lime, 32.5 kg of rutile, 19.2 kg of magnesium oxide and 84 kg of quartz sand were fed into an electric arc furnace and melted by an oxidizing atmosphere (normal pressure) at 2300°C. This received melt was cast into forms and therein cooled for 5 days by using heat insulators.

The formed parts having the dimensions of at least 600 x 300 x 200 mm had, with the exception of the differences of 2 mm, the nominal values of the dimensions expected. A jointing therefore was not necessary. The compound consisted, among others, also of Fe_2O_3, Na_2O, and K_2O. The latter oxides partially came from the calcined alumina, and the iron oxide from the walls of the electric furnace and the forms. The formed parts were characterized by a considerably high resistance to pressure, by abrasiveness and resistance to temperature changes.

HIGH CARBON REFRACTORY WITH SILICON

Carbon and graphite refractories and mineral refractories containing elemental carbon are exceptionally suitable for applications requiring resistance to slags and metals at very high temperatures. Unfortunately, the high oxidation rate of carbon limits its usefulness to areas covered by molten materials where exposure to oxygen is minimized or to situations where the burning rate is tolerated with periodic replacement or frequent maintenance patching. Replacement requires cooling of the installation, whether a furnace, trough or spout, and consequent downtime. Patching has always left much to be desired as the performance of the applied material is much dependent on preparation of the eroded surface, quality of the ramming or gunning placement techniques, care taken in drying or curing, heatup schedule, etc.

The use of brick linings in ladles, furnaces, and metal and slag runners is advantageous in providing refractories with maximum slag and metal resistance, superior in density to rammed or gunned materials and not subject to the dangers of a dry-out or cure, as with wet rams and plastics or with materials deriving their bond from pitches. A carbon brick or block lining, however, requires constant maintenance to prevent their burning by overcoating with a variety of carbonaceous patching formulations, and their manufacture requires high-temperature treatments under totally oxygen-free atmospheres at considerable expense in comparison with firing of clay-based brick.

Accordingly *J.R. Parsons and H.L. Rechter; U.S. Patents 3,842,760; October 22, 1974; and 3,810,768; May 14, 1974; both assigned to Chicago Fire Brick Company* describe both shaped articles in the form of blocks such as brick or other shapes consisting of a matrix of clay, grog, alumina, bentonite, bauxite, zircon, or mixtures thereof with 20 to 80% amorphous carbon or graphite, preferably 40 to 60%, and 5 to 30% of silicon or ferrosilicon, preferably 10%, which are fired in a kiln to provide extraordinary oxidation resistance and strength. Firing the brick in a conventional kiln at ordinary temperatures for fired clay brick gives these desired properties. Suitable temperatures are 2400° to 3200°F, near or above the melting point of silicon (2570°F), to achieve the oxidation protection.

The shaped products should be brought through the range of oxidizing temperatures of 800° to 2500°F in no more than 4 hours. Firing is preferably continued at above 2500°F for at least 3 hours. The fired brick can then be employed in high-temperature service under oxidizing conditions, with a very low rate of carbon burning and high strength for withstanding the impact of molten metals and slags.

Instead of firing the composition, the composition may include bonding agents such as sodium silicate, phosphoric acid, ammonium phosphates (preferably mono- and diammonium phosphates), aluminum phosphates (preferably monoaluminum dihydrogen phosphate and others used in the art) in bonding amount such as 4 to 10% by weight. These shaped compositions are then dried to remove water, such as at above 212° to 500°F.

PROTECTING LININGS FROM SKULL AND SKULL REMOVAL

Refractory-lined metallurgical vessels such as ladles or other items are subject to skulling. Heat transfer, which always occurs to some extent through the refractory lining of such vessels, results in gradual chilling and solidification of a layer of metal adjacent the refractory surface. Eventually a solidified metal skull builds up to a point where it is necessary to take the vessel out of service to remove the skull. Because the interior surfaces of the refractory lining of the vessel are always to some extent uneven, and because cracks and fissures develop in the refractory lining, skull removal has been attached by considerable damage to the refractory lining. Molten metal finds its way into cracks and fissures in the lining.

Chunks of refractory are commonly broken and removed with skull. Moreover, sliding of the uneven surface of the skull across the uneven surface of the refractory lining results in abrasion and breakage of the refractories. The result is that the refractory lining employed in a vessel such as a tundish must ordinarily be replaced after seven to 10 skull-removing operations. Such refractory linings are commonly fabricated of alumina or other acidic brick, but may also be made from other refractory materials, either bricks or as a monolithic lining.

Accordingly *M.E. Harnish and M.A. Peters; U.S. Patent 3,830,653; August 20, 1974; assigned to International Minerals & Chemical Corporation* describe a method for protecting the refractory lining of metallurgical vessels from the destructive effects of skulling and skull removal by applying to the molten-metal-adjacent interior surfaces of the refractory lining a thin layer of a special refractory parting composition. The layer is applied by conventional means such as trowelling, gunning or brushing in place while the vessel is at room temperature, but can be applied at a slightly elevated temperature, 225° to 300°F. A thin parting layer, which may be ¼ inch to 1 inch in thickness, and preferably ½ inch in thickness, is applied uniformly to the interior surfaces of the refractory lining.

The special refractory parting composition must be capable of application at slightly elevated temperatures; must be sufficiently adherent to remain in place without slumping until it hardens; and must be capable of setting without blistering as the vessel is heated prior to returning it to service. Moreover, the special parting composition must set up to sufficient strength so that it will withstand the pouring of metal into the vessel. It must be nonbonding to the refractory lining of the vessel throughout the temperature range at which it will be employed, namely, 225° to about 3000°F, but should bond securely to the solidified skull at skull-forming temperatures. Moreover, the special parting composition must crumble freely during skull removal to provide minimum abrasion against, and damage to, the refractory lining.

All of the foregoing criteria can be met by employing as the refractory parting composition an aqueous mortar comprising a mixture of a dry aggregate and 10 to 20 parts by weight of water per 100 parts of dry ingredients. The dry aggregate comprises (a) a first grog portion which is –4+50 mesh and provides at least 45%, and preferably 50%, by weight of said dry aggregate; (b) an intermediate fraction which is –50+100 mesh and comprises 2 to 10% by weight of said dry aggregate; and (c) a fines fraction which is –100 mesh and comprises at least 40%, and preferably 45%, by weight of said dry aggregate.

The grog, intermediate, and fine fractions consist essentially of chrome ore, periclase, or mixtures of chrome ore and periclase. The grog and fines fractions are further characterized in meeting one of two conditions, viz, (1) the grog fraction contains chromite ore in the amount of at least 40% by weight of the total dry aggregate and the fines fraction contains periclase (calculated as MgO) in the amount of at least 20% by weight of the total dry aggregate; or (2) said grog fraction contains periclase (calculated as MgO) in the amount of at least 25% by weight of said total dry aggregate and said fines fraction contains chromite ore in the amount of at least 35% by weight of said total dry aggregate.

The aggregate will include 2 to 6%, preferably 4%, of a magnesium sulfate salt. It also preferably includes a clay, especially bentonite, in the amount of ½ to 3%, more preferably 1% by weight, and a plasticizing agent such as methylcellulose in the amount of 0.1 to 0.5%, but more preferably in the amount of 0.2% by weight. The mixture should be substantially free of nonchromite-form iron.

A dry aggregate is prepared using 50 parts by weight of Grootboom chrome ore. The ore is in the form of a grog which is substantially 100% –5 mesh and 98% +100 mesh, 94% of the grog being +50 mesh. The periclase constituent is prepared from fired bodies which contain periclase and some chrome ore and are ground to substantially 100% –100 mesh. The periclase is incorporated in the amount of 44.8% by weight. Epsom salt in the amount of 4%, bentonite in the amount of 1% and methylcellulose in the amount of 0.2% are added and the total is mixed together to form a uniform aggregate. The approximate chemical and screen analyses of the dry mix are as follows.

Chemical Analyses

	Percent
MgO	36.0
Cr_2O_3	26.0
Fe_2O_3	18.0
Al_2O_3	10.0
SiO_2	7.0
CaO	0.7
SO_3	1.3
LOI	1.3

Screen Analyses

Total on 5 mesh	0.0
Total on 10 mesh	26.0
Total on 20 mesh	41.0
Total on 100 mesh	54.0
Minus 200 mesh	37.0

PATCHING MIX FOR BLAST FURNACE FLOOR

G.F. Paolini; U.S. Patent 3,826,662; July 30, 1974; assigned to Bethlehem Steel Corporation describes a patching mix which when mixed with water is trowellable, and which contains a high percentage of carbon. The mix is suitable for repairing the refractory lining in the network of troughs, runners, dams, cinder falls and spouts in a blast furnace casting floor. The patching mix consists of 60 to 80% carbon which can be in the form of particles of coke breeze, 10 to 30% fireclay and 1 to 20% pitch. The patching mix is blended with sufficient water to produce the consistency and the plasticity needed to hand-trowel the patching mix into the area being repaired and to adhere to the surrounding refractory lining until it has been burned-in.

The pitch can be coal pitch or petroleum pitch having a softening point of 300° to 340°F. The fireclay can be any one of several well-known fireclays having typical chemical compositions listed below:

	Percent
Alumina	15
Silica	68
Iron Oxide	1.5
Titania	1.0
Lime	0.2
Magnesia	0.2
Alkalies	2.0

It has been found that a fireclay having a PCE (Pyrometric Cone Equivalent) of 19 to 24 has sufficient refractoriness, bonding and resistance to erosion to be used. Carbon in the form of particles of coke breeze is used as the main constituent of the patching mix. Coke breeze can have a carbon content of 70 to 92%, 1.0 to 3.0% volatile matter, 15 to 25% ash and 10 to 20% moisture. The coke breeze, which is obtained by screening coke, is crushed to reduce the particle size sufficiently so that 90% of the particles will pass a 1/8 of an inch screen (USS).

Example 1: A mix comprising 1,400 pounds of coke breeze, all particles of which passed a 1/8 of an inch sieve (USS), 400 pounds of fireclay having a PCE of 21 and 200 pounds of coal pitch having a softening point of 310°F, was blended with 25 gallons of water per ton of dry mix. The mix was used to line a cinder fall. The bottom of the cinder fall was covered with a coating of 4 inches to 5 inches and the sides of the fall with a coating of 2 inches to 3 inches by hand trowelling the mix. The time to coat the cinder fall was ½ hour. Then iron sheet was placed over the trowelled mix.

The mix was fired at 1200° to 1800°F for 1 hour. The lining was in place for 90 flushes, i.e., 90 passes of molten slag, and did not require any maintenance other than routine cleaning during this time.

Example 2: A mix comprising 1,400 pounds of coke breeze which was crushed so that 100% of the particles passed a 1/8 of an inch sieve (USS), 300 pounds of fireclay having a PCE of 22 and 300 pounds of coal pitch having a softening point of 324°F, was blended with 20 gallons of water to a consistency of stiff cement. The mix was trowelled into an iron trough to protect the carbon blocks lining the trough. The coating of the mix in the trough was 5 inches in thickness. An iron sheet was placed over the mix which was burned-in at 1200° to 1800°F for 1½ hours. The mix was still in place and giving good service one month after installation.

PREPARATION OF SINTERABLE ALUMINUM TITANATE POWDER

Aluminum titanate is known for its high refractoriness (melting temperature), its near-zero coefficient of linear thermal expansion which results in a high thermal shock resistance, and its high negative temperature coefficient of electrical resistivity. Such physical properties are desirable for products used in high-temperature applications such as required of thermal insulators, ablators, furnace liners, and crucibles.

Accordingly, *W.K. Duerksen, C.E. Holcombe, Jr., and M.K. Morrow; U.S. Patent 3,825,653; July 23, 1974; assigned to U.S. Atomic Energy Commission* describe a method for preparing sinterable aluminum titanate (Al_2TiO_5 or $Al_2O_3 \cdot TiO_2$) powder and products therefrom characterized by an average linear coefficient of thermal expansion of less than 1×10^{-6} in/in °C over a temperature range of about 25° to 1000°C. Such products in thin-walled configurations such as crucibles have an average compressive strength of 5,000 psi and a density greater than 85% of theoretical (3.73 gm/cc) when isostatically pressed and sintered, and a density of 70% of theoretical when slip-cast and sintered.

Generally, the method for preparing the aluminum titanate powder comprises mixing a solution of an aluminum compound at a plus-3 valence state and a titanium compound at a plus-4 valence state, coprecipitating the aluminum and titanium from solution as aluminum titanium hydroxide, filtering the precipitate, drying the precipitate, and thereafter calcining the aluminum titanium hydroxide at a temperature adequate for removing volatiles and increasing the average size of the particulates. The resulting calcined Al_2TiO_5 powder may then be formed into the desired product configuration by isostatic pressing and sintering, by slip casting and sintering, or by any other suitable powder metallurgical technique.

Example 1: Sinterable aluminum titanate was prepared under an argon atmosphere by dissolving one kilogram of aluminum isopropylate in two liters of benzene, and adding to the resulting solution 736 milliliters of titanium isopropylate. The benzene solution was refluxed for 3 hours to insure homogeneity of the solution and then allowed to cool. The aluminum and titanium isopropylates were hydrolyzed and coprecipitated from the solution by adding demineralized water dropwise to the solution as it was being stirred.

The hydroxides thus formed were collected by filtration and vacuum dried in an oven at 50°C for 48 hours. The dried hydroxides were calcined at 800°C in air for 40 hours. The aluminum titanate hydrate powder had an average particle size of 50 angstroms before calcining and an average particle size of 650 angstroms after calcining. The calcined product was compacted into a cylinder 1 inch in length by 0.25 inch in diameter at an isostatic pressure of 30,000 psi and sintered at 1650°C for 2 hours in an argon atmosphere.

Analytical data indicated the cylinder had a bulk density of 3.16 g/cc which is 85% of theoretical, an average compressive strength of 5,000 psi and an average coefficient of linear thermal expansion of 0.5×10^{-6} in/in/°C over 25° to 850°C.

PERICLASE-OIL-WAX READY-TO-USE RAMMING MIX

Refractory ramming mixes are used for forming monolithic refractory masses, for example the bottoms of furnaces such as electric steelmaking furnaces or open-hearth furnaces.

Accordingly, *N. Cassens, Jr.; U.S. Patent 3,816,146; June 11, 1974; assigned to Kaiser Aluminum & Chemical Corporation* describes a ready-to-use refractory ramming mix which can be rapidly burned in immediately after forming. It is made from a composition consisting of 0.5 to 3% of a nonsetting oil having a pour point no higher than 0°C, a flash point of at least 100°C, a viscosity at 20°C of from 10 to 100 centipoises, a viscosity at 0°C of not over 1,000 centipoises, and a viscosity at 100°C of not less than 1 centipoise; and from 0.5 to 3% of a wax having a congealing point of at least 40°C; the remainder of the composition being refractory grain together with up to 3% bonding and sintering additives, at least 10% of the composition being hydratable refractory grain passing a 100 mesh screen.

The refractory grain used may be any such material, but the advantages of the nonaqueous bond are achieved when the composition contains periclase, particularly fine periclase, or some other hydratable material. The oil used may be any such material having the specified rheological properties, and can be a petroleum oil.

The wax can be any such material having the specified congealing characteristics. Petroleum-based waxes are a preferred form of wax, particularly when the oil used is a petroleum oil.

In making compositions, the wax is added to the oil and the admixture heated until the wax is completely dissolved in the oil. The heated oil-wax mixture is then added to the grain, and the two mixed in suitable equipment such as a muller mixer. Alternatively, the wax can be added to the grain, or a portion of the grain, for example in a ball mill when the fine fraction of grain is being milled, and the oil then added to the wax and grain admixture.

The composition so formulated is then generally bagged and either stored or shipped directly to the user. The latter has only to open the containers of ramming mix, dump it out, and ram it into place in a furnace bottom or wherever desired.

After the composition is rammed in place, it can immediately be burned in. It has been found that the rammed composition can be heated to 500°C at a rate as high as 200°C per hour, and then heated to the operating temperature of the furnace at a rate as high as 120°C per hour.

It can be calculated that, following the above heating schedule, a rammed lining of this composition can be heated to 1650°C in 12 hours. This burn-in time can be compared with that required for a widely used refractory ramming mix which is mixed with water for ramming. This latter mix requires 6 hours drying at room temperature prior to any heating, and must then be heated at a rate such that 1650°C is not reached for 42 hours, a total of 48 hours before the furnace can be used. Thus, this composition can be burned in some four times as fast as a conventional ramming mix using water.

A composition of 96% periclase, 1.5% Chevron base oil 120, 1.5% Chevron slack wax 140, 0.75% pigment grade Cr_2O_3 and 0.25% volatilized silica was made by mixing the wax and oil, heating the combination to dissolve the wax, and adding the combination to the solid ingredients in a muller mixer. The periclase had the following typical chemical analysis: 1.1% CaO, 2.1% SiO_2, 0.3% Al_2O_3, 0.4% Fe_2O_3, 0.3% Cr_2O_3 and (by difference) 95.8% MgO, and was sized so that all passed a 4 mesh screen and about 35% passed a 100 mesh screen. Volatilized silica and pigment-grade Cr_2O_3 both substantially all passed a 325 mesh screen. Slack wax is made by chilling straight crude petroleum to precipitate the wax, which is then separated by filtering. Chevron slack wax 140 has a congealing point of 140°F, a flash point, COC, of 540°F, a specific gravity of 0.883, an oil content of 21% by weight, and is free of phenolic or unusual odor. Chevron base oil 120 has a dark color, a pour point of -40°F (-40°C), a flash point of 335°F (168°C), and a viscosity of 54 centipoises at 22°C.

After ramming at room temperature, this composition had a density of over 175 pounds/cubic foot (pcf), and when tested for cold crushing strength at room temperature after being heated to 850°C, had a strength of over 2,000 psi.

When this composition was rammed into 1-gallon cans and placed in a furnace and heated to 650°C in 10 minutes, and to 1000°C in an additional hour, the mix began to burn with occasional flames after 17 minutes heating (at 730°C), showed low continuous burning after 40 minutes (at 800°C), the burning being ended after 75 minutes (by the time 1000°C was reached).

This very rapid burn-in of the composition can be compared with the same rapid burn-in of a similar composition containing 3% Chevron base oil 120 and no wax. In this case, the composition began burning in less than 12 minutes (before 700°C was reached), and within 33 minutes (at 840°C) continuous long flames were bursting from the specimen. These long continuous flames continued beyond 44 minutes heating (until a temperature of over 900°C was reached), and intermittent burning continued until after 70 minutes (1000°C). The specimen of this comparison mix without wax was severely ruptured after the burn-in test.

COKE OVEN PATCHING MATERIAL

D.H. Hubble and J.G. Yount, Jr.; U.S. Patent 3,814,613; June 4, 1974; assigned to United States Steel Corporation describe a refractory composition for patching coke oven walls. Such a composition can be applied to either hot or cold silica brick walls. It is comprised of 50 to 93% siliceous aggregate, 3 to 15% plastic clay, 2 to 12% chemical binder and 1 to 25% of a suitable source of manganese oxide. The total manganese oxide content of the patching material as MnO should be 0.5% to 15%. The preferred patching material is 72 to 84% siliceous aggregate, 8 to 12% plastic clay, 5 to 8% chemical binder and 3 to 12% manganese oxide source.

The siliceous aggregate is preferably quartzite or other ground silica brick, but may also be some other siliceous material such as sands or clays. The siliceous aggregate is sized minus 3 mesh and preferably minus 10 mesh to allow penetration of the material into the cracks being patched. 5 to 25% of the siliceous aggregate should be sized minus 100 mesh. All screen sizes are Tyler Standard.

The plastic clay is a fire clay product sized -14 mesh, preferably -100 mesh. The pyrometric cone equivalent (PCE) (ASTM) of this clay should be from 26 to 33. The chemical binder is preferably sodium silicate, but may be another chemical binder, such as chromic acid, boric acid, sodium sulfate, magnesium sulfate, sodium phosphate or a suitable organic binder such as lignin sulfates or dextrose. The binder may be added in either solid or liquid form. The manganese oxide source is preferably a manganese ore containing more than 50% manganese (as MnO). Other sources of manganese oxide may be used such as ferro-manganese flue dusts or chemically pure manganese oxides. The manganese oxide source should be sized minus 48 mesh and preferably minus 100 mesh.

The material is mixed in dry form and may be premixed with sufficient water to obtain the desired workability prior to shipment to the patching location, or may be shipped in dry form and mixed with water at the site. 10 to 20% water must be added to obtain a workable mixture when the mix is to be trowelled. Higher amounts of water may be used when pumping or gunning with nutrients.

The patching material may be applied by trowelling or plastering over a cracked area on either a cold or hot wall, by pumping or injecting the material into cracks in a cold or hot wall, or by pneumatic gunning or other suitable application technique. Examples of the composition are given below.

Component and Mesh Size	Trowelling Mix A	Gunning Mix B	C
Quartzite:			
-4	- -	40	0
-10	61	36	0
-100	15	0	0
Siliceous clay, -10	0	0	80
Plastic clay, -100	10	10	10
Sodium silicate	6	6	6
Manganese ore (68% MnO), -100	8	8	4
Water added	13 - 14	0	0

Tests showed excellent bonding of all mixes to used silica brick from a 15-year-old coke oven at 1500 to 2800°F in both oxidizing and reducing atmospheres. Tests using the same mixtures without manganese oxide showed no bonding under the same conditions. Little or no bonding was obtained from experimental mixtures containing a variety of other additives such as iron oxide, titania, magnesia, iron and steel slags, chromic oxide, sodium phosphate, fluorspar, alumina, volatilized silica, chrome ore, sodium-lime-silica glasses, etc.

Mixes A, B and C were applied to both cold and hot coke oven walls. In all cases, the bonding of the patching material to the used silica brick walls was superior to that obtained from any other patching material that had been heretofore employed. Petrographic examination seems to indicate that the excellent bonding results from the formation of a liquid which is subsequently absorbed into the silica brick.

REFRACTORY WITH EXPANDED PERLITE AND ALUMINUM PHOSPHATE

L.J. Jacobs, R.E. Fisher and F.T. Felice; U.S. Patent 3,793,042; February 19, 1974; assigned to Combustion Engineering, Inc. describe moldable, monolithic refractory materials having high insulating values which can be readily and inexpensively installed as furnace linings.

The refractory material comprises a base of calcined diatomaceous earth, bentonite or other plastic clay, expanded perlite, aluminum phosphate and water according to the following composition table:

	Typical Composition Parts by Weight	Range Parts by Weight
Calcined diatomaceous earth	40.0	20 - 70
Bentonite	7.5	3 - 12
Expanded perlite	2.5	0 - 10
Aluminum phosphate	0.0	0 - 10
Water	50.0	20 - 70

The calcined diatomaceous earth is a coarse-graded calcined diatomaceous silica aggregate that has been converted to the crystobalite form by calcining at not lower than 2100°F. This calcining gives the diatomaceous earth maximum volume stability which prevents swelling during the heating cycles.

The maximum bulk density of the calcined diatomaceous earth tapped in place is 35 lb/ft^3 and it will absorb 90 to 110% water by weight. The diatomaceous earth is a relatively low-cost material and it contains microsize pores which are not readily crushed in the normal manufacturing operation and in the installation techniques used to install monolithic materials.

Preferably, expanded perlite is included in the composition to increase the insulating qualities and to improve the strength characteristics. The perlite is extremely light-weight and it begins to soften or melt at 1500°F and 1800°F. This softening or melting characteristic has been found to enhance the strength of the refractory above 1500°F thus reducing the friability of the product. The expanded perlite is produced from raw perlite ore, which is a siliceous glass containing 2 to 5% combined water, by expanding 1600° to 2100°F.

The refractory composition also includes a finely ground clay mineral which becomes plastic when mixed with water. This material is preferably bentonite, but other minerals may be used such as kaolinite, halloysite, illite and attapulgite. Bentonite is a very finely ground air-floated montmorillonite clay which becomes very sticky and plastic when mixed with water. The bentonite-water mixture swells considerably and has the ability to coat and bind the extremely large surface area of the calcined diatomaceous earth.

Although the perlite and bentonite may provide sufficient bonding strength, aluminum orthophosphate may be added to provide additional bonding strength in the quantities indicated hereinbefore.

The refractory material is manufactured by first charging the desired amounts of the calcined diatomaceous earth and water into a muller-type mixer. It is preferred that the muller clearance be on the order of 2 inches above the mixer floor in order to obtain the desired mixing effect and to avoid crushing the diatomaceous earth particles. After the water and diatomaceous earth have been mixed, the perlite is added to the muller-type mixer and mixed after which the bentonite and aluminum phosphate are added. The material is then discharged from the mixer, formed into blocks, and packaged in watertight containers. At the job site the material is removed from the packages and rammed into place in the furnace structure. No mixing or forming is required other than the ramming operation.

Other monolithic refractory materials may be readily laminated thereto in order to provide different refractory materials in different portions of the furnace.

For example, a monolithic refractory material which will withstand higher temperatures may be laminated over the refractory material of the present method in selected furnace locations at which higher temperatures might occur. Also, other refractory materials which are resistant to slag attack may be laminated in locations where such attack is likely to take place.

LOOSE GRANULAR MIX FOR FURNACE LININGS

According to *O.M. Burrows; U.S. Patent 3,793,040; February 19, 1974; assigned to Norton Company* a loose granular mix for producing heat-refractory linings for furnaces is characterized by including as at least a substantial portion of the aggregate material, a coarse crystalline fused refractory metal oxide of mixed sizes having a fused stable glass coating thereon. Fused alumina grit, having a grit size of 100 microns to 6,000 microns (6 to 150 grit), is the generally preferred aggregate material.

The mix consists of the coated aggreagate in a graded mixture of sizes to achieve maximum density (minimum porosity) of the fired lining, plus uncoated aggregate and includes at least 5% of refractory oxide having a particle size of 30 microns and finer; and it may include plasticizers such as clay for use where a wet mix is desired, and chemical additives to aid in the sintering process and/or react with the aggregate to produce crystalline compounds of lower specific gravity, thus producing volume growth in the body to maximize elimination of porosity. The total glass coating content in the mix should be 2 and 15% by volume of total solids. The employment of graded sizes of refractory oxide is important in providing maximum porosity of the tamped mix of 30% or less.

In the case of alumina aggregate such chemical additives include clay and silica, which may desirably be in the form of ground flint. The silica in the flint or clay will react on firing to produce mullite. Another useful additive is magnesia, preferably calcined or fused, relatively coarse (capable of passing a 100 mesh screen, but not an impalpable powder). Magnesia, upon firing with alumina, will form spinel, to prevent shrinkage.

Plasticizers such as clay, or organic plasticizers such as starch or methylcellulose may be employed if it is desired to employ a wet mix. Although the use of organic plasticizers tends to induce porosity in the mix, the beneficial effect of the use of glass-coated aggregate inherently produces a fired body of low porosity as compared to the use of conventional mixes.

It should be noted that not all of the grit is coated. It has been found that preferred results are achieved when at least the major portion of the glass in the mix is present on grits of 100 to 200 microns in size (as coated). Preferably at least 15% of the mix consists of uncoated refractory oxide particles. The total glass content of the mix should generally not exceed 15% by volume, to preserve the refractory nature of the product and should not be less than 2% to preserve the strength and low porosity of the fired body.

When clay is employed to plasticize a wet mix, the raw clay content should not exceed 40%, by weight of the mix. Potentially reactive materials for producing growth are effective when employed in 1 to 20%, by volume.

SEMISILICA PLASTIC RAMMING MIXTURE

According to *G.F. Carini, G.H. Williams and R.O. Burhans; U.S. Patent 3,791,835; February 12, 1974; assigned to Dresser Industries, Inc.* semisilica plastic refractories comprised of 10 to 40% crude flint clay, 48 to 78% quartz or quartzite, and up to 15% bond clay, preferably, 5 to 12%, are suitable for ramming into place to form monolithic linings in high-temperature furnaces.

Preferably, the overall batch analyzes, on a dry basis, 72 to 90% SiO_2, 7 to 22% Al_2O_3, less than 0.2% alkalies ($Na_2O + K_2O + Li_2O$) and the balance incidental impurities. The crude flint clay is typically sized minus 3 mesh, while the quartzite is partially added as minus 3 mesh and partially as minus 200 mesh.

Numerous bond clays are suitable so long as the alkali oxide content is less than 0.4%. Although the presence of a bond clay increases the plasticity of the mix, it is not required as a plasticizer as in a high calcined mix. The primary functions of the bond clay are to reduce the total alkali content of the mix and to balance, along with the crude flint clay, the permanent expansion resulting from the conversion of quartz to beta-cristobalite on firing. Both the crude flint and bond clays exhibit a permanent linear shrinkage on firing. Preferably, the crude flint clay and bond clay analyze:

	Crude Flint Clay, %	Bond Clay, %
Silica (SiO_2)	40–45	58.0
Alumina (Al_2O_3)	38–43	27.0
Titania (TiO_2)	2–3	1.5
Iron oxide (Fe_2O_3)	<1.5	1.4
Lime (CaO)	<0.3	0.1
Magnesia (MgO)	<0.4	0.04
Alkalies ($Na_2O + K_2O + Li_2O$)	<0.6	0.4
Ignition loss	Balance	–

Examples 1 through 4: Materials were weighed and thoroughly mixed and tempered with 5 to 15% water in a muller-type mixer and pressed into 9" x 4.5" x 2.5" specimens at about 1,000 psi. Thereafter, the shapes were dried at 250°F and subjected to the typical tests for plastic refractories. Compositions and the results of these tests are shown in the table below.

Example	1	2	3	4
Crude flint clay, %	28	14	–	32
Calcined flint clay	–	14	28	–
Quartzite	60	60	60	60
Ball Clay	12	12	12	5
Bulk density, pcf				
After drying 18 hours at 250°F	130	133	136	130
After heating 15 hours to 1500°F	122	126	130	–
After heating 15 hours to 2550°F	114	117	121	
After heating 5 hours at 2550°F	–	–	–	116
Modulus of Rupture, psi (ASTM Test C-133-55)				
After drying at 230°F	300	310	380	
After drying at 250°F	–	–	–	280
After heating to 1500°F	44	40	50	–
After heating to 2550°F	45	140	50	–
After heating at 2550°F	–	–	–	50
Linear change from mold size, %				
After heating 5 hours to 2550°F	+0.9	+1.9	+2.5	–

(continued)

Example	1	2	3	4
After heating 5 hours at 2550°F	–	–	–	+1.1
Load subsidence (25 psi for 90 min) at 2460°F	1.2	0.2	-0.3	-0.8

The chemical analysis of Example 1 is given in percentages as follows: silica, 81.6; alumina, 16.6; titania, 1.0; iron oxide, 0.4; lime, 0.1; magnesia, 0.1; and alkalies, 0.2.

As can be seen, the properties of the mixes are similar except that Example 1 has a lower, yet still acceptable strength while exhibiting less dimensional change after heating to 2550°F.

In the preferred composition, the 60% free silica addition comes from two sources. Approximately 40% was added as minus 3 mesh novaculite (a polycrystallized quartz) and the remaining 20% was added as minus 200 mesh potters flint. Both of these materials are composed of over 99% plus free silica in the form of quartz and have an extremely low (i.e., 0.03%) total alkali content. The use of this filler portion of quartz is believed important in obtaining a better packing density leading to improved properties.

A permanent linear expansion results as quartz progressively converts to the high-temperature crystalline form (beta-cristobalite) on firing. It compensates for the shrinkage of the crude clay materials, thus achieving a dimensionally stable composition.

REFRACTORY LINING FOR PIG IRON TROUGHS

R. Visser, J.M.J. Bormans, W.P.J. Van Haaren, and D.M. Homan; U.S. Patent 3,775,140; November 27, 1973; assigned to Koninklijke Nederlandsche Hoogovens En Staalfabrieken N.V., Netherlands describe a refractory lining for pig iron troughs and the like which contains (1) granular refractory material preferably 30% of which is a 1 to 3 mm sieve fraction, the remainder being finer, (2) tar 8 to 15%, preferably containing over 70% pitch and 9 to 11% C_2 resins, and preferably having a viscosity of less than 8,000 cp at 20°C; and (3) has at least 75% of the mass consisting of magnesite. It (4) preferably contains Cr_2O_3, and (5) may contain ground coal (anthracite and/or bituminous coal and/or coke) from 0 to 15%.

MOLTEN NONFERROUS METAL-RESISTANT REFRACTORY

The predominant cause of failure and shortness of life of refractory furnace linings in nonferrous melting and processing furnaces is attack and penetration of the lining by metal and/or metal oxides. The refractory aggregate of the refractory body, is generally more resistant to metal and metal oxide penetration and subsequent degradation, than is the bond portion. Therefore, if the bond is fortified in a way that renders it resistant to wetting by nonferrous metals and oxides, and highly stable towards metal and metal oxides, a superior refractory material is obtained.

According to *E.S. Gamble and W.S. Peterson; U.S. Patent 3,775,139; November 27, 1973; assigned to Olin Corporation* inorganic refractory bonding material is provided for bonding refractory oxides or compounds to form refractory bodies having improved resistance to attack by molten nonferrous metals and slags.

The first component of the bonding composition is 30 to 55 parts by weight of a nonbasic and nonaggregate metallic oxide or compound having a particle size such that 97% will pass through a 200 mesh U.S. standard sieve and having a melting point above 2500°F.

A preferred material is tricalcium phosphate, for example, commercial bone ash. Typical other materials which may be readily employed are: silica; mullite; beryllium oxide, titanium oxide; fire clay; kaolinite; aluminum nitride; chromium oxide; bauxite; spinel; silicon carbide; titanium diboride; zirconium diboride; and so forth. This component acts as a filler of high refractoriness, and is generally resistant to attack by molten nonferrous metals including copper and aluminum and their alloys and the slags commonly encountered in their processing. In general, the component resists wetting by nonferrous metals. The refractory material is also compatible with the bonding agent.

The second component is a finely divided nonaggregate material selected from aluminum, aluminum oxide and/or zirconium oxide. This component is utilized in an amount up to 30 parts by weight based on a dry basis parts. The aluminum powder is preferably utilized where the refractory is to contact copper or copper-base alloys and the aluminum oxide is preferably utilized when the refractory is to contact aluminum or aluminum-base alloys.

The aluminum powder is preferably the nonleafing, nonlubricated type of aluminum or aluminum alloy containing a minimum of 95% by weight of elemental aluminum. The aluminum powder particle size as well as the particle size of the aluminum oxide or zirconium oxide is such that 97% will pass through a 200 mesh U.S. standard sieve and the shape is preferably spherical.

The third component is a bonding agent such as aluminum phosphate or zirconium phosphate. In addition, the bonding agents may contain small amounts of organic bonding agents. The refractory bonding agent is utilized in an amount from 15 to 50 parts by weight.

In addition to the foregoing constituents, one may optionally employ 1 to 5% by weight of a suspending or thickening agent, such as kaolin. The purpose of kaolin is to keep the ingredients of the bonding composition in suspension. Otherwise, the ingredients would tend to settle on standing. The suspending agent should preferably be an acidic or neutral refractory material. For example, fire clay may be utilized. Alternatively, one might use an organic suspending agent, but it is less preferred.

Another optional component is silica flour of at least −325 mesh U.S. standard sieve in an amount from 1 to 5% by weight. Alternatively, one may use any thixotropic or gelling agent preferably refractory in nature. This material acts as a control over the thixotropic or gelling properties of the bonding composition.

The preferred bonding composition is as follows: (a) bone ash, $Ca_3 (PO_4)_2$, 30 to 55 parts by weight; (b) aluminum powder, up to 30 parts by weight, part or all of which may be replaced by aluminum oxide, Al_2O_3, for applications in processing aluminum alloys; (c) kaolin clay, 1 to 6 parts by weight; (d) silica flour, SiO_2, up to 5 parts by weight; (e) aluminum phosphate, the useful range of which bond former is 15 to 50 parts by weight.

The bonding composition is used in the proportion of 1 part of the sum of the aforementioned requisite three components, as for example, items (a), (b) and (e) of the above-prepared bonding composition, to 3 to 7 parts by weight of graded refractory grain, as, for example, –4+170, and preferably –4+100, mesh U.S. standard sieve fused aluminum oxide. The preferred ratio is 1 part of the bonding composition to 4 parts of refractory grain, on a dry basis.

POURABLE, NONHYDRATABLE PREMIXED REFRACTORY

According to *R.N. McFadgen; U.S. Patent 3,763,085; October 2, 1973; assigned to The J.E. Baker Company* a pourable, nonhydratable, premixed, refractory ramming, patching or lining material for high temperature apparatus is provided by mixing nonacid refractory aggregate such as particulate dolomite, magnesite, etc., with 2 to 5% of a water-resistant, normally liquid or plastic polymeric material selected from the lower molecular weight petroleum hydrocarbon resins, α-methylstyrene polymers, polybutenes and coumarone-indene resins.

As mixed, the material is free-flowing and can be stored in this manner until used. It can be applied by pouring into any desired aperture. Upon the application of slight pressure such as trowelling, ramming, etc. this material becomes self-adherent so that it forms a cohesive monolithic mass adherent to itself and to most of the types of surfaces which are normally found in refractory applications, such as metal or refractories. After application, it remains in place as a monolithic mass in any position, even in an inverted position.

Upon heating, a strong carbonaceous bond replaces the original bond at moderate temperatures. This carbonaceous bond can last indefinitely at these temperatures under reducing conditions and then last until a ceramic bond is formed at the appropriate temperatures under oxidizing conditions. The fact that there is a relatively small amount of carbonaceous material facilitates the formation of a very dense ceramic structure in the area in which the premix has been applied when the apparatus is burned in use.

The premix can be used as packing in the repiping of tap holes of electric steel making furnaces, to build up the bottom of ladles for molten steel, as a buffer lining between the inner lining and safety lining in basic oxygen furnaces as well as in patching and relining work. Denser backup linings are formed between the linings of the basic oxygen furnace than can be realized with the ordinary solvent-tar bonded ramming mixtures of the art. It will be apparent that the mixture would also have utility in the preparation of shaped articles such as bricks, and other molded refractory articles.

A batch of premix material was prepared from dead-burned dolomite of a particle size 30% through No. 4 and retained on No. 12, 50% through No. 12 and retained on No. 50 and the remaining 20% passing through No. 50. This aggregate was thoroughly mixed in a blade-type mixer with 3.1 lb of liquid α-methylstyrene polymer and 1.7 lb of No. 4 fuel oil, per 100 lb of dolomite, so as to substantially coat the dolomite granules. The premix was poured into containers and transported to the location of a rotary furnace. It remained pourable during transit and was trowelled and tamped into a two-inch-wide joint extending 360° around the interior of a 10-foot-diameter rotary furnace between old and new sections of the furnace lining. The material had excellent adhesion around the entire 360° and formed a permanent and successful ceramic bond between the brick courses when the furnace was put back in normal operation. In a subsequent shut-down for repairs on another section of the furnace, inspection of the subject joint showed that the material had burned in well around the full circle, had kept the bricks of the joined linings properly spaced and had become coated in the same manner as the adjacent linings.

POROUS PLUG FOR REFRACTORY LINING

O. Koerner; U.S. Patent 3,753,746; August 21, 1973; describes a porous refractory body, primarily a porous plug for the refractory lining of a vessel containing molten metal through which plug gas can be blown into the metal. This body has a cold crushing strength of not less than 200 kg/cm^2 and consists of a coarse-grained refractory material with a minimum quantity of a suitable binding agent, and has permeable pores of a cross section not less than 0.05 mm and a gas permeability of at least 100, and preferably more than 500 nanoperm (npm).

The hot bending strength or modulus of rupture of the body at 1400°C is 20 to 60 kg/cm^2. The refractory material is mixed with the binding agent, and the mixture is compressed with a pressure of at least 300 kg/cm^2 before firing to bring substantially all of the grains into contact with adjacent grains.

The oxide-containing refractory body includes dead-burned (or fused) and coarse-grained neutral refractory substances, such as mullite and corundum, which are rich in aluminum oxide (alumina), or basic refractory materials, such as magnesia (magnesite).

The oxide-containing binding agent is a material having a refractory component which is used to join the refractory particles at their points or areas of contact. Binding agents preferably used in conjunction with the refractory granular materials rich in aluminum oxide (e.g., mullite and corundum) are refractory clays having a high aluminum oxide content. Finely dispersed alumina and water-soluble alumina, with or without the addition of clay, finely dispersed ceramic slurry of mullite, along with additions of a bond-activating agent, such as monoaluminum phosphate or chromium aluminum phosphate, also may be used.

The latter two materials are preferably used together with alumina. When a basic refractory material, such as sintered magnesia, is used to form the refractory body, finely dispersed magnesia or caustic-burned magnesia is the binding agent with additions of small amounts of a bond-activating agent, such as finely grained chrome ore, the sesquioxides (e.g., Fe_2O_3, Al_2O_3, and Cr_2O_3), chrome salts, chromic acid or B_2O_3.

The ratio of refractory part to bond-activating agent is at least 2:1. In other words, at least 30% of the binding agent is the same oxide as that mainly contained in the refractory material.

Example 1: Porous Plug for Steel Casting Ladles — Raw Material: Sintered mullite having the following granulation:

2.5 to 3.0 mm	20% by wt
2.0 to 2.5 mm	20% by wt
1.5 to 2.0 mm	18% by wt
1.0 to 1.5 mm	17% by wt
0.5 to 1.0 mm	15% by wt
0.0 to 0.5 mm	10% by wt

Binding Agents: (a) Clay containing not less than 43% Al_2O_3, 3% by wt, and (b) aluminum monophosphate (50% aqueous solution), 1.2% by wt. The granulated mullite is mixed with the two binding agents which are added in a known manner as a slurry. The mixture is pressed into the desired shape under a pressure of 600 kg/cm^2 and fired for 4 hours at 1550°C. The resulting physical properties of the plug are:

Permeability	400 npm
Open pore volume	10% to 15%
Cold crushing strength	400 kg/cm^2
Hot bending strength	
(or modulus of	63 kg/cm^2 at
rupture)	1200°C and
	56 kg/cm^2 at
	1400°C

Example 2: Permeable Block for Use in Converters — Raw Material: Sintered magnesia (0.3% Fe_2O_3) of the following granulation:

3.0 to 4.0 mm	4% by wt
2.0 to 3.0 mm	24% by wt
1.0 to 2.0 mm	34% by wt
0.5 to 1.0 mm	27% by wt
0.0 to 0.5 mm	11% by wt

Binding Agents: (a) Caustic magnesia, 3% by wt, (b) boric acid, 0.2 to 0.5% by wt and/or (c) iron oxide, 0.2 to 1.0% by wt. The magnesia is mixed with the sintering agents in the conventional manner. The mixture is formed into blocks under a pressure of 900 kg/cm^2 and fired at 1750°C. The resulting physical properties of the block are:

Permeability	1,080 npm
Cold crushing strength	400 kg/cm^2

The hot bending strength of the permeable block made without the addition of boric acid is comparable to the strength of the plug manufactured according to Example 1.

Example 3: Porous Brick for Casting Ladles — Raw Material: Corundum with a grain size range of either 0.5 to 3 mm or 1 to 3 mm. Binding Agents: (a) Clay containing not less than 43% Al_2O_3 (grain size of up to 0.25 mm), 5% by wt, (b) aluminum monophosphate (50% aqueous solution), 1.5% by wt. The mixture is pressed at a pressure of 500 to 600 kg/cm^2 and fired for not less than 4 hours at 1600°C. Tbe resulting physical properties are:

Permeability   500 to 700 npm
Cold crushing strength 250 to 350 kg/cm^2

The hot bending strength is comparable to that of the plug manufactured according to Example 2. Porous plugs according to Example 1 are suitable for flushing steel with argon in casting ladles prior to casting. One or more plugs are arranged within the refractory lining at the bottom of the ladle. The ladles contain up to 300 tons of steel, and 2 to 6 m^3 of argon per ton may be used in the process which should not last longer than 10 minutes. This short time for the process is possible because a large volume of small gas bubbles can flush the liquid steel due to the highly permeable plugs. The agitation is strong enough to homogenize the molten metal in the ladle as well as equalize the temperature at various points in the molten metal. On the other hand, the upward momentum of the small gas bubbles is not strong enough to prevent a clean separation of slag and metal.

Porous blocks according to Example 2 are suitable for blowing oxygen and other gases into the molten pig iron of converters. Due to the high permeability of the block, oxygen can be blown at a sufficiently high volumetric rate so that a high quality output is maintained. The metallurgical reactions are much more efficient due to the very large surface of contact between the gas and metal and to the longer period of contact between them. Due to the very high refractoriness and mechanical strength of the blocks as well as the great evenness of the gas flow, the life of such blocks is very satisfactory.

PLASTIC ALUMINA-SILICA FOR STEEL REHEATING FURNACES

W.T. Bakker; U.S. Patent 3,751,274; August 7, 1973; assigned to General Refractories describes a plastic refractory for use at intermediate surface temperatures up to 2750°F which is suitable for use in steel reheating furnaces.

The plastic refractory comprises 45 to 65% calcined kaolin; 20 to 40% silica; and 10 to 25% plastic refractory clay, based on the total weight of the composition. The composition generally possesses a relatively low concentration of alkalies which contributes to its low deformation under load.

The alkali impurities of the refractory brick composition are kept to a minimum preferably 1.0% or below, based on the weight of the composition. To provide a relatively high refractory capability under load, it is advantageous to reduce the amount of impurities in general, especially alkali impurities present in the raw material, to a value of 1.0% or under. For this reason, calcined kaolin is preferred as the coarse fraction of the refractory composition since this material generally is low in alkali content, namely 0.5% and below. The silica used is preferably a super-duty silica or a glass sand, in either case containing less than about 0.1% alkali impurities.

Suitable type clays to be used as binder and plasticizer clays include ball clays, from Kentucky, Tennessee or Texas, having an alkali impurity content of 0.5 to 1.5% by weight. The total composition generally will have an alkali content of 1.0 and below, preferably 0.5%, calculated by weight.

Examples 1 through 4: Plastic refractories were produced having the following compositions:

	Example 1	Example 2	Example 3	Example 4
Calcined-Georgia kaolin (-3+28 mesh), %	52.0	52.5	59.0	47.5
High-purity silica gravel (-28+150 mesh), %	20.0	-	18.0	29.5
High-purity silica sand (-48+200 mesh), %	-	20.0	-	-
High-purity ground silica sand (-200 mesh), %	10.0	10.0	5.0	5.0
Kentucky Ball Clay, % (air-floated)	18.0	17.5	18.0	19.0

The above materials were tempered in a pug mill or a wet pan using 7.5% water and 0.5% of a lignosulphonate solution as tempering liquids and extruded in an auger-type extruder, with a deairing chamber operating at 26-inch mercury pressure. The mixture was extruded into a continuous column having a cross section of 9 x 6 inches and cut into 13 ½-inch-long blocks. The plastic mix may also be vibrated into dense slabs measuring 8 x 12 x 2½ inches. The specimens prepared were tested by repressing in a hydraulic press at 1,000 psi and the following results were noted:

	Example 1	Example 2	Example 3	Example 4
Deformation under load (%) 1½ hr at 2460°F	1.5-2.0	1.8	nd	nd
Deformation under load (%) 1½ hr at 2640°F	2.5-3.1	2.6	3.9	4.1
Permanent linear drying change (%)	0.3S	0.3S	0.3S	0.3S
Permanent linear drying and firing change (%):				
at 2000°F	0.3E	nd	0.1S	0
at 2550°F	0.4E	0.2S	0.1S	0.2S

SEMIRIGID INSULATING MATERIAL

J.-P. Kiehl and V. Jost; U.S. Patent 3,752,684; August 14, 1973; assigned to Société Générale des Produits Refractaires, France describe composite insulating materials which have a melting point of at least 1500°C and a use-limit temperature of at least 1200°C. The materials having a density of less than 0.9 g/cc, preferably 0.4 to 0.6 g/cc, are prepared from a batch comprising, by weight: neutral magnesium phosphate, 12 to 25%; alkaline earth oxides in excess, up to 5%; silica, 50 to 80%; other acid oxides, up to 20%, and mineral fibers, 2 to 20%.

The neutral magnesium phosphate is an element essential to the fabrication process; the alkaline earth oxides are present only to ensure absolute neutrality of the phosphate employed; the silica and the acid oxides constitute a refractory charge. The mineral fibers form a reinforcement against shrinkage and also add to the lightness and flexibility of the material.

These refractory materials have a resistance to compression which increases with the rate of crushing (percentage deformation) at least equal to the following values:

Rate of Crushing, percent:	Resistance to Compression, kg/cm^2
5	2
15	5
30	10

Their insulating power is due to their low heat conductivity which at 400°C is no greater than 0.25 kcal/m/m^2/°C/hr and at 800°C is no greater than 0.35 kcal/m/m^2/°C/hr. These products may be formed as blocks or panels of standard dimensions or shaped as required. They may be cut and machined. They may, for example, be used to advantage for: insulating the cowpers (a type of regenerator) of blast furnaces, between the sheet work and the interior refractory chimney, as well as insulating the hot air ducts that connect the cowper apparatuses to the blast furnaces; insulating the side walls and arched roofs of metallurgical heat-treating furnaces; insulating the primary furnaces of the petroleum and petrochemical industry (cracking furnaces, reforming furnaces); insulating aluminum and nonferrous metals electrolysis tanks; insulating furnaces with glassmaking tanks; and fabricating insulating feedheads for steel casting.

A fabricating process consists of (a) forming an aqueous mixture of silica, magnesia, possibly alkaline earth oxides, mineral fibers and at least one acid magnesium phosphate, (b) pouring it into a mold, (c) permitting it to set and harden at ambient temperature by formation of hydrate neutral phosphates, and (d) oven drying it between 100 and 350°C in order to remove at least 75% of the water it contains.

The following are mixed in 125 liters of water: 85 kg of siliceous sand crushed into 75 micron size; 3 kg of long-fiber amosite asbestos; 4 kg of short-fiber amosite asbestos; and 8 kg of dead-burned magnesia crushed into 75 micron size.

The obtained slip is homogenized and then 14.5 kg of an 85% solution of phosphoric acid is added. It is homogenized again for less than 2 minutes and poured into a 200-liter mold. The reaction of the phosphoric acid begins to set at once and produces hard gelatin of the mixture in 20 minutes. The piece is removed from the mold and is dried for 48 hours in a kiln at 200°C. A shrinkage of 0.5% is ascertained. The semirigid insulating refractory thus obtained has the following composition by weight:

	Percent
Neutral magnesium phosphate	14.5
Magnesia	0.3
Silica	74.7
Fibers	10.5

It has the following properties:

Density at 20°C	0.56 g/cc
Density at 1000°C	0.58 g/cc
Overall porosity	77 percent

Resistance to compression when cold:

Percent	Kg/cm^2
5	2.7
15	6.0
30	14.0

Reheat—shrinkage:

	Percent
24 hr at 1200°C	1.3
24 hr at 1400°C	1.5

Melting point: 1680°C

Heat conductivity:

°C	Kcal/m/m^2/°C/hr
200	0.110
400	0.124
600	0.162
800	0.178
1000	0.190

LUBRICANT-STEARATE BINDER FOR RAMMING MIX

In certain metallurgical processes, the basic furnace structure is comprised of a metal skin or shell having a refractory lining interiorly thereof to define the furnace space in which the process is carried out. Generally, the refractory lining of oxygen steelmaking vessels is comprised of an inner working lining of such as tar bonded, chemically bonded, or burned basic brick, an outer lining adjacent the inner wall of the metal shell usually of a burned magnesite brick, and an intermediate layer.

The intermediate layer is usually monolithic, and is formed by such as ramming of a refractory composition in situ. The intermediate layer can vary in thickness, depending on the vessel being lined and the operating parameters to which the lining is to be subjected.

M. Morris and G. R. Henry; U.S. Patent 3,729,329; April 24, 1973; assigned to Dresser Industries, Inc. describe a basic refractory ramming mix or brick made from a batch comprised of size-graded basic refractory aggregate and a nonaqueous, bonding system. This bonding system is comprised of a lubricating oil in combination with a metallic stearate. Preferably, the total bonding system amounts to 3 to 8 parts, by weight, for each 100 parts, by weight, of basic refractory aggregate, although 2 to 12 parts is workable. The preferred basic refractory aggregate is dead-burned dolomite, dead-burned magnesite and/or mixtures thereof. In addition, hard-burned lime or chrome ore can be used in combination with dead-burned dolomite and dead-burned magnesite.

The oil and stearate mixture serve a dual purpose—the oil provides lubrication necessary for achieving required rammed densities, while the stearate functions as a binder. Mixes containing this binder system have excellent ramming and storage properties. In addition, this lubricant-binder system is nonaqueous. It is well known that water is an undesirable constituent with mixtures of basic refractory materials because of hydration and subsequent danger of cracking.

Examples: Mix A, 100 parts magnesite, 3½ parts of partially neutralized soybean pitch; Mix B, 100 parts magnesite, 3 parts of nondetergent 40 weight motor oil and ½ part magnesium stearate powder. Both mixes had excellent ramming properties made up for storage tests at room temperature and at 140°F.

Appearance (as stored in polyethylene-lined sacks)—After storage, 140°F, 5 days: Mix A, Slightly dried and stiffer, partially compacted but easily broken up; Mix B, No change. After storage at room temperature for 8 weeks: Mix A, Tight compaction, difficult to loosen; Mix B, No change.

While the mix above illustrates the use of magnesite grain, equally satisfactory are mixes of dead-burned dolomite and magnesite, such as: Mix C, 60 parts dolomite, 40 parts magnesite, 2 parts carbon black, 3 parts of nondetergent 40 weight motor oil and ½ part magnesium stearate powder.

This mix had superior strength properties compared to similar mixes containing a partially neutralized soybean pitch binder. In addition, the mix exhibited excellent storage life, being very workable after seven weeks storage at room temperature. In preparing refractory batches using this lubricant-binding system, the lubricating oil is first mixed with the refractory aggregate and then the magnesium stearate is added to the mix.

While the use of a magnesium stearate $[Mg(C_{18}H_{35}O_2)_2]$ has been illustrated in the foregoing examples, other metallic stearates such as aluminum, calcium, sodium and zinc were evaluated and found to be equally effective. The oil, which provides the lubrication necessary for achieving required rammed densities, can be any oil which provides adequate lubrication, with economics and availability being a determining factor as to which particular oil is to be used.

ALKALI-RESISTANT LINING BY IMPREGNATING WITH PHOSPHORUS COMPOUND

According to *A.J. Owen, R. Visser, and J. Van Laar; U.S. Patent 3,708,317; January 2, 1973; assigned to Koninklyke Nederlandsche Horgovens En Staalfabrieken NV, Netherlands,* a refractory body, such as bricks for lining metallurgical equipment such as blast furnaces having a porosity of at least 10% and the grains of such body being held together by ceramic bondings, is subjected to an impregnation treatment with an aqueous solution of a phosphorus compound and the impregnated material is then heated above 100°C, preferably 400° to 500°C for at least 1 hour, preferably 5 to 6 hours.

The treatment results in a product in which the impregnated phosphorus compound is only present in the top layer of the inside pore surface whereas the ceramic bonding of the refractory product is substantially free from the phosphorus compounds. Such a product is highly resistant to alkali attack.

The ceramic bond is obtained by firing above 900°C so that from the refractory grains sufficient glassy liquid is formed to bind the more refractory parts of the grains, mainly mullite, together. Many refractory clays satisfy the general formula $(K_2O)_x(Al_2O_3)_y(SiO_2)_z \cdot (H_2O)_n$. By firing such a clay a transformation takes place into: $3Al_2O_3 \cdot 2SiO_2$ (mullite), SiO_2 (cristobalite) and $K_2O \cdot SiO_2$ (glass phase). The mullite and cristobalite grains are molten together with the glass plate.

Preferably the amount of phosphorus compound which is impregnated is such that the content of the phosphorus compound, calculated as P_2O_5, is 2 to 8% by weight of the refractory material. The amount of the phosphorus compound which is impregnated depends on the porosity and the pore-size of the refractory material.

The higher the porosity and the bigger the pores of the refractory material the higher also the amount of phosphorus compound which is impregnated. Suitable phosphorus compounds for use are orthophosphoric acid (H_3PO_4) and ammonium phosphate [$(NH_4)_3PO_4$]. Aluminum phosphates such as monoaluminum phosphate [$Al(H_2PO_4)_3$] (aluminum dihydrogen orthophosphate) and aluminum metaphosphate [$Al(PO_3)_3$] are preferred.

An aluminosilicate refractory brick of quality commonly used as lining material for iron-making blast furnaces having an apparent porosity of 11.1%, a cold crushing strength of 840 kg/cm^2 and being produced by firing aluminosilicate refractory material having the desired shape at a temperature above the vitrification temperature of this material so as to vitrify the refractory grains together forming ceramic bonds in the brick, was found to be subject to alkali attack. The brick had the following chemical analysis:

SiO_2	51.80% by weight
Al_2O_3	44.21% by weight
TiO_2	1.27% by weight
Fe_2O_3	0.93% by weight
CaO	0.34% by weight
MgO	0.21% by weight
K_2O	0.23% by weight
Na_2O	0.61% by weight
Other elements	0.40% by weight

One-half of the brick was subjected to an impregnation treatment as described below in order to introduce a phosphorus-containing compound into the pores of the brick; the other half of the brick was left untreated. The impregnation treatment on one-half of the brick was carried out as follows:

The brick-half was placed in a vessel and the vessel was sealed. A vacuum pump was then attached to the vessel and the pressure reduced to 2 inches water gauge. An aqueous solution of orthophosphoric acid having a concentration of 50% was then introduced into the vessel until the brick half was completely covered. This situation was maintained for 30 minutes. Then atmospheric pressure was reintroduced into the vessel above the orthophosphoric acid solution. The orthophosphoric acid solution was removed from the vessel and the brick was found to have their pores completely filled up with phosphoric acid solution. The impregnated brick was transferred to an electrically heated furnace and the temper-

ature was raised to 450°C, this being maintained for 5 hours. The furnace was
then allowed to cool and the brick was removed. In the following table, the
cold crushing strength and resistance to alkali attack (expressed in the index
figures according to crucible test) are given for an unimpregnated brick and for
a brick after the above-mentioned impregnation treatment.

	Cold Crushing Strength	Crucible Test Index Figure
Unimpregnated	860	4 to 5
Impregnated	860	1

The index figures before and after the impregnation treatment show that the
untreated brick is prone to complete disintegration or at least to light cracking
and that such a brick is made wholly resistant to alkali attack or prone to no
more than light cracking when it is treated in the above-mentioned way. The
treated brick is thus eminently suitable for parts of the lining of an iron-making
blast furnace, by virtue of their greatly improved resistance to attack by alkali
metal compounds.

After the above-mentioned impregnation treatment, the brick was found to have
an apparent porosity of 6.82% and a P_2O_5 content of 2.5%. Specimens of the
above-mentioned treated brick were boiled in water but it was found that the
impregnated phosphorus compound was completely insoluble.

REFRACTORY CEMENTS

FIRE RESISTANT MOLDINGS

M. Matsuo and H. Yamakita; U.S. Patent 3,957,522; May 18, 1976; assigned to Nippon Gohsei Kagaku Kogyo KK, Japan describe a process for preparing fire resisting moldings having an improved mechanical strength by admixing gypsum with crystalline calcium silicate in an amount of 5 to 100% by weight to the gypsum and water to give a slurry, molding the slurry under pressure and drying the raw articles. The water resistance of the moldings can be improved by the further addition of cement.

As crystalline calcium silicate, there may be employed known crystalline calcium silicates obtained by subjecting silica component such as siliceous sand or diatomaceous earth, calcium source such as quick lime or slaked lime, and water to hydrothermal reaction at a high temperature under a high pressure. Crystalline calcium silicates useful in the process include xonotlite ($6CaO \cdot 6SiO_2 \cdot H_2O$) and tobermorite ($5CaO \cdot 6SiO_2 \cdot 5H_2O$), which can be optionally prepared by selecting conditions of the hydrothermal reaction. Particularly, tobermorite has the advantages that starting materials can be selected from a wide range and the hydrothermal reaction can be carried out under moderate conditions.

Examples of gypsum employed are natural gypsum, calcined gypsum of α-type and β-type, and soluble anhydrous gypsum having a hydraulic property. For the purpose of reinforcing mechanical strength, there may be preferably added a filler such as cellulose fiber, asbestos, glass fiber, perlite or clay besides crystalline calcium silicate.

In case of further incorporating cement into gypsum together with crystalline calcium silicate and water, the obtained moldings are applicable to not only building materials for interiors but also building materials for exteriors since water resistance of the moldings is extremely increased. In that case, cement is employed at the ratio of 10 to 30% by weight of the total amount of gypsum and crystalline calcium silicate, preferably 10 to 20% by weight of the total amount. As a cement employed in this process, there can be employed known cements such as Portland cement, blast furnace cement and alumina cement.

Example: Powdery siliceous sand was admixed with quick lime at the molar ratio ratio of 1:1.05 as SiO_2:CaO, and then to the obtained mixture was added water in an amount of 12 times the total weight of the mixture and blended to give a slurry. An autoclave was charged with the slurry and the hydrothermal reaction was carried out at 220°C for 4 hours to give xonotlite.

Thus obtained xonotlite was admixed with powders of commercial gypsum at the ratios shown in Table 1. To the obtained mixture was further added asbestos and water in an amount of 5% by weight of the mixture and of 300% by weight of the mixture, respectively, and the mixture was agitated to give a slurry. The slurry was then poured into a mold made from mild steel and pressed, removing water under the pressure shown in Table 1. Then, the raw shaped article was taken out from the mold and dried at 100°C for 24 hours to give a board having a thickness of 2 cm. The results of the measurement of mechanical strength of the boards by testing machine (Autograph) are shown in Table 1.

Bending strength was measured on the test piece of $2 \times 2 \times 10$ cm under the conditions of a span distance of 7 cm and a rate of loading of crosshead of 0.25 mm/min. Compressive strength was measured on the test piece of $2 \times 2 \times 2$ cm under the condition of a rate of loading of crosshead of 1.0 mm/min.

TABLE 1

Xonotlite:Gypsum (weight ratio)	Molding Pressure kg/cm^2	Bulk Density g/cc	Bending Strength kg/cm^2	Specific Bending Strength*	Compressive Strength kg/cm^2	Specific Compressive Strength**
1:9	13	0.91	57.0	62.6	62.5	68.6
	35	1.12	92.7	82.8	116	104
	50	1.23	117	95.1	147	120
2:8	13	0.83	64.6	77.8	64.9	78.2
	35	1.06	105	99.1	110	104
	50	1.18	139	118	150	127
3:7	13	0.68	37.3	54.9	46.5	68.4
	35	0.87	62.5	71.8	84.8	97.5
	50	0.98	78.4	80.0	108	110
5:5	13	0.56	29.6	52.9	27.2	48.6
	35	0.77	59.9	77.8	68.9	89.5
	50	0.88	78.8	89.5	84.6	96.1
0:10	13	1.37	89.6	65.4	76.4	55.8
	35	1.57	162	103	350	223
	50	1.72	193	112	485	282
7:3	13	0.46	19.7	42.8	21.1	45.9
	35	0.60	38.0	63.3	43.6	72.7
	50	0.67	48.6	70.3	59.5	88.8
10:0	13	0.30	7.1	23.7	13.8	46.0
	35	0.35	28.1	80.3	31.0	88.6
	50	0.45	41.0	91.1	48.2	107

*Bending strength/bulk density.
**Compressive strength/bulk density.

Further, fire resisting property was estimated by the following manner. The board cut into the size of $2 \times 2 \times 10$ cm was put in an electric furnace and heated. The temperature was elevated to 750° to 800°C from room temperature for the first hour, elevated to 1000°C for the next hour, and then maintained at 1000° to 1050°C for the next 2 hours. After the conclusion of the heating, the board was taken out and the physical properties thereof were measured. The results were shown in Table 2.

TABLE 2

Xonotlite:Gypsum (by weight)	Bulk Density g/cc	Linear Shrinkage, %	Weight Loss, %	Drop of Bending Strength, %	Appearance
2:8	0.83	1.2	17.8	83.1	No crack, no deformation
	1.06	1.2	17.2	86.8	No crack, no deformation
	1.18	1.1	16.2	85.0	No crack, no deformation
3:7	0.68	1.3	13.2	80.4	No crack, no deformation
	0.95	1.3	11.8	80.6	No crack, no deformation
	0.93	1.2	11.0	78.6	No crack, no deformation
0:10	1.24	-	-	-	Breakdown
	1.57	-	-	-	Generation of crack
	1.72	2.5	20.0	98.3	Generation of crack
7:3	0.46	1.2	10.6	45.8	No crack, no deformation
	0.60	1.1	10.0	50.9	No crack, no deformation
	0.67	1.3	11.3	49.2	No crack, no deformation

FIREPROOF FOAMED MAGNESIUM OXYCHLORIDE CEMENT

H.C. Thompson; U.S. Patent 3,951,885; April 20, 1976; assigned to Thompson Chemicals, Inc. describes a method of making a foamed fireproof product of magnesium oxychloride cement. A porous substrate is impregnated with a foaming mixture of magnesium chloride, magnesium oxide, and frothing agent in water. The mixture hardens with small voids throughout the porous substrate, thus providing a fireproof product of low density.

The cement mixture may be cofoamed with polyurethane foam. The fireproof products of relatively low density are particularly valuable for building and construction purposes. Magnesium powder is the preferred frothing agent which, in combination with a surfactant, induces a large volume of small bubbles that remain in the composition as it is set.

Magnesium oxychloride cement (sorel cement) is prepared by the reaction of magnesium chloride, magnesium oxide and water. Magnesium oxychloride cement, before it has set, may be made to froth with the addition of a frothing agent. By adding a frothing agent during the initial reaction period of the oxychloride cement a continuing froth is obtained which may be impregnated in porous materials. The frothing agent continues to form bubbles in the magnesium oxychloride cement within the porous material until the initial set of the cement takes place, at which time the bubbles are frozen into a low density fireproof product.

In preparing magnesium oxychloride cement water is added to $MgCl_2 \cdot 6H_2O$ to form a mixture having a specific gravity of 22° Baume. In practice, this is a 1:1 ratio with water. The specific gravity is important and should not exceed the range of 21.5° and 22.5° Baume where $MgCl_2 \cdot 6H_2O$ is used.

About 4¼ pounds of magnesium chloride is added to water to make a gallon of liquid at 22° Baume. It follows that the range of 21.5° and 22.5° Baume is achieved by roughly the addition of 4 to 5 pounds of magnesium chloride with water to make a gallon. In practice, it is convenient to purchase solutions of magnesium chloride which are commercially available at 36° Baume.

Magnesium oxide is added to the mixture of magnesium chloride in water to commence the reaction to form magnesium oxychloride cement. It is essential to have between 5 and 6½ pounds of magnesium oxide added per gallon of magnesium chloride solution to form a satisfactory cement. The ratio of magnesium oxide to magnesium chloride is critical.

The preferred embodiment is a ratio of 5¼ pounds magnesium oxide per 4¼ pounds of magnesium chloride, or a ratio of 1.23. The preferred range of magnesium oxide to magnesium chloride is 1.23 to 1.42.

In practice, it is convenient to maintain a 22° Baume solution of magnesium chloride and water as a liquid component separate from the dry powder which is added to the liquid to commence formation of the cement. The dry powder may include appropriate fillers or solid additives as desired. Substantial amounts of fillers, may be included such as powdered limestone, marble dust, silica flour, or magnesium carbonate.

The preferred frothing agent is powdered magnesium, which reacts with the oxychloride cement to continuously produce small bubbles. In order to control the size of the bubbles generated by the reaction of the foaming agent and the magnesium oxychloride cement, a small amount of surfactant is used. Inasmuch as magnesium oxychloride cement system is basic, the surfactant should be anionic, although it is believed that certain nonionic surfactants are also suitable. The preferred detergent is isooctylphenoxypolyethoxyethanol containing 10 mols ethylene oxide.

While the surfactant controls the size and amount of bubbles, it does not in itself contribute gas bubbles. In order to increase frothing, the addition of a small amount of lactic acid renders the froth much more reactive.

When the frothing combination of magnesium chloride, magnesium oxide, powdered magnesium, detergent and water is prepared, it is ready for coating, impregnating, laminating, or molding. These steps must take place before the initial set commences. In the absence of any additives, magnesium oxychloride cement sets 120 minutes after the ingredients are mixed when ordinary commercial grades of calcined magnesium oxide are used.

However, where calcining of the magnesium oxide takes place at a lower temperature and under appropriate conditions, the initial cure of the cement may be commenced much sooner. For most purposes, a 40 minute cure is highly satisfactory since it permits thorough impregnation and working of the product before the initial set and still does not require undue periods of waiting for curing during the processing step.

Magnesium oxychloride cement is at least four times as strong as Portland cement so that products fireproofed with oxychloride cement retain substantial strength along with low density. Moreover, Portland cement is moisture absorbent, while magnesium oxychloride is not. These properties permit utility in aircraft, for example, for soundproofing and insulation.

The density of the cofoam may be easily controlled by varying the ratio of polyurethane to cement. The lightest practical weight, while still retaining fireproof characteristics is 50% polyurethane and 50% cement. The preferred ratio of cement to urethane is between 1.0 and 9.0.

MAGNESIA SETTLING AGENT IN COLD-SETTING REFRACTORIES

J.E. Cassidy; U.S. Patent 3,923,534; December 2, 1975; assigned to Imperial Chemical Industries Limited, England describes a cold setting refractory composition comprising a water-soluble aluminum phosphate binding agent, refractory filler and, as setting agent, magnesia of low reactivity.

The use of magnesia of low reactivity permits a controlled set giving satisfactory initial strength (green strength) in the set refractory product, and good strength in the final product which may be obtained on subsequent heating to high temperatures. By selecting magnesia of low reactivity, the setting time of the composition may be extended to a period of hours or even days. The extended setting time allows the material to be shaped into the desired form, for example, by casting.

The reactivity of the magnesia depends on its surface area. Thus a convenient empirical way of expressing the reactivity of magnesia is to express it in terms of the surface area of the magnesia. Magnesia of low reactivity generally has a surface area of less than 5 square meters per gram. The particular grade of magnesia selected will depend upon the setting time required, but it is generally preferred to use magnesia having a surface area below 2 square meters per gram. Where it is desired to increase the setting time still further, the use of magnesia having a surface area of less than 0.5 square meter per gram is advantageous. Examples of forms of magnesia that may be used include the so-called fused magnesia, and hard-burnt natural or precipitated magnesia.

The magnesia used is conveniently in finely-divided form, and consequently, if fused magnesia is used, it will generally be ground before use. It is preferred that the median particle size of the magnesia be less than 200 μm and especially 20 to 100 μm. The magnesia used preferably is at least 90% pure and it is especially preferred that the magnesia be at least 98% pure.

The setting time of the composition is also influenced by pH; the higher the pH, the shorter the setting time. Chloride ion, if present, increases the setting time. Any water-soluble aluminum phosphate binding agent may be used, for example the acid orthophosphates $Al_2(HPO_4)_3$ and $Al(H_2PO_4)_3$, and mixtures containing them.

The water-soluble complex phosphates containing aluminum and phosphorus in a ratio of 1:1 (for example 0.8:1 to 1.2:1) give especially good results. The complexes contain the anion of a mineral acid (other than an oxyphosphorus acid) or a carboxylic acid, for example, citric acid. The solid complexes generally also contain chemically bound water and/or chemically bound alcohol.

When the anion is a halogen, it is preferably chlorine. Examples of other preferred anions which may be present are the anions of mineral oxyacids especially monobasic oxyacids, for example nitric acid. The solid complex phosphates generally contain three to five molecules of the hydroxy compound per phosphorus atom, for example, the water-containing complexes having an empirical formula corresponding to $AlPO_4 \cdot HCl \cdot (H_2O)_x$ where x is in the range 3 to 5.

The complex phosphate may be prepared by, for example, reacting an aluminum salt containing the desired anion, for example the halide or nitrate, with water or an alcohol and phosphoric acid; by mixing aluminum phosphate with an aqueous acid containing the desired anion or by treating aluminum phosphate hydrate with gaseous hydrogen chloride or nitrogen dioxide. When the complex phosphates are prepared in solution they may be separated in solid form by spray drying.

The amounts of the components will generally be chosen to give a ratio of magnesia to aluminum phosphate binding agent of 1:200 to 1:1, preferably 1:50 to 1:3, and usually 1:10 to 1:2, on a weight:weight basis.

The refractory filler is preferably an acid or neutral refractory filler. Silica, alumina (for example calcined alumina, tabular alumina and bubble alumina) and zirconia are especially useful as refractory fillers, as are aluminum and zirconium silicates.

In the production of refractory cements and castables, only a small proportion of binding agent, relative to the refractory filler, is generally necessary. For example, the binding agent may used in an amount of 1 to 25% by weight, especially 2 to 10% by weight, based on the weight of the refractory filler.

The amount of water in the composition depends on the consistency required which, in turn, depends on the use for which the composition is intended. For example, a thinner mix will generally be used for a mortar than for a concrete. Usually, there will be sufficient water present to dissolve at least a major proportion (and preferably all) the aluminum phosphate binding agent. For example, the composition will generally contain 1 and 30%, preferably 4 and 20%, by weight of water based on the weight of refractory filler.

These compositions may be used, for example, in concrete mixes, as a mortar or grouting or as a castable composition, for example in the production of refractory bricks. Examples of areas in which cements and/or concretes may be useful are torpedo ladles, blast furnace runners, desulfurizing systems, cement kilns, lime kilns, sliding gate nozzles and petrochemical plants.

REFRACTORY SILICEOUS BARIUM CEMENTS

L. Walter; U.S. Patent 3,897,258; July 29, 1975 describes the preparation of refractory siliceous barium cements based on dibarium silicate by baking to vitrification or fusion (1500° to 1700°C) a homogeneous mixture of very fine powders of raw materials containing mainly barium, silicon and oxygen. Their clinkers have a chemical composition, expressed in oxides, of 75 to 82% BaO, 14 to 17% SiO_2, less than 3.5% Al_2O_3, less than 1.5% Fe_2O_3, less than 2% CaO, less than 1% MgO, less than 1% $Na_2O + K_2O$, for a total of 100%, all percentages being by weight.

The cement clinkers have a mineralogical constitution, expressed in hydraulic constituents of 70 to 90% dibarium silicate, $2BaO \cdot SiO_2$ as hydraulic main-constituent and of less than 8% monobarium silicate, $BaO \cdot SiO_2$; less than 8% monobarium aluminate, $BaO \cdot Al_2O_3$; less than 8% tetrabarium aluminateferrite, $4BaO \cdot Al_2O_3 \cdot Fe_2O_3$; less than 12% calcium-barium orthosilicate, $CaO \cdot BaO \cdot SiO_2$; together with magnesium-barium orthosilicate, $MgO \cdot BaO \cdot SiO_2$, as hydraulic secondary constituents.

To the cement clinker, there is added, during the very fine grinding, an additional compound intended to slow down the setting of this cement, i.e., calcium sulfate or gypsum in quantities of 3 to 5% or 3 to 5% gypsum together with 5% bentonite.

The refractory siliceous barium cements based on dibarium silicate present the following advantages in comparison with the refractory aluminous cements based on calcium monoaluminate and calcium dialuminate:

   (1)   They are cheaper, the barytes having a lower cost price than the artificial alumina or the very pure natural hydrated alumina.
   (2)   They have higher mechanical strengths 24 hours after mixing with water.
   (3)   Their hydration takes place with lower water quantities, thus leading to lower porosities after hardening and smaller contractions after calcination.
   (4)   They do not expand during setting and consequently can be rapidly removed from the shuttering.

As hydraulic binders for refractory mortars and concretes with refractory granulates made from crushed alumino-silicate bricks containing less than 40% Al_2O_3, with sillimanite or mullite granulates and for insulating concretes with very porous siliceous granulates of expanded fireclays for instance, they can be used at higher temperatures of 100° to 150°C because of their superior resistance to softening under load at high temperature.

Ground and homogenized with 20 to 70% by weight, of basic or acid high furnace slag, poor in Al_2O_3, they can be used as unequalled sulfate-resisting cements or sea water cements. Ground and homogenized with 30 to 60% barytes, they can be used as shielding cements of great value, absorbing x- and gamma-radiations.

Example: 91 parts by weight of siliceous barytes and 9 parts by weight of siliceous clay are very finely ground, then admixed and homogenized. The raw materials used have the following chemical composition.

Siliceous Barytes	Percent	Siliceous Clay	Percent
SiO_2	4.52	SiO_2	73.95
Al_2O_3	0.78	Al_2O_3	12.58
Fe_2O_3	0.42	Fe_2O_3	5.81
CaO	0.56	CaO	0.92
BaO	61.08	MgO	0.47
MgO	0.35	Alkalis	0.90
Alkalis	0.26	Loss on ignition	5.66
SO_3	31.84		
CO_2	0.41		

The above mixture is baked until clinkers are formed (at 1500° to 1600°C) in a usual cement furnace. After cooling, the resulting cement clinker is ground as finely as artificial Portland cement together with 4.5% gypsum as setting retarder. The raw material mixture contains 23% SO_2, which emanates during formation of clinkers and is recuperated to be then used according to known methods. The refractory siliceous cement clinker resulting from the above method of production has the following chemical composition.

SiO_2	15.30%	BaO	78.99%
Al_2O_3	2.62%	MgO	0.51%
Fe_2O_3	1.29%	Alkalis	0.45%
CaO	0.84%		

Its mineralogical constitution is: 77.10% dibarium silicate, $2BaO \cdot SiO_2$, as main constituent and 3.63% monobarium silicate, $BaO \cdot SiO_2$, 4.51% monobarium aluminate, $BaO \cdot Al_2O_3$, 7.06% tetrabarium aluminate ferrite, $4BaO \cdot Al_2O_3 \cdot FeO_3$, 4.04% calcium barium orthosilicate, $CaO \cdot BaO \cdot SiO_2$, 3.21% magnesium barium orthosilicate, $MgO \cdot BaO \cdot SiO_2$, as secondary constituents and 0.45% alkalis, free and included in the barium belite (mineralized dibarium silicate).

Without gypsum, the setting of this cement is very rapid. With the very finely ground and homogenized addition it becomes normal. Its hardening is very rapid (high mechanical strengths 24 hours after mixing with water).

It has a refractoriness of 1670°C, and its compressive strengths, after 12 hours in humid air and then in water, are the following: 450 kg/cm² after 24 hours, 600 kg/cm² after 3 days, 660 kg/cm² after 7 days, and 730 kg/cm² after 28 days.

The refractory concrete prepared from 15% of cement and 85% of refractory granulates made from crushed aluminosilicate brick containing 30.2% of Al_2O_3, having 16% of the total weight of the concrete granules of less than 0.2 mm, 32% granules between 0.2 and 2 mm and 32% granules between 2 and 5 mm, has a refractoriness of 1630°C. The refractoriness under load of 2 kg/cm² of the refractory concretes realized with this refractory siliceous barium cement together with refractory granulates made from crushed aluminosilicate fire clay bricks containing 30.2% Al_2O_3, having the above granulation, is 1330°C.

This is 120°C higher than the refractoriness under load of the corresponding refractory concretes realized with the refractory aluminous cements based on calcium monoaluminate and calcium dialuminate, which is only of 1210°C. This results in much wider use for these less refractory concretes and in a lower cost.

MAGNESIA-CHROME REFRACTORY MORTAR

One problem with mortars made using binders such as sodium silicate and suspending agents such as clay, is that they are less refractory than desired since both sodium silicate and clay, as well as other binder and suspension ingredients which might be used, are low melting and thus reduce the refractoriness of the overall composition compared to that of the refractory material itself.

Other desirable characteristics of a refractory mortar are that it be plastic (i.e., easily worked with a trowel) after mixing with water, that it not settle out of suspension when diluted with water to a dipping consistency, that it have relatively little shrinkage upon heating to its use temperature, and that it resist erosion of the joints between refractory brick.

Plasticity in a mortar is generally achieved by relatively fine grinding of the refractory and other ingredients. Thus, for example, refractory mortars are generally all finer than 28 mesh, often finer than 48 mesh, and sometimes substantially all finer than 100 mesh. In addition, plasticity is obtained by adding plasticizing agents such as clay or methylcellulose. Fine grinding of a mortar also aids in preventing settling when it is diluted to dipping consistency. Addition of suspending agents such as dextrin assist in preventing settling.

However, in general, the finer the refractory aggregate in a mortar, the greater its shrinkage upon heating to a elevated temperature, other things being equal. Thus, one problem in designing a refractory mortar is to obtain the desired non-settling characteristics without introducing undue shrinkage.

Accordingly *N. Cassens, Jr. and J.E. Neely; U.S. Patent 3,868,261; February 25, 1975; assigned to Kaiser Aluminum & Chemical Corporation* have found a heat setting refractory mortar with good suspension properties and low shrinkage. It is made from a composition consisting of: (1) 85 to 99.4% prereacted magnesia-chrome grain containing 40 to 80% MgO, the grain being substantially all finer than 35 mesh and containing 25 to 75% –325 mesh material; (2) up to 13% Cr_2O_3, substantially all of which passes a 325 mesh screen; (3) 0.5 to 1.5% dextrin; and (4) 0.1 to 0.5% methylcellulose; the amount of Cr_2O_3 and the percentage of prereacted grain finer than 325 mesh being such that the linear shrinkage of the mortar upon heating to 1650°C is not over 5%.

Example: Prereacted grain was made by admixing chrome ore concentrates, 95% of which passed a 325 mesh screen, with magnesium hydroxide produced from sea water, all of the hydroxide passing a 325 mesh screen, and firing the admixture to 2000°C in a rotary kiln. The grain so produced was ground so that 35% of it passed a 325 mesh screen, substantially all of it passing a 48 mesh screen.

To 88.9 parts of this prereacted grain were added 9.8 parts Cr_2O_3, one part dextrin, and 0.3 part Methocel 65 HG. This mortar was mixed with 26% water, based on the weight of the dry ingredients, to a trowelling consistency and troweled into molds 3" long, ¾" wide, and ½" deep. The bars thus formed were dried at 105°C for 12 hours and then fired to 1650°C in six hours with a five hour soak. After cooling to room temperature, the bars showed an average linear shrinkage of 1.0%.

The same composition was mixed with 37.5% water, again on the basis of the dry ingredients, to a dipping consistency having a viscosity of 1,980 centipoises. The slurry was then poured into a 1,000 ml beaker and allowed to sit for 24 hours; it showed 10.5% settling, determined by dividing the volume of clear supernatant liquid after 24 hours by the total volume of the suspension. A value less then 20% is considered acceptable in a mortar at dipping consistency.

A slag test was carried out by mortaring together two 4½ x 2¼ x 3 inch pieces of direct bonded 60% MgO magnesia-chrome brick to form a 4½ x 4½ x 3 inch base. On top of this base, oriented with the mortar joint vertical, was mortared a 4½ x 4½ x 1½ inch piece of the same brick with a 2½" diameter hole cut through the 1½" dimension. In the cup provided by this structure were placed about 200 grams of a slag containing 30% CaO, 50% SiO_2, 10% Al_2O_3 and 10% FeO. The assemblage was fired to 1700°C for 1½ hours.

After firing, the samples were cut vertically down the middle, the cut being at right angles to the vertical joint in the base portion. The top, cup-forming, portion of the specimen was then removed from the base so that the horizontal joint could be examined. Also, the two pieces of the base were also separated so that the vertical joint could be examined. For comparison purposes specimens of the same brick were tested in the same way at the same time but using as comparison mortars two commercially available products. It could be seen that they underwent markedly more shrinkage and cracking than the mortar of the process.

The mortar exhibits better refractoriness than mortars made with binders and suspending agents such as sodium silicate and clay, since its only nonrefractory ingredients are organic materials which burn out at relatively low temperatures and do not lower the softening temperature of the refractory ingredients. It exhibits better resistance to erosion than a mortar of similar chemical composition made from an admixture of chrome ore and periclase because it uses prereacted grain and Cr_2O_3. Finally, because of the careful control of the degree of fineness of the prereacted grain and the amount of Cr_2O_3, it combines volume stability upon firing with the ability to remain in suspension when diluted to dipping consistency.

CaSO₄-GROG-PRIMARY FLUXING AGENT CEMENT

According to *H.W. Burr; U.S. Patent 3,841,886; October 15, 1974; assigned to Motus Chemical, Inc.* refractory properties are imparted to a cement prepared from nonrefractory $CaSO_4$ by incorporation of a primary fluxing component and grog.

Adding a primary flux to gypsum or plaster of Paris does not control shrinkage under heat anymore than it does when added to a clay or refractory cement, but it does enable the gypsum or plaster of Paris to become very hard and dense and resistant to moisture degradation when fired to a stabilized temperature. Shrinkage control is facilitated by the incorporation of a grog.

Depending upon the relative amounts of the component, the compositions can be used to form insulating fire bricks for furnaces, wallboards, structural members having greater fire resistance than presently obtainable, oil well casing cement, hot tooling and welding fixtures and dies, lightweight floor fill, roof decking, ceiling panels, fire door cores, insulation for safes, pipes, and beams.

Useful primary fluxes include inorganic metal compounds having a fusion point below 1450°C when fired with calcium sulfate. Certain compounds which have melting points above 1450°C form a eutectic below 1450°C with calcium sulfate when fired and are therefore suitable as primary fluxes. Specifically, fluxes useful herein are compounds selected from the oxides, hydroxides, and carbonates of a metal from Groups I-a, I-b, II-a, IV-a, IV-b, VII-b and VIII and any other inorganic compound of a metal from Groups I-a, I-b, II-a, II-b, IV-a, IV-b, VII-b and VIII provided such compound has a fusion point below 1450°C when fired with calcium sulfate.

The calcium sulfate, both alpha and beta crystalline form can be in any of its anhydride or hydrous forms such as anhydrite ($CaSO_4$), plaster of Paris (calcium hemihydrate, $CaSO_4 \cdot \frac{1}{2}H_2O$) and gypsum (calcium sulfate dihydrate, $CaSO_4 \cdot 2H_2O$).

Examples of grogs are fired clay, silica (quartz or flint), alumina, zirconia, expanded perlite, pumice, silicon carbide and artificial and natural aluminum silicates such as expanded vermiculite, analcite, and andalusite, kyanite, orthoclase, spodumene, microcline, nepheline and syenite. Particularly useful grogs imparting low or no shrinkage and high refractory properties include expanded vermiculite and other forms of aluminum silicate.

In some cases certain natural minerals, such as feldspar, have both flux and grog properties and can be used as both. The alkali metal component of such materials apparently selectively leach at the operating temperature, leaving a supporting matrix.

The compositions, exclusive of H_2O whether crystalline or added, comprise 10 to 90 weight percent calcium sulfate and 90 to 10 weight percent of flux and grog material, the flux and grog material comprising 0.2 to 50 weight percent primary fluxing component and 99.8 to about 50 weight percent grog component, provided the flux component is at least 2 weight percent based on the calcium sulfate content. The relative amounts depend on the use for the finished product.

With plasterboards having, e.g., 65% or more calcium sulfate, one would use relatively higher proportions of flux. With castables, or even with wallboards and the like having a low calcium sulfate content (e.g., 10% plus a filler such as fly ash) one would use less flux in the total composition.

Example 1: A test sample was made by intimately mixing and pressing together 100 parts of calcium sulfate dihydrate, 5 parts of sodium hydroxide and 15 parts of water into a cylindrical form 1⅛ inch in diameter and 2⅔ inches in length. The cylinder was fired for three hours to 1000°C and then observed for any apparent deformation, densification, improved surface hardness, moisture degradation and weight change. It was found that the sample lost about 9.5 parts in weight and had a good structural change in that there was a glassy surface and excellent densification, with no cracking observable, nor was there any moisture degradation and the sample displayed improved surface hardness.

Examples 2 through 16: The procedure of Example 1 was repeated except that test samples were prepared using the following materials as flux in place of the sodium hydroxide:

Lithium chloride	Potassium hydroxide
Sodium chloride	Potassium tripolyphosphate
Sodium sulfate	Calcium chloride
Sodium hydroxide	Calcium phosphate
Sodium metaphosphate	Strontium chloride
Potassium chloride	Lead oxide
Potassium sulfate	Lead sulfate
Potassium carbonate	

In each case, after firing at 1000°C the structural changes were determined as good as defined in Example 1.

FIREPROOF ALUMINOUS CEMENT

M. Grylicki, F. Nadachowski and S. Pawlowski; U.S. Patent 3,826,664; July 30, 1974; assigned to Instytut Materialow Ogniotrwalych, Poland describe a method of manufacturing a highly fire resistant aluminous cement by sintering raw nate-

rials containing CaO, Al_2O_3 and 3 to 50% addition of $CaCl_2$, preferably 20 to
35%, at 800° to 1500°C, preferably 900° to 1300°C. Instead of $CaCl_2$ other
chlorides can be used, for example, $MgCl_2$, NaCl, KCl, $AlCl_3$ and others.

Calcium chloride melts at 772°C forming a liquid phase which considerably accel-
erates the reaction between the lime and aluminous components owing to the
exchange of ions through the liquid.

The reaction of the sintered lime-aluminous mixtures containing calcium chloride
proceeds as follows:

$$3CaCl_2 + Fe_2O_3 \longrightarrow 3CaO + 2FeCl_3$$

The ferric chloride product is a volatile compound which escapes from the material during sintering, thereby removing from the resultant cement ferruginous
materials, expressed herein as Fe_2O_3. The second product of the reaction, i.e.,
calcium oxide, remains in the material in an active form, enabling the accelera-
tion of the reaction with the aluminous compound of the mixture. Also, since
it introduces additional CaO into the mass, the calcium chloride can partly replace
raw materials usually used to provide CaO, such as limestones and hydrated lime.

The $CaCl_2$ content in the mixture of raw materials should exceed by weight the
Fe_2O_3 content by at least 1.5 times or preferably 2 times. The sintering process
is carried out until the main part of the ferruginous admixture is expelled from
the material in the form of $FeCl_3$.

Preferably, where a bauxite of high iron content is used as a raw material, the
process consists in separately sintering such bauxite with an appropriate quantity
of calcium chloride in order to remove Fe_2O_3 before the basic synthesis process
of cement.

Sometimes it is more convenient to introduce water vapor or gas mixtures containing water vapor and an excess of oxygen to the kiln during the sintering,
with the resultant conversion of calcium chloride into CaO. In this way calcium
chloride is removed from cement after it fulfils its function of accelerating basic
reactions through the liquid phase. At the same time, reactive lime produced as
a product is able to bind easily with aluminum oxide.

Cement manufactured according to the process may contain certain quantities
of $CaCl_2$ remaining after the reaction besides the main compounds, i.e., calcium
aluminates. For special purposes, when it is desired to obtain cement free of
chlorides, $CaCl_2$ is also removed by means of washing the granules of cement,
e.g., with water or alcohol, such washing being based on the solubility of calcium
chloride in water or alcohol.

Example: A mixture composed of 20% by weight of technical aluminum oxide,
30% by weight of bauxite, 30% by weight of raw calcium carbonate and 20%
by weight of calcium chloride is wet ground in a tube mill until it comes below
0.2 mm in particle size and then it is fed into the rotary kiln in the form of a
slip and sintered up to the maximum temperature of 1200°C while applying
gaseous fuel. Substantially the entire iron content of the admixture evolves
from the material during sintering in the form of volatile $FeCl_3$. The $FeCl_3$ is
precipitated from the cooled waste gases.

HYDRAULIC CONCRETE

L. Prost and A. Pauillac; U.S. Patent 3,802,894; April 9, 1974; assigned to Societe Prost, France describe hydraulically setting refractory compositions comprising 5 to 8 parts by weight of at least one hydraulic alumina cement; 2.4 to 4 parts by weight of at least one pulverulent refractory material of very high specific surface area and of high water absorbency; 0.01 to 0.30 parts by weight of at least one fluidizing and/or deflocculating agent; and 86 to 92 parts by weight of at least one refractory aggregate.

Such compositions may be made into concrete in conventional manner by mixing with the requisite amount of water for good workability. The hydraulic alumina cement or cements used are employed at the rate of 5 to 8 parts per 100 parts by weight of the total mixture.

The pulverulent refractory material of very large specific surface area and of high water absorbency may be, for example: clay, kaolin, micronized silica, micronized alumina, micronized magnesia, micronized chromite or micronized forsterite.

Such materials should be in the form of a fine powder of particle size less than 50 microns and preferably less than one micron. The fluidizing and/or deflocculating agent may be an alkali metal phosphate, alkali metal polyphosphate, alkali metal carbonate, alkali metal carboxylate, alkali metal humate, or an organic material similar to these salts. Any of these compounds may be used alone excepting for sodium silicate which is in admixture with at least one of the noted compounds.

The refractory aggregate may be: calcined refractory clay, bauxite, cyanite, sillimanite, andalusite, corundum, tabular alumina, silicon carbide, magnesia, chromite and zircon. These aggregates should be used in the form of particles of which the largest dimension is less than 30 mm. However, it is preferable to use aggregates of which all particles pass through a sieve of 10 mm mesh, and which comprise 25% of particles passing through an 0.5 mm mesh sieve.

Example: A mixture was prepared which contained, in parts by weight: 90 parts aggregate containing 40% of alumina, particle size 0 to 5 mm; 6 parts alumina cement; 4 parts clay; and 0.12 part of fluidizing agent.

This mixture was mixed with 7 parts of water to prepare a concrete. The properties of this concrete are reported in the table below, in which they are compared with the properties of a conventional type of concrete obtained by mixing 80 parts by weight of the same aggregate with 20 parts by weight of the same cement and making up the mixture with 12 parts by weight of water.

They are also compared with the properties of a supercompressed refractory brick obtained from a mixture of 93 parts by weight of the same aggregate and 7 parts by weight of the same clay, by treating the mixture with 7 parts by weight of water, molding under high specific pressure and subsequently baking at 1400°C.

	Conventional Concrete	Moldable Refractory Material	Supercompressed Refractory Brick
Ingredients (parts by wt):			
Aggregate	80	90	93
Alumina cement	20	6	–
Clay	–	4	7
Fluidizing agent	–	0.12	–
Water (additional)	12	7	7
Raw dried product:			
Apparent density (g/cc)	2.00	2.25	–
Open porosity (vol %)	23	15.2	–
Compressive strength when cold (bars)	350	350	–
After baking (°C) at:	1250	1350	1400
Apparent density (g/cc)	1.90	2.20	2.15
Open porosity (vol %)	27	14.4	20
Compressive strength when cold (bars)	250	850	250
Dimensional change (%)	–0.4	–0.4	–
Pyroscopic resistance (°C)	1480	1710	1720
Temp (°C) for 0.5% settling under a load of 2 kg/cm²	1250	1390	1400
Flexural strength (bars)* at: Temp (°C).			
20	48	65.3	54.8
400	27	54.8	57.3
600	28	57	62.2
800	29	58.6	57.3
1000	26	55.5	58.4
1200	18	43.5	44
1300	13	33.4	33.6

*Flexural strengths for the conventional concrete and moldable refractory material are for the raw dried product; for the supercompressed brick for the baked product.

CONCRETE FROM CALCIUM-ALUMINATE CEMENT, SLAG AND BORON PHOSPHATE

R.W. Wallouch; U.S. Patent 3,798,043; March 19, 1974; assigned to Airco, Inc. describes a refractory concrete of high fired strength, comprising an admixture of graded ferrochromium slag and a calcium-aluminate cement. Addition of several parts per hundred by weight of powdered boron phosphate, furthermore, have been found to considerably increase the strength of the fired refractory concrete, most particularly at 1000° to 1200°C. The ferrochromium slag utilized as the refractory aggregate is a composition resulting from production of ferrochrome alloys comprising 25 to 40% MgO; 20 to 50% SiO_2; and 10 to 40% Al_2O_3; together with less than 15% by weight of such constituents as Cr, Cr_2O_3, CaO, FeO, C, and S.

In order to illustrate the effectiveness of compositions, refractory products comprising a calcium-aluminate cement and the ferrochromium slags, were prepared and compared to refractory products based upon conventional blast furnace slag and calcium-aluminate cements. The typical blast furnace slags so utilized comprised: 40 to 50% CaO; 30 to 40% SiO_2; 8 to 18% Al_2O_3.

All aggregates were crushed and sized to form graded grog and combined with calcium-aluminate cement as hydraulic bond material. Concrete samples were then formed and fired in 100°C increments up to 1200°C and the fired compressive strengths determined. As may be seen by examining the following table, the incorporation of ferrochromium slag into the mixed configuration greatly improved the high temperature properties of the concrete as compared to a formulation using blast furnace slag as aggregate.

Specifically it will be seen by comparing in the table the results for Mix 1 with those yielded by Mixes 2 and 3, that the ferrochromium slag in combination with the calcium-aluminate cement greatly out performs blast furnace slag where used as an aggregate. The aggregate used in all instances to prepare Mixes 1, 2 and 3 was well graded from coarse to fines and contained at least 50% of sized slag which would pass the 14 mesh screen. The cement-to-aggregate ratio in the mix in all instances was 1:4, and the water to cement ratio, W/C, was equal to 0.60. In each instance the compressive strength test was run on refractory cylindrical samples which had been heat treated up to 1200°C in 100° increments, with 48 hours hold at each of the stated incremental temperatures.

It may be noted that Mix 2 differed from Mix 3 in that the former was derived from production of ferrochrome silicon alloy, while the latter was generated in the production of so-called high carbon ferrochrome. The former slags generally include by weight percent 30 to 40% MgO; 30 to 50% SiO_2; and 10 to 25% Al_2O_3; the latter slags (i.e., similar to Mix 3) generally include by weight percent 25 to 35% MgO; 20 to 35% SiO_2, and 20 to 40% Al_2O_3.

Compressive Strength Test

Heat Treatment Temp (°C)	Cumulated Days	- - - Compressive Strength (psi)- - -			
		Mix 1	Mix 2	Mix 3	Mix 4
125	2	4,207	4,514	6,490	5,514
200	4	4,349	4,869	6,493	5,133
300	6	2,830	4,417	4,978	4,760
400	8	2,813	4,808	6,110	4,463
500	10	3,437	4,256	5,630	4,511
600	12	2,737	4,582	4,827	4,125
700	14	1,731	4,447	4,814	3,802
800	16	2,134	3,313	4,159	3,377
900	18	1,828	2,712	2,740	3,080
1000	20	1,907	2,503	2,825	4,756
1100	22	1,754	2,470	2,321	6,870
1200	24	1,713	3,069	3,659	9,343

Legend to mixes:

Mix 1: Graded blast furnace slag and calcium-aluminate cement.

Mixes 2 & 3: Graded ferochromium slags and calcium-aluminate cement.

Mix 4: Same as Mix 2 with 3.85 parts of boron phosphate (BPO_4) per hundred parts of dry ingredients.

It will be noted from examination of the table that Mixes 2 and 3, while clearly in all respects superior to Mix 1, do not have exceedingly high strength characteristics at the higher temperature range of 900° to 1200°C. The addition of small quantities by weight of powdered boron phosphate (BPO_4) to the mix, as specified in the table, results in outstanding strength properties in the fired refractory, particularly at the higher temperature range of use.

In particular, as may be seen from the table, the Mix 4 differs from Mix 2 only in that 3.85 parts per hundred by weight of the dry ingredients of the additive boron phosphate is now present. As may be seen from the table, compressive strengths as high as 9,343 psi have been obtained where the concrete was fired up to 1200°C over 24 days. Generally, less than about 5 pph by weight of the dry ingredients are found to be a suitable addition range for the BPO$_4$, with a preferred range of addition being for about 3 to 4 pph.

It appears that some sort of ceramic bond is formed transforming the concrete quite rapidly into a dense ceramic body. The resulting refractory material when used, for example, in sideblock applications is found to have outstanding non-spalling and noncracking characteristics even after repeated heating to 1200°C and subsequent cooling.

REFRACTORY GAS CONCRETE

K.D. Nekrasov, A.P. Tarasova, A.A. Bljusin, T.P. Avdeeva, V.A. Elin, P.A. Roizman, and A.P. Denisenko; U.S. Patent 3,784,385; January 8, 1974 describe a method of preparing a mix for producing refractory gas concrete, which can find application as a material for heat insulation and lining of heat-treating, annealing and open-hearth furnaces where a temperature of up to 1200°C is to be maintained.

When preparing a mix for producing refractory gas concrete on the basis of a binder containing sodium silicate and nepheline slurry or ferrochrome slag, by mixing the components of the binder with water and subsequently adding a gas-developing agent, use is made as a filler of powdered chrome-alumina slag taken in an amount of 22 to 32% of the total weight of the mix and of a high alumina refractory material taken in the same amount. An average chemical composition of the slag in percent by weight is presented in the table below.

Slag Oxide Content

Al_2O_3	CaO + MgO	Cr_2O_3	Na_2O	SiO_2
75–80	4–10	5–10	3.5	0.7–1.0

A high-alumina powder prepared from wastes of broken high-alumina articles must contain not less than 62% of aluminum oxide. Both the high-alumina refractory material and chrome-alumina slag must be ground to such a degree of fineness, that not less than 70% of a sample should pass through a sieve with a mesh of 4,900 apertures per cm^2.

A gas concrete mixer is started and then charged with water preheated to 65° or 70°C, an aqueous solution of sodium silicate and sodium hydroxide; then powdered materials are introduced into the mixer: chrome-alumina slag, high-alumina refractory, finely ground sodium silicate and nepheline slurry or ferrochrome slag. The mix having been thoroughly mixed, aluminum powder mixed with a small quantity of water is introduced into the mix, and the resulting composition is thoroughly stirred again so as to preclude the commencement of gas evolution in the mixer.

On completion of the stirring the gas concrete mix is poured into metal molds preheated to 38° to 42°C, and the mix is allowed to stay in the molds at this temperature for 3 to 5 hours. After preliminary hardening of the articles, the hump is cut off from them, and the shaped articles are subjected to autoclave treatment by self-curing techniques.

With the help of electric heaters the temperature in the autoclave is maintained within 170° to 180°C. The steam which evolves in the autoclave builds up a pressure which during three hours reaches 8 gauge atmospheres and is maintained at this level for 4 hours. Then the pressure is relieved to 0 during a period of three hours, and the gas concrete articles are removed from the autoclave. After that the articles are kept under shop conditions for three days at 20°C to complete readiness.

The refractory gas concrete produced has the following properties: operation temperature, up to 1200°C; ultimate compression strength after the maximum operating temperature, not less than 12 to 20 kg/cm²; additional shrinkage at operating temperature, not higher than 1%; and volume, not less than 500 to 800 kg/m³ and over.

CALCIUM ALUMINATE-SPINEL CEMENT

A.I. Braniski, T.D. Ionescu and N.D. Deica; U.S. Patent 3,748,158; July 24, 1973; assigned to Institutul de Cercetari Metalurgice, Rumania describe a refractory cement composition consisting of a hydraulic component in the form of 75 to 40% by weight of calcium monoaluminate (CaO·Al$_2$O$_3$) or calcium dialuminate (CaO·2Al$_2$O$_3$) or a combination of both, and 20 to 55% by weight of magnesium oxide-alumina spinel (MgO·Al$_2$O$_3$), the latter being a nonhydraulic constituent which, does not destroy the binder characteristics of the composition and in fact constitutes a superrefractory. Preferably, the cement consists of 75 to 50% by weight of calcium aluminate (i.e., the monoaluminate or dialuminate or both) and 25 to 45% by weight of spinel (MgO·Al$_2$O$_3$).

Example 1: 9.5% by weight magnesia, 23.5% by weight limestone and 67.0% by weight calcined alumina, are fineground and mixed to make them homogeneous. The raw materials have the following compositions (all by weight). Magnesia: 0.55% SiO$_2$, 0.87% Al$_2$O$_3$, 0.60% Fe$_2$O$_3$, 2.15% CaO, 95.83% MgO, together 100%. Limestone: 0.32% SiO$_2$, 0.14% Al$_2$O$_3$, 0.11% Fe$_2$O$_3$, 54.63% CaO, 0.20% MgO, 44.20 less on ignition, together 99.60%. Calcined alumina: 0.04% SiO$_2$, 98.2% Al$_2$O$_3$, 0.2% Fe$_2$O$_3$, 0.3% CaO, 0.7% MgO, 0.62% Na$_2$O + K$_2$O, together 100.06%.

The abovementioned mixture is burned in a conventional cement kiln until sintering (1580°C). After slow cooling, the resulting clinker is ground at the Portland cement fineness.

The resulting refractory calcium and magnesium aluminous cement is an hydraulic binder with normal setting and rapid hardening. Its refractoriness is 1770°C, its compressive strength after 3 days 412 kg/cm² and after 7 days 506 kg/cm².

The mineralogical composition (by weight) of the refractory cement is 34.3% spinel MgO·Al$_2$O$_3$ (melting point 2135°C) superrefractory constituent; 62.7% calcium dialuminate CaO·2Al$_2$O$_3$ (melting point 1750°C) hydraulic constituent; 1.5% calcium monoaluminate CaO·Al$_2$O$_3$ (melting point 1600°C) hydraulic constituent; 0.7% brownmillerite 4CaO·Al$_2$O$_3$·Fe$_2$O$_3$ (melting point 1415°C) secondary hydraulic constituent and 0.8% gehlenite 2CaO·Al$_2$O$_3$·SiO$_2$ (melting point 1590°C) secondary nonhydraulic constituent.

The refractory concrete resulting from 20% refractory calcium and magnesium aluminous cement and 80% refractory magnesite grog, with 16% <0.2 mm, 32% 0.5 to 2 mm and 32% 2 to 5 mm diameter grading, has the refractoriness of 1960°C (SK 41/42).

Example 2: 47.4% of dolomite and 52.6% of calcined alumina are very finely ground, then admixed and homogenized. The raw materials have the following chemical composition. Dolomite: SiO_2 0.69%; Al_2O_3 0.73%; Fe_2O_3 0.55%; CaO 29.43%; MgO 21.46%; $Na_2O + K_2O$ 0.21%; loss on ignition 46.54%. Calcined alumina: SiO_2 0.04%; Al_2O_3 98.2%; Fe_2O_3 0.2%; CaO 0.3%; MgO 0.7%; $Na_2O + K_2O$ 0.62%.

The above mixture is fired up to sintering or melting (1530° to 1630°C) in a usual cement furnace. After slow cooling, the resulting clinker is ground as finely as a Portland cement.

The refractory calcium and magnesium aluminous cement obtained is a hydraulic bonding agent (it is not disaggregated under water). It has a normal set, a rapid hardening and a refractoriness of 1630°C. It has a compressive strength of 358 kgf/cm² after 3 days, and of 507 kgf/cm² after 28 days.

The basic hydraulic constituent of this refractory cement is the calcium mono-aluminate, and its refractory constituent spinel. The refractory concrete prepared from 20% of the above refractory calcium and magnesium aluminous cement and 80% of white electrocast corundum as refractory aggregate, having 16% of the granules of diameter <0.2 mm, 32% between 0.5 and 2 mm, and 32% between 2 and 5 mm, has a refractoriness of 1865°C.

GENERAL REFRACTORIES

REFRACTORIES WITH IMPROVED STABILITY UNDER GRAVITATIONAL LOADS

A property of refractories which is of particular importance is their stability under gravitational loads when held at elevated temperatures for long periods, i.e., their refractoriness under load. Heretofore, in order to improve this property it has been usual to calcine the raw materials at very high temperatures and to include a briquetting or pelletizing step in the process prior to the calcination of the raw material.

N.O. Clark; U.S. Patent 3,912,526; October 14, 1975; assigned to English Clays Lovering Pochin & Company Limited, England have found that if the initial filtration process is carried out at pressures substantially higher than those conventionally used and the dewatered product thus obtained then subjected to calcination there can be obtained a product with superior refractory properties. Alternatively, if improved refractoriness under load is not required, the calcination temperature can be reduced with consequent savings in cost.

More particularly, there is provided in a process for producing a refractory material in which an aqueous suspension of a raw material is dewatered and the dewatered solid is calcined to form the desired refractory, the improvement which comprises dewatering the aqueous suspension of the raw material at a pressure greater than 700 psig (4.8 MN/m^2).

The dewatering of the raw material at a pressure greater than 700 psig (4.8 MN/m^2) can be conveniently achieved by means of a tube pressure filter. When operating at 700 to 1,500 psig (4.8 MN/m^2 to 10.4 MN/m^2), a plate filter press may be employed. The dewatering of the raw material is advantageously effected at a pressure greater than 1,500 psig (10.4 MN/m^2). The product of the high pressure dewatering step can be calcined directly, after briquetting or pelletizing or after shaping in a mold. Examples of refractories include calcined kaolin clays, other calcined aluminosilicates, aluminas, magnesias, dolomites, magnesium aluminate spinels, zirconia, beryllia, etc.

PUMPABLE REFRACTORY INSULATING COMPOSITIONS

B.J. Harvey; U.S. Patent 3,883,359; May 13, 1975; assigned to The Carborundum Company describes a pumpable refractory insulating composition comprising 20 to 50% of hydraulic setting cement, 10 to 25% of finely divided refractory material, 1 to 5% of a particulate synthetic organic polymer, 0.15 to 0.3% of a surface active agent and 20 to 70% water.

The surface active agent may be either ionic or nonionic. An example of an ionic agent is an aqueous solution of a sodium alkylnaphthalenesulfonate, modified with glue and foam stabilizers such as glycerol, urea and ethanol. An example of a nonionic agent is an aqueous solution of an alkylated phenol ethylene oxide condensate. The synthetic organic polymer must be inert to the other constituents of the slurry and insoluble in water.

The particulate synthetic organic polymer has the function of a lightweight pore former, the particles being spherical or bead-like in shape to aid the flow of the pumpable insulating composition. Examples of polymeric materials which may be used are those such as polystyrene, polyamides, polyethylenes, polypropylenes, polyurethanes and the like; of these, polystyrene is preferred. The beads or particles may be cellular and should have average diameters of 0.3 to 2 mm, a preferred diameter being 0.6 to 1.5 mm.

The hydraulic cement comprises a calcium aluminate hydraulic setting cement having an analysis of 54 to 56% Al_2O_3, 30 to 33% CaO, 7 to 9% Fe_2O_3 and 0.5 to 1% TiO_2. The refractory materials included in the composition may be in particulate or fibrous form, and may comprise refractory materials such as alumina, magnesia, titania, silica, aluminum silicate compositions, and the like. It is preferable to use particulate materials having average diameters of 1.4 mm or finer.

The most preferred composition comprises 30 to 35% of hydraulic setting cement, 14 to 18% of finely divided refractory material, 1.7 to 2% of a particulate synthetic organic polymer, 0.1 to 0.15% of a surface active agent, and 50 to 60% water. The viscosities of these mixtures range from 140,000 to 80,000 seconds. Up to 4% of colloidal silica may be added, if desired, to enhance the strength of the finished refractory coatings. The additions of the particulate organic polymer and the surface active agent are essential to achieve the desired pumpable characteristics of the refractory insulating compositions.

The advantages of the pumpable compositions may be summarized as savings in labor costs, reduction of down time of an installation under repair, improved insulation effectiveness by virtue of the lower density of the insulation, greater resistance to thermal shock, and the feasibility of pumping insulation into normally inaccessible positions. The compositions are easily prepared and applied without the use of special equipment, the resulting insulation setting up and being ready for use within 24 to 36 hours after application.

Example 1: The composition (percent by weight) is as follows: 10.0 alumina-silicate ceramic fiber, 29.6 calcium aluminate hydraulic setting cement, 4.0 colloidal silica, 1.7 synthetic organic polymer (spherical, cellular polystyrene particles), 0.1 surface active agent (alkylated phenol ethylene oxide condensate) and 54.6 water. A batch weighing 68 kg, sufficient for a dried insulation volume of 0.085 m^3 was prepared by mixing the dry ingredients first and then mixing with the water in a small concrete mixer for two minutes, after which the mix was transferred to the feed hopper of a pump having a double internal helix stator with helical rotor, specially designed for plastic viscous materials.

The mix had a cream consistency and could be pumped to a height of 9 meters. It was found that the composition, as pumped, would bridge gaps of up to 6.5 mm, provided that the hydrostatic head imposed while still in the fluid condition was not more than 1½ meters. Once the initial hydraulic set had taken place (20 minutes), additional amounts of mix could be added. A 100 mm thickness of insulation pumped into position had a linear shrinkage, wet to fired, of 0 at 600°C and 0.5% at 800°C. Final density was 0.4 g/cc.

Example 2: The composition (percent by weight) was as follows: 61.8 refractory aggregate (420 micrometers and finer), 15.5 calcium aluminate hydraulic setting cement, 3.2 synthetic organic polymer (spherical, cellular polystyrene particles), 0.1 surface active agent (alkylated phenol ethylene oxide condensate), and 19.4 water. The above mix was prepared and pumped as in Example 1. A 100 mm thickness of insulation pumped into position had a wet to fired shrinkage of 0 at 600°C and 0.2% at 1000°C. Final density was 0.48 g/cc.

URANIA-FUELED BERYLLIA COMPACT

J.B. Cahoon, Jr.; U.S. Patent 3,849,329; November 19, 1974; assigned to the U.S. Atomic Energy Commission has produced a high density fueled beryllia compact thermally stable above 1500°C, comprising:

(1) dissolving 4% by weight uranium chloride in ethyl alcohol,

(2) slurrying an intermediate grade purity beryllia having a mean particle size of less than 10 microns in ethyl alcohol solution,

(3) evaporating the solution at room temperature by vacuum pressure, whereby free-flowing powders are produced,

(4) compacting the powders isostatically above 20,000 psi,

(5) heating in a furnace through which air is freely circulating for at least 10 hours and to at least 1000°F, and thereafter

(6) sintering for at least 1 hour in hydrogen at a temperature at least as high as 1500°C.

Specifically the process relates to a method for producing a urania-beryllia fuel-moderator composition which is thermally stable above 1500°C.

Example: 11.8 grams of BeO were slurried in a quantity of ethyl alcohol in which 0.5 gram of camphor and UCl_4 had been dissolved. The BeO particles had a mean particle size of less than 10 microns and contained 1% impurities listed by the manufacturer as follows: 500 ppm Al, 320 ppm Ca, 4,750 ppm Mg, 548 ppm Cr, 580 ppm Si, 3,800 ppm Na, 100 ppm Fe, 1.30 ppm B, 210 ppm Mn, 26 ppm Ni, and 0.35 ppm loss on ignition. The alcohol was completely evaporated by maintaining the slurry under a vacuum of a few microns. The particles were mechanically agitated, resulting in a thin, uniform coating of camphor and uranium tetrachloride.

The dry particles were placed in an isostatic water-operated die and cold molded at 30,000 psi. The molded compact was placed in an electric furnace through which air freely circulated and the temperature was slowly raised, in 250°F increments, over 15 hours, to 1000°F to burn out the camphor and completely oxidize and convert the uranium chloride to uranium oxide, U_3O_8. The compact was finally fired in a tungsten resistance furnace in a hydrogen atmosphere for 1 hour at 1560°C. Final weight was 11.7 grams.

The density measurements indicated a slightly higher density than theoretical, the theoretical density being 3.085 g/cc and the actual density being 3.095 g/cc. The density of a sample of pure beryllia and camphor run as a control without any uranium containing material being added was 3.03 theoretical and 2.86 actual. A sample in which $Be(OH)_2$ was used instead of BeO as the initial constituent, the hydroxide decomposing to form the oxide in the oxidation step, had a theoretical density of 3.085 and an actual density of 3.10.

The fueled compact was heated in air to 2750°F without loss of weight or mechanical strength. It was further thermally cycled 6 times between 1600° and 2290°F, two cycles being run during each of succeeding days to the higher temperature which was maintained each time for 1 hour. The specimen was maintained at the lower temperature overnight. There was no weight change or loss of mechanical strength and the density at the conclusion of the six cycles was 99.38% theoretical. X-ray diffraction showed the sample to consist of U_3O_8, BeO and two unknown phases, thought to be uranium compounds and alloys. The mechanical properties of the fueled beryllia compact were observed to be comparable with the unfueled beryllia compacts of the prior art.

The experiment was repeated a number of times using comparable amounts of materials and virtually identical techniques and conditions, except that the camphor was omitted completely. In every instance the properties of the resulting compact, including density, mechanical properties and retention of uranium at high temperatures, were observed from measurements to be of the same quality as the compacts in which camphor was used.

SYNTACTIC CARBON FOAM WITH RESIN-STARCH BINDER

Carbon foams exhibit desirable thermal insulating characteristics, high strength-to-weight ratios, and are particularly useful at relatively high temperatures, particularly in nonoxidizing atmospheres. Of the various types of carbon foams available syntactic carbon foam is especially suitable for some applications where relatively high compressive strengths at low densities are desired. Syntactic carbon foam is formed of hollow carbon spheres or spheroids bound together in a carbon matrix which is provided by a carbonized resin binder.

According to *W.B. Malthouse and D.R. Masters; U.S. Patent 3,832,426; Aug. 27, 1974; assigned to the U.S. Atomic Energy Commission* syntactic carbon foam comprising hollow carbon microspheres in a carbon matrix is prepared by mixing hollow phenolic resin microspheres with a carbonizable binder consisting of resin and starch particulates, compacting the mixture, gelatinizing the starch in the binder while maintaining the mixture under a load corresponding to a pressure greater than ambient pressure, and then heating the mixture to convert the mixture of spheres and binder to carbon. The use of the resin-starch binder signifi-

cantly increases the plasticity of the foam during the carbonization step to minimize deleterious internal stressing.

Depending on variables such as density and wall thickness of the resin spheres,
sphere size, pressure loading during molding and curing, quantity of binder, the
compressive strength of the carbonized foam product is 1,300 psi at 0.260 g/cc
to 1,500 psi at 0.280 g/cc. The modulus of elasticity of the carbonized foam
product is 852.4×10^3 psi at the low density to 1136.6×10^3 psi at the high
density.

Satisfactory syntactic carbon foam has been prepared by using commercially
available hollow microspheres of 5 to 300 microns diameter with the best results
being achieved by using microspheres in a narrow size or diameter range wherein
the microspheres vary in diameter from one another by a maximum of 50 microns.
The use of microspheres in a narrow size range provides a stronger product and
uniform density. The wall thickness of the microspheres, which has a direct
bearing on the strength and density of the product, is preferably 2 to 5 microns.

The binder is a mixture consisting of a water-soluble starch and an organic resin,
both in particulate form, with the quantity of binder providing 30 to 40 wt %
of the microsphere-binder-water mixture. The starch may be any of the naturally
occurring plant starches such as cereals, corn, potato, tapioca, etc. with a particle
size of 15 to 50 microns. The binder contains 30 to 70 wt % starch.

Upon forming the microsphere-water-binder mixture with the particulate binder
being uniformly dispersed throughout the mixture, the mixture is molded under
a pressure of 10 to 200 psi and then heated, while under pressure, to a temperature adequate to gelatinize the starch and then to a higher temperature, if desired, to thermoset the gelatinized starch and promote crosslinking of the polymer materials, i.e., the resin in the binder and the resin microspheres.

The gelatinized starch serves to hold the resin particles and microspheres in place
prior to and during carbonization. Satisfactory results have been achieved at 80°
to 100°C for gelatinizing the starch and 130° to 170°C to thermoset the gelatinized starch and initiate crosslinking of the polymers. The quantity of water employed in the mixture is preferably no more than that which will promote the
pressing of the foam mixture during the molding step and effect the gelatinization of the starch. Excess water is driven off during the heating step under
pressure by a conventional venting mechanism in the pressing mechanism.

The particulate resin binder may be of any suitable high-carbon yielding organic
resins such as epoxies; aldehydes; partially or fully polymerized alcohols; pitches
of coal tar, vegetable, or petroleum origin; pyrrolidones; polyphenylenes; poly-
acrylonitrile and copolymers of vinylidene chloride-acrylonitrile; decacyclene;
and derivatives of indene. The resin particulates are preferably 100 to 200 mesh
(Tyler).

Upon completing the gelatinization of the starch and, if desired, the thermosetting of the starch and the curing of the resins under pressure, the pressure is released and the composite is then carbonized by heating to 750° to 1100°C in an
inert atmosphere. The starch-resin binder system introduces sufficient plasticity
in the microsphere-binder composite during essentially the entire carbonizing
step so as to allow the attendant shrinkage during heating to occur without gen-

erating deleterious stresses in the carbonized foam.

The mixture of microspheres and binder is preferably both molded and heated to a temperature adequate to at least gelatinize the starch under the influence of pressure loadings. These pressure loadings effect an increase in the compressive strength of the syntactic carbon foam product. The pressure employed during the molding step is preferably 100 to 200 psi while the pressure employed in the starch gelatinizing and polymer curing step is preferably 10 to 50 psi.

Example: Two syntactic foam discs, each having a diameter of 2.6" and a thickness of 0.9", were prepared from two mixtures, each containing 60 grams of carbon microspheres (average diameter of 200 microns), 24 grams of phenolic resin having an average size of 150 mesh, 24 grams of tapioca starch particles having an average size of 17 microns, and 100 cc of water. The carbon microspheres, phenolic resin, and tapioca starch were blended and combined with the 100 cc of water to form a mixture in which the microspheres and binder particles were uniformly dispersed.

Each mixture was cold pressed in a carbon die at 200 psi to form a molded composite. The pressure in the die was reduced to 50 psi and the composite placed in an oven at 90°C for 24 hours for gelatinizing the starch. After the curing cycle the pressure was released and the composite was carbonized at 850°C in an argon atmosphere while still in the die. A 60 hour heat cycle having a temperature increase rate of 15°C/hr was used to carbonize the discs. The physical properties of the disc are shown below in the table.

Disc No.	Density, g/cm³	Compressive Strength, psi	Modulus of Elasticity × 10³, psi
1	0.265	1,385	852.4
2	0.280	1,455	1,136.6

IMPREGNATING POROUS REFRACTORY WITH CHROMIUM COMPOUNDS

P.K. Church and O.J. Knutson; U.S. Patents 3,789,096; January 29, 1974 and 3,734,767; May 22, 1973; both assigned to Kaman Sciences Corporation have found that underfired or so-called machinable grade refractory ceramics can be shaped while in the relatively soft state and then impregnated and heat treated to produce a ceramic having all the desirable characteristics of a vitrified ceramic without the usual change in dimensions. The process appears to be useful in the treatment of such refractory ceramic materials as the oxides of aluminum, beryllium, zirconium, titanium, magnesium and the like.

These materials in the commercially available machinable grade are quite soft and easily broken. Also, in the soft state, they can be readily cut with carbide cutting tools, drilled, filed, sanded and otherwise formed to practically any desired shape. When the machinable ceramics are treated they become very hard, approximating highly vitrified ceramic and, in addition, will retain the original machined and pretreated dimensions. The treated material becomes so hard that the only practical method to do further machining is with diamond cutting wheels or by using lapping techniques.

The process comprises the steps of impregnating the interstices of the porous body with a chromium compound capable of being converted to an oxide in situ at relatively low temperatures, heating the body at a temperature well below normal vitrification to at least 600°F to convert the compound impregnated to an oxide and repeating the impregnation and heating steps until the desired degree of hardness is obtained.

STABILIZATION OF POSITIVE SOL-REFRACTORY GRAIN SLURRIES

Slurries of refractory material suspended in aqueous sols of positively charged silica particles coated with alumina are useful in a rapid process for making precision investment casting molds and other ceramic laminates. Briefly, laminates are built up on a pattern by alternately dipping the pattern in reagents containing oppositely charged materials such as a sol of positively charged particles followed by a sol of negatively charged particles.

To increase the rate of buildup and impart the desired properties to the ceramic article, refractory material is slurried in the reagent containing the charged particles. Slurries of refractory grain and refractory fibrous material suspended in aqueous sols of positively charged alumina coated colloidal silica are particularly useful in this process. However, these slurries tend to thicken on standing and their maximum working life is only about six to eight days. *E.P. Moore, Jr.; U.S. Patent 3,764,355; October 9, 1973; assigned to E.I. du Pont de Nemours and Company* has found stabilizing agents with which the working life of these slurries can be extended to 25 days or more.

The working life of slurries of up to 80% by weight particulate or fibrous refractory material suspended in aqueous sols of positively charged alumina coated colloidal silica can be extended by the addition of water-soluble hydroxy-substituted aliphatic and aromatic carboxylic acids as well as some mineral acids such as phosphoric acid and certain other organic acids such as ethylenediaminetetraacetic acid and nitrilotriacetic acid. Addition of 0.00834 to 0.0834 part by weight of stabilizer per part of colloidal solids provides the improved stability in the slurries.

Specific stabilizers include hydroxyacetic acid, citric acid, maleic acid, tartaric acid, gluconic acid, diglycolic acid, phosphoric acid, ethylenediaminetetraacetic acid and nitrilotriacetic acid. The last two stabilizing agents named, polyfunctional carboxylic acids, must be partially converted to salts to adjust their water solubility so that useful amounts may be dissolved in the slurries of this process.

Any finely divided particulate or fibrous refractory material may be used provided that it does not react with the positive sol. Among the suitable granular refractory materials are zircon, alumino-silicates such as mullite, sillimanite and molochite, fused silica and alumina. Useful refractory fibers are volcanic rock fibers, glass fibers, and alumino-silicate fibers.

MONOLITHIC INORGANIC STRUCTURES

E.P. Moore, Jr. and D.M. Sowards; U.S. Patent 3,758,317; September 11, 1973; assigned to E.I. du Pont de Nemours and Company describe a process for forming monolithic inorganic structures, particularly suited to the production of adhesives, coatings, shaped refractories and refractory molds.

Two oppositely charged agents which have high binding efficiency are combined in homogeneous fashion in compositions. These agents can be combined to give a broad spectrum of mixture properties and working lives: rapid forming, still gels to slow curing, fluid-to-plastic compositions to noncuring, long-lived compositions are possible.

These homogeneous compositions contain negatively charged colloidal silica particles and positively charged colloidal particles. The ratio of positively charged colloidal particles to negatively charged colloidal silica particles is 1:2 to 6.5:1. These compositions are formed by combining a sol of positively charged colloidal particles and a sol of negatively charged colloidal silica particles. These dispersions vary in consistency from fluid to gel. Fluid to stiff mixtures, long-lived to very short-lived mixtures can readily be formulated to fit requirements. Preferably, the positively charged colloid is alumina-coated colloidal silica.

These matrices can contain particulate or fibrous refractory material or metal as monolithic structures. The matrices serve as suspension media for these materials and contain binders for these materials.

ALUMINUM HYDROXIDE-BORATE COMPLEX HARDENING AGENT

E. Shimazaki, S. Araki and H. Akazawa; U.S. Patent 3,746,557; July 17, 1973; assigned to Taki Fertilizer Manufacturing Co., Ltd., Japan describe a hardening agent for refractory use in combination with a phosphate or phosphoric acid-type binder, consisting of active aluminum hydroxide-borate complex where the molar ratio of Al_2O_3/B_2O_3 is 1 to 24.

This hardening agent is prepared by reacting in the presence of a proper amount of water, aluminum alcoholates or aluminum phenolate with boric acid, ammonium borate, alkali borates or boric esters in such a proportion as to bring the molar ratio, Al_2O_3/B_2O_3, to 1 to 24, and, if necessary, drying the reaction mixture. Active aluminum hydroxide-borate complex may also be produced starting from active aluminum hydroxide, or so-called fresh alumina gel as the aluminum compound.

The phosphoric acid and phosphate type binders include phosphoric acid, polyphosphoric acid, monoaluminum phosphate, aluminum-chromium primary phosphate, monomagnesium phosphate, monocalcium phosphate, monobarium phosphate and other soluble acid phosphate binders.

As for the aggregates, any known kind that is ordinary bound by the above mentioned phosphoric acid and phosphate type binders may be used. For instance, chamotte, quartz sand, zircon sand, fused silica, fired bauxite, corundum, silicon carbide, glass fibers, etc. are employed according to the refractoriness desired.

In general, the binder is used in an amount ranging from 10 to 30% by weight based on aggregates. The amount of the hardening agent is 2 to 15% by weight based on the total amount of the binder.

RUBBING SEAL FOR CERAMIC REGENERATOR

V.D. Rao and Y.P. Telang; U.S. Patent 3,746,352; July 17, 1973 describe a seal having good oxidation resistance and a low coefficient of friction and low wear when rubbing against a ceramic regenerator. The seal comprises a metal substrate having good oxidation resistance at the anticipated temperatures and a coefficient of thermal expansion matched as closely as possible to the thermal expansion properties of the coatings applied thereto.

A typical steel substrate is made from nickel-chromium stainless steels or high temperature alloys. A surface layer consisting of a glazing material and a matrix material nonabradable to the ceramic is applied to one side of the substrate where the surface layer will contact the ceramic regenerator. Glazing materials consist of lithium fluoride, sodium fluoride, potassium fluoride, or calcium fluoride. Other fluorides or chlorides can be added to the glazing materials to reduce the glazing temperature and thereby lower the operating range.

An intermediate layer of nickel oxide and calcium fluoride can be placed in contact with the surface layer and a bonding layer of nickel aluminide can be located between the substrate and the intermediate layer to improve bonding and prevent oxidation of the substrate. A bonding layer of nickel chromium alloy can be substituted for the nickel aluminide layer and generally permits eliminating the intermediate nickel oxide-calcium fluoride layer.

In general, glazing materials of lithium fluoride produce a low friction, low wear glaze at 500°F to 900°F. Similarly, sodium and potassium fluoride materials produce useful glazes at 650° to 1100°F, and calcium fluoride materials produce useful glazes at 800° to 1800°F. Adding up to 20 wt % of other fluorides or chlorides of lithium, sodium, potassium and calcium to the glazing materials reduces both the minimum and maximum temperatures by as much at 150° to 500°F. Such additions preferably are fluorides since the fluorides produce glazes having better overall properties.

Matrix materials useful in the surface layers include zinc oxide, cuprous oxide and stannous oxide for the lithium, sodium, and potassium fluoride glazing materials. A higher temperature glazing material also can be used as a matrix material; for example, fluorides of sodium, potassium, and calcium can be used as matrix materials for lithium fluoride glazing materials, fluorides of potassium and calcium can be used as matrix materials for sodium fluoride, and calcium fluoride matrix material can be used with potassium fluoride glazing materials. Mixtures of these fluorides and oxides compatible with the ceramic of the regenerator also can be used as matrix materials.

Zinc oxide, cuprous oxide, stannous oxide, nickel oxide, strontium zirconate, barium zirconate or barium titanate are used as matrix materials for the high temperature calcium fluoride glazing materials. Brass or bronze powders can be used to supply matrix materials containing both cuprous oxide and zinc oxide or cuprous oxide and stannous oxide. Small amounts of carbon preferably are added to matrix materials containing cuprous oxide to prevent complete conversion of cuprous oxide to cupric oxide which increases friction considerably. The carbon generally is added as a metallic carbon composite such as a nickel carbon composite. The nickel carbon composite contains 25 wt % carbon and is used in proportions ranging up to 20% of the surface layer.

5 to 90 wt % of glazing material with the balance matrix material produces surface layers having good friction, low wear, and good adhesion to the bonding or intermediate layers. The minimum temperature at which the glazing material produces the glaze can be lowered by using eutectic mixtures of the glazing material and the matrix material.

Each of the layers can be applied by a plasma spraying technique. The bonding layer is prepared by mixing powders of the ingredients and spraying the powder onto a fully annealed or aged substrate. An intermediate layer is applied by ball milling nickel oxide and calcium fluoride powders free of iron compounds, sintering the mixture at 2500° to 3000°F in an inert or slightly oxidizing atmosphere of nitrogen, argon, xenon, or helium that prevents any reduction of nickel oxide, dry grinding the sintered product, and spraying the resulting powder on the bonding layer.

An inert atmosphere prevents any formation of calcium oxide and is preferred. Plasma spraying with an inert gas also is preferred. Surface layers are prepared in much the same manner from powders of the appropriate components. Intermediate and surface layers also can be prepared by mixing powders with a suitable binder such as gum arabic or PVA in a water base slurry and agglomerating the mixture in a spray drier. The resulting mixture is applied to a substrate and heated to burn off the binder.

At the operating temperatures and pressures encountered in the regenerator environment of a gas turbine engine, the glazing materials develop a glaze surface on the seal member that has satisfactory friction and wear when rubbing against a ceramic regenerator. Friction coefficients below 0.45 are considered to be satisfactory for use in the gas turbine engine. Small amounts of the glazing materials can transfer onto the rubbing surface of the regenerator to produce a similar glaze there which reduces wear rates.

Materials such as boric oxide that are detrimental to the ceramic regenerator core are avoided in any of the layers of the seal. The best combination of friction, wear and seal integrity is provided by surface layers comprising lithium fluoride glazing material and zinc oxide matrix material for temperatures at 450° to 900°F and by calcium fluoride-zinc oxide surface layers for higher temperatures.

ALUMINUM PHOSPHATE BONDED REFRACTORY

P.J. Yavorsky; U.S. Patent 3,730,744; May 1, 1973; assigned to Basic Ceramics Incorporated provides a composition containing an aluminum phosphate bonding solution for making refractory ceramic shapes, and also a process by which such composition can be cast and hardened in situ without external heating.

Accordingly, a ceramic article is prepared by mixing a ceramic aggregate having a distribution of particle sizes necessary for such an article with aluminum dihydrogen orthophosphate, $Al(H_2PO_4)_3$, bonding solution and a curing agent capable of supplying alkalinizing ions to the solution to gel them. The resulting slurry is cast into a mold, and hardens in situ without the application of heat. The finished ceramic product is obtained by removal of the article from the mold (hardening being sufficient to permit this), and relatively intense heating to produce a polyphosphate bonded product.

The ceramic aggregate material employed is substantially insoluble and inert in the aluminum phosphate bonding solution. Suitable ceramic materials are alumina, mullite, zirconia, magnesium aluminates, beryllia, cordierite, dumortierite, and pyrophyllite. Other aggregate materials which can be employed are calcined clay, silicate aggregates, carbides, silicates, nitrides, borides, and some metal powders and grits. Preferred ceramic materials are alumina, mullite and zirconia.

The aggregate has a particle size distribution suitable for the production of a ceramic article. Preferably, at least 10% by weight of the aggregate material should have a particle size of –325 mesh or smaller (U.S. Standard Sieve Size) and preferably 10 to 30% by weight should be of such mesh. This material is frequently termed flour. The use of so-called flour achieves optimum density and crushing strength in the resultant ceramic product.

The preferred concentration of the phosphate solution is that to produce a specific gravity of 1.12 to 1.5, with corresponding P_2O_5 and Al_2O_3 contents as follows:

Phosphate Solution	Specific Gravity of		
	1.12	1.25	1.50
P_2O_5 content			
Wt % (typical)	8.0	16.0	32.6
g/cc (typical)	0.103	0.203	0.406
Al_2O_3 content			
Wt % (typical)	2.2	4.3	8.6
g/cc (typical)	0.027	0.054	0.108

The specific concentration selected is determined by the fluidity of the solution versus the final bond strength desired and is limited at the maximum concentration by the tackiness which can be tolerated in the casting. It is essential that the curing agent be alkaline and that the alkaline ions be released sufficiently slowly that a gradual reaction is effected. A preferred curing agent is magnesium oxide.

Other suitable curing agents are lithium silicate, lithium aluminate, magnesium silicate, and calcium silicate. Generally, any oxide, carbonate, hydroxide, silicate, aluminosilicate and phosphate of magnesium, calcium, barium, strontium, lanthanum, yttrium, and lithium can be employed.

Examples 1 through 3: Fused mullite having a distribution of particle sizes necessary for a ceramic castable refractory was dry tumbled for 1 hour to insure uniform distribution of agglomerates. Three 100 gram samples of the dry aggregate were then wetted with equal amounts of an aqueous aluminum dihydrogen orthophosphate bonding solution (SG 1.25) having a concentration of 0.203 g/cc P_2O_5 and 0.054 g/cc Al_2O_3. Fused magnesia curing agent of a weight percent based on the weight of the aggregate had been dry tumbled with the fused mullite.

The slurry was cast into a rubber mold with a cavity 1¼" wide by 3" long and by 1" deep. The following table indicates the variation in setting time which resulted.

| | - - - - - - - Example - - - - - - - | | |
	1	2	3
Batch ceramic aggregate, wt %			
-4+8 mesh fused mullite	10	10	10
-8+14 mesh fused mullite	10	10	10
-14+30 mesh fused mullite	13	13	13
-30+50 mesh fused mullite	13	13	13
-50+100 mesh fused mullite	18	18	18
-100 mesh fused mullite	18	18	18
-325 mesh fused mullite	9	9	9
-325 mesh calcined china clay	9	9	9
Total	100	100	100
Fused MgO curing agent (wt % based on			
weight of aggregate)	0.75	0.75	1.00
Sizing MgO, mesh	-200+325	-325	-325
Aluminum phosphate bonding solution per			
100 grams of aggregate (SG 1.25), cc	12	12	12
Setting period, hours	4	1	0.2

The set period is that interval from batching to a hardness that would not permit penetration by a sharp probe without chipping. At such hardness, the cast article has sufficient wet strength and no tackiness to permit removal from the mold.

The curative agent for most examples was magnesia to simplify the comparisons. The Example 2 batch contains the same amount of magnesia as Example 1, but it is more finely sized in Example 2. The finer magnesia reacts more readily, and thus, reduces the setting period. Example 3 bearing an increased amount of the fine magnesia still further reduces the setting period.

COMPANY INDEX

The company names listed below are given exactly as they appear in the patents, despite name changes, mergers and acquisitions which have, at times, resulted in the revision of a company name.

INVENTOR INDEX

U. S. PATENT NUMBER INDEX

3,811,899 - 139	3,844,802 - 66	3,887,387 - 40
3,811,900 - 163	3,844,803 - 205	3,892,579 - 38
3,813,252 - 162	3,846,144 - 203	3,892,583 - 117
3,814,613 - 212	3,846,145 - 177	3,892,584 - 200
3,814,782 - 111	3,847,629 - 23	3,893,867 - 64
3,816,145 - 180	3,849,329 - 249	3,897,256 - 199
3,816,146 - 210	3,852,078 - 155	3,897,258 - 234
3,816,163 - 6	3,852,099 - 137	3,899,341 - 37
3,817,765 - 90	3,853,566 - 136	3,901,721 - 83
3,819,786 - 161	3,853,567 - 117	3,903,230 - 153
3,821,005 - 160	3,854,965 - 23	3,904,427 - 116
3,824,105 - 4	3,854,966 - 176	3,909,278 - 119
3,825,653 - 210	3,854,967 - 154	3,912,526 - 247
3,826,658 - 179	3,856,538 - 199	3,920,464 - 62
3,826,662 - 208	3,859,399 - 170	3,923,531 - 175
3,826,664 - 239	3,859,426 - 43	3,923,532 - 61
3,830,652 - 159	3,859,427 - 22	3,923,534 - 233
3,830,653 - 207	3,860,529 - 129	3,928,244 - 168
3,832,193 - 24	3,861,947 - 128	3,930,874 - 80
3,832,194 - 26	3,862,283 - 41	3,932,681 - 12
3,832,426 - 250	3,862,845 - 52	3,933,513 - 174
3,833,389 - 158	3,864,136 - 87	3,943,216 - 59
3,833,390 - 68	3,865,599 - 108	3,945,816 - 3
3,833,391 - 68	3,868,241 - 51	3,945,839 - 54
3,834,981 - 15	3,868,261 - 237	3,948,670 - 52
3,835,054 - 121	3,875,296 - 132	3,948,671 - 80
3,836,374 - 158	3,879,208 - 201	3,950,177 - 183
3,837,871 - 157	3,879,210 - 21	3,950,464 - 152
3,839,054 - 178	3,879,211 - 14	3,951,885 - 231
3,839,057 - 89	3,881,911 - 134	3,953,563 - 20
3,839,540 - 156	3,883,359 - 248	3,957,522 - 229
3,841,886 - 238	3,885,977 - 14 Appl. (Pub.)	B 554,655 - 11
3,842,760 - 206	3,887,384 - 41	

NOTICE

Nothing contained in this Review shall be construed to constitute a permission or recommendation to practice any invention covered by any patent without a license from the patent owners. Further, neither the author nor the publisher assumes any liability with respect to the use of, or for damages resulting from the use of, any information, apparatus, method or process described in this Review.

GLASS TECHNOLOGY 1976
Recent Developments

by G. B. Rothenberg

Chemical Technology Review No. 63

Glass is one of man's most valuable and versatile materials. About 700 different glass compositions are in actual use. By all standards commercial glassware consists of high quality articles made to precise dimensions seldom equalled in other technologies. Profitable mass production of millions of window panes, plate glass mirrors, no-return bottles, jars, electric light bulbs, etc. has been made possible by comparatively recent advances in continuous melting, high speed-forming and tempering.

During manufacture the most important factor controlling the workability of glass is its viscosity. Small changes in chemical composition may significantly affect the viscosity at certain temperatures. Strain point, annealing point, and working point, all of which are viscosity measures, must be known for a given glass before it can be blown, molded, or cast into a rigid product.

This book describes the large scale synthesis and manufacture of glass. Special attention is paid to the basic raw materials and the many chemical additives that go into the melt to make a specific glass product. Formulations are given for many specialty glasses, glass fibers, lasers and glass ceramics, to mention only a few. In all, close to 300 processes are presented as depicted in the U.S. patent literature from 1973 to mid-1975. In this book the various glasses are discussed under their best known major application. It is recognized that many of these glasses may be used for other purposes as well as in a variety of complex formulations. A condensed table of contents follows with numbers of processes in ( ).

ISBN 0-8155-0609-0

270 pages

POLLUTION CONTROL 1975
IN THE ASBESTOS, CEMENT, GLASS
AND ALLIED MINERAL INDUSTRIES

by Marshall Sittig

Pollution Technology Review No. 19

The mineral or silicate industries are engaged in the processing or manufacture of asbestos, bricks, cement, clay, glass, fritted glass and ceramics, fiber glass, mineral wool and stone, sand and gravel, etc. They generate serious air and water pollution aggravated by disposal problems of heavy solid wastes which fortunately are biologically and chemically inert for the most part.

Here dust is the major problem and dust reclamation at the source is an economic necessity. Submicron particles are barely influenced by gravity. Once they enter the atmosphere, they stay there for a long time.

The asbestos industry, in particular, is faced with the twin threats of asbestosis and lung cancer as a result of airborne asbestos dust. Now, in addition, waterborne asbestos from processing of the mineral is being assessed as a potential hazard in drinking water.

There are, of course, the usual industrial pollution problems as well, such as stack gas emissions from glass-producing furnaces, phenolic emissions from fiber glass manufacture, and many more.

Reports of accomplishments and proposed solutions to these problems are afforded by this book which is based on various government-sponsored surveys with practical examples detailed from late U.S. patents. A partial and condensed table of contents is given here. Chapter headings are enumerated and followed by examples of important subtitles.

ISBN 0-8155-0578-7

333 pages

POLLUTION CONTROL
IN METAL FINISHING 1973

by Michael R. Watson

Pollution Technology Review No. 5

Electroplating and other metal finishing waste streams are significant contributors to water pollution, either directly by their content of toxic and corrosive materials, such as cyanide, acids, and metals, or indirectly through the deleterious effect these effluents exert on sewage treatment systems.

Federal, state and municipal regulations fixing the allowable concentrations have been established and enforcement may be expected to become increasingly strict.

There is ample technology available for treating chromium and cyanide rinse waters to any required degree of detoxification. The problems facing the plater are:

1. Volume and composition of the rinse waters.
2. EPA requirements for an acceptable waste effluent.
3. Capital outlay for treatment facilities.
4. Operating costs for the treatment processes.
5. Disposal costs of unusable recovered materials.
6. Credits from the recovery of metals and decrease in sewage fees.

Solutions to these problems are afforded by this book which is based on various government surveys and reports with practical examples detailed from late U.S. patents. A partial and condensed table of contents follows. Chapter headings are given here, followed by examples of important subtitles.

ISBN 0-8155-0505-1

295 pages

METAL-BASED LUBRICANT COMPOSITIONS 1975

by Henry M. Drew

Chemical Technology Review No. 48

Lubricants are designed to reduce friction between moving parts and are usually of the fluid film type, made from petroleum or increasingly nowadays by total synthesis (silicones etc.). Lubricants are getting better all the time and last longer, still the demand for them is on the increase. It is estimated that by 1980 the petroleum-lubricant market will grow 25%, and that of synthetic lubricants by more than 40% over the present volume.

The search for compounds which minimize friction and wear under boundary lubrication conditions has provided a wide variety of additives based on metal salts of sulfur and phosphorus. Lead, zinc, molybdenum and the fatty acid soaps of lithium, calcium, sodium, aluminum and barium are formulated for efficient multifunctional lubricants made from petroleum and synthetic oils. Ashless dispersants of the polybutenyl succinic anhydridepolyamine type are often used in combination with metal salts or may be reacted directly with alkali metals leading to overbased additives.

The synthetic greases such as silicones, organic esters and the polyphenyl ethers are extremely well suited for the addition of metal-based compositions yielding products with excellent performance over wide temperature ranges. About 255 processes are described. A partial and condensed table of contents follows. Numbers in parentheses indicate the number of processes per topic.

ISBN 0-8155-0577-9

356 pages

HOW TO DISPOSE
OF TOXIC SUBSTANCES
AND INDUSTRIAL WASTES 1976

by Philip W. Powers

Environmental Technology Handbook No. 4

This book discusses all recognized ultimate disposal methods in detail and contains a long list of specific recommendations for specific substances plus alternative disposal or recovery methods.

Ultimate waste disposal implies the final disposition of non-degradable, persistent, harmful, and cumulative wastes that may be solid, liquid, or gaseous. Workable solutions to ultimate disposal problems as described in this book include conversion to harmless end products, subsurface storage in ponds or landfills, and disposal in the ocean. This condensed information will enable anyone to establish a sound background for action towards disposal of toxic and hazardous materials with safety.

The five general categories of hazardous wastes are:

1. Toxic Chemical
2. Radioactive
3. Flammable
4. Explosive
5. Toxic Biological

Many toxic or hazardous wastes contain valuable materials. Whenever this is the case, recovery and reuse is one of the most desirable methods of hazardous waste avoidance.

The book is based on government-sponsored reports (often elaborated by highly qualified personnel from industrial companies), U.S. patents, and on pertinent articles in authoritative journals. A partial and condensed table of contents follows here. Chapter headings are given in full whenever possible.

ISBN 0-8155-0615-5

500 pages

CORROSION INHIBITORS 1976
Manufacture and Technology

by Maurice William Ranney

Chemical Technology Review No. 60

Modern corrosion inhibitors with anti-degradant properties not only inhibit the corrosion of metals and other materials, but also stabilize and readjust the medium or environment in which they are applied.

A customary corrosion inhibitor, for instance, will prevent acid attack in an automotive crank case by building up a protective barrier on the surface of the metal substrate by physical adsorption, chemisorption or other reaction with the substrate, but a modern antidegradant corrosion inhibitor will try to forestall the formation of acids in the crank case oil in the first place, while forming a protective coating on the metal as a second step.

Naturally, such corrosion inhibitors must be designed individually for given applications: Aluminum is attacked by alkalis, salts, and certain acids. Concrete is attacked by acids. Iron and its various alloys are very sensitive toward oxygen-containing compounds.

How to design, manufacture and apply such antidegradant corrosion inhibitors is taught in this book which describes over 260 processes relating to the protection of materials susceptible to corrosive chemical action.

A partial and very condensed table of contents follows here. Numbers in parentheses indicate the number of processes per topic or chapter. Chapter headings and some of the more important subtitles are given here.

ISBN 0-8155-0606-6

338 pages

CORROSION RESISTANT MATERIALS HANDBOOK 1976

Third Edition

by Ibert Mellan

This famous book, first published in 1966 and now already in its third, greatly enlarged and completely revised edition, will help you cut losses due to corrosion by enabling you to choose the proper *commercially available* corrosion resistant materials for your particular purpose.

The great value of this outstanding reference work lies in the extensive cross-indexing of thousands of substances. The 151 tables are arranged by types of **corrosion resistant materials.** The **Corrosive Materials Index** is organized by *corrosive chemicals* and other *corrosive substances:* it refers you to specific recommendations in the tables. New in this edition are a separate **Trade Names Index** and a listing of **Company Names and Addresses.**

For the first time there appears also a group of 13 tables comparing the respective anticorrosive merits of commercial engineering and construction materials essential to industry.

The tables in this book represent selections from manufacturers' literature made by the author at no cost to, nor influence from, the makers or distributors of these materials.

Contents:

ISBN 0-8155-0628-7

665 pages

SYNTHETIC OILS & GREASES FOR LUBRICANTS 1976
Recent Developments

by M. William Ranney

Chemical Technology Review No. 72

Fully synthetic motor oil is now being test marketed at select locations all over the world. Such synthesized engine lubricants are said to improve gasoline mileage up to 5% and require oil changes only every 40,000 miles or so, compared with the present 6,000 mile average for premium motor oils from the best petroleum crudes.

The organic esters represent the largest single segment of the synthetic lubricant market and are becoming widely used in crank case oils.

Silicone fluids are well known for their thermal stability, inertness and relatively slight change in physical properties over extremely wide temperature ranges. Recently organic phosphates, fluorocarbons, silicate esters and polyphenyl ethers have found their place. Polyglycols have low pour points, high viscosity indexes and show very little deleterious effects on rubber compounds and gasket materials.

This book describes over 290 processes relating to all phases of synthesis, formulation and evaluation of synthetic fluids and greases. Additionally olefin polymerization provides the opportunity to tailor-make hydrocarbon polymers for various applications.

The advent of the "economical" automotive engine, with its complexity increased by antipollution devices, has resulted in a substantial increase of oil additives, also described in this book, yet the formulator is reminded that the primary mission of the lubricant is to lubricate. A partial and condensed table of contents follows. Numbers in parentheses indicate numbers of processes per topic.

ISBN 0-8155-0624-4

367 pages